"This is a highly timely volume that takes a valuable step towards a greater understanding of the inequalities inherent in the climate crisis. Bringing together a range of important perspectives, this edited collection is a call to address the need for epistemic justice in the climate justice literature, and itself adds knowledge that is of use to a range of audiences – academic, policy, and practice-based – and a significant addition to the literature on climate justice."

Ali Watson, *OBE, Managing Director, Third Generation Project and Professor of International Relations, University of St Andrews, UK*

"An important book on climate justice where the majority of the world lives. It brings vital insights often lost in doomsday tales on climate change – on how people find ways to support each other in times of climate crisis, as in Bangladesh; digital advocacy of social movements in Brazil that shows us how to hold institutions accountable for disasters and how one may grieve, organize and resist in times of crisis. Across the world, these strategies are just as crucial in India where 'climate friendly projects' on the age-old-trope of 'empty wasteland' continue to threaten the lives of local populations."

Seema Arora-Jonsson, *Professor in the Division of Rural Development, Department of Urban and Rural Development, Swedish University of Agricultural Sciences, Sweden*

"A wonderfully illuminating collection that diagnoses the structural drivers of climate injustice – by authors mostly from/of the Majority World. The book intricately weaves together diverse voices, identities, and sites of knowledge production to document the profound consequences of climate change on peoples, cultures, and communities. It asks us to confront the lived realities of neoliberal development, socioeconomic inequalities, and the ongoing trauma of colonization. The book offers an important advancement in our collective effort to decolonize climate justice research and to elevate the contestations, movements, and narratives from communities living on the frontlines of climate change."

Eric Chu, *Assistant Professor, Department of Human Ecology, University of California, Davis, USA*

"*Climate Justice in the Majority World* opens new ground in the growing, and increasingly urgent literature on climate justice. Indeed, we cannot truly know what climate justice means without hearing from authors and places such as those represented, for the first time at such a broad scope, in this volume."

Brandon Barclay Derman, *Assistant Professor in Environmental Studies, University of Illinois Springfield, USA*. Author of Struggles for Climate Justice: Uneven Geographies and the Politics of Connection

"*Climate Justice in the Majority World* comprises an eclectic collection of chapters that foreground scholarship from the Majority World. This volume challenges the perceived neutrality of climate change and disaster events, and analyses them through intersections with (neo)colonial extractivist models, neo-liberal development policies, institutional structures, social norms and social identities. It emphasizes the often muted voices in the struggle for climate justice through a depiction of the so-called alternate knowledge frames and innovative movements of resistance and struggles from the Majority World. This volume is an essential read for academics, students, policy makers and practitioners committed to decolonizing climate change and climate justice scholarship."

Saleemul Huq, *OBE, Director, International Centre for Climate Change and Development (ICCCAD) and Professor, Independent University, Bangladesh*

"*Climate Justice in the Majority World* calls out the continuity of the colonial legacy in defining climate injustice in the majority world. The book builds on an interesting set of case studies with an intersectional perspective, representing multiple voices in the discourse – social movement organisers, climate activists, vulnerable communities, among others. The book provides a compelling call for recognition of diverse knowledge frames in the Majority World that are often sidelined in the conversations around transitions to sustainable futures."

John Paul Jose, *Youth Environment and Climate Activist, India*

"Reading this book illuminated my understanding of climate justice in a fresh, new way. I found myself agreeing loudly to statements that resonated deeply as I read *Climate Justice in the Majority World*. It is my hope that this knowledge will challenge readers into acting better for climate justice."

Susan Nanduddu, *Executive Director, African Centre for Trade and Development, Uganda*

"*Climate Justice in the Majority World* is an important read for anyone seeking to better understand climate change as a social and political issue. By putting inequalities and injustices at the center of the analysis, it offers a powerful critique of the conviction that 'we're all in this together'."

Diana Ojeda, *Associate Professor, Interdisciplinary Center for Development Studies (Cider), Universidad de los Andes, Colombia*

"A must read book for those interested in climate justice and decolonising climate change. This book presents stories from various places all over the globe, diverse topics in climate change and is written by people of various backgrounds, which provides different points of view. It also highlights important elements in knowledge production processes that perpetuate the gap between the Minority and Majority World. I found myself saying out-loud 'yes, yes, that's the problem, tell me more!'"

Desy Ayu Pirmasari, *Research Fellow, University of Leeds, UK*

"*Climate Justice in the Majority World* puts people at the centre of the climate justice debate. Without explicitly claiming so, it illustrates what peoples' science perspective on climate change looks like. It unsettles the narratives of equity coming out from the power structures embedded in the Assessment Reports and Gap Reports, and alerts us that the discourse on climate justice must go beyond the macroscientific and economic facts of historical responsibility. The world needs to understand the microcosm of non-linear relationships between scientific facts and social values of communities, and that economic calculations and political institutions must be adjunct to these relationships, not the other way round."

Manish Kumar Shrivastava, *Senior Fellow, TERI–The Energy and Resources Institute, India*

CLIMATE JUSTICE IN THE MAJORITY WORLD

This edited collection explores a diverse range of climate (in)justice case studies from the Majority World – where most of humans and non-humans live. It is also the site of the most severe impacts of climate change and home to some of the key solutions for the climate crisis. The collection brings together 12 chapters featuring the work of over 30 authors from around the globe.

The impacts of climate change are disproportionately affecting individuals, communities, and countries in the Majority World who historically have contributed little to rising global temperatures. The 12 chapters focus on a range of cross-cutting themes, demonstrating both individual and collective experiences of climate change and struggles for achieving climate justice from the Majority World. This includes activism, resistance, and social movement organizing in India and Brazil; lived experiences and understandings of frontline communities in Bangladesh and South Africa; consequences of and responses to disasters in Mozambique and Puerto Rico; and contested accounts, narratives, and futures in the Maldives and Pakistan, among other topics.

By adopting a decolonial lens, this book provides rich empirical content, insightful comparisons, and novel conceptual interventions. It foregrounds climate justice from an intersectional perspective and contributes to the ongoing efforts by scholars and activists to address epistemic injustice in climate change research, policy, and practice. It will appeal to undergraduate and graduate-level students, academics, activists, policymakers, and members of the public concerned with the impacts and inequalities of climate change in the Majority World.

Neil J. W. Crawford is a research fellow in climate action and member of the Priestley Centre for Climate Futures, University of Leeds, UK. Their research focuses on forced migration and displacement, refugee rights, climate and environmental justice, gender and sexuality, and urbanism.

Kavya Michael is an environmental-social scientist at Chalmers University of Technology, Sweden. Her research broadly examines environmental change and energy-related issues through a human rights and justice lens. She studies the multiple intersections of climate change, urban inequality, informality, and inter- and intraregional migration with a focus on the Global South.

Michael Mikulewicz is an assistant professor at the Department of Environmental Studies at the State University of New York College of Environmental Science and Forestry (SUNY ESF). Michael is a human geographer who applies critical theory to study issues of climate and environmental (in)justice, adaptation, and urban sustainability.

Routledge Advances in Climate Change Research

Governing Climate Change in Southeast Asia
Critical Perspectives
Edited by Jens Marquardt, Laurence L. Delina, and Mattijs Smits

Kick-Starting Government Action against Climate Change
Effective Political Strategies
Ian Budge

The Social Aspects of Environmental and Climate Change
Institutional Dynamics beyond a Linear Model
E. C. H. Keskitalo

Climate Change and Tourism in Southern Africa
Jarkko Saarinen, Jennifer Fitchett, and Gijsbert Hoogendoorn

Climate Cultures in Europe and North America
New Formations of Environmental Knowledge and Action
Edited by Thorsten Heimann, Jamie Sommer, Margarethe Kusenbach, and Gabriela Christmann

Urban Planning for Climate Change
Barbara Norman

For more information about this series, please visit: www.routledge.com/Routledge-Advances-in-Climate-Change-Research/book-series/RACCR

CLIMATE JUSTICE IN THE MAJORITY WORLD

Vulnerability, Resistance, and Diverse Knowledges

Edited by Neil J. W. Crawford, Kavya Michael, and Michael Mikulewicz

Routledge
Taylor & Francis Group

LONDON AND NEW YORK

from Routledge

Designed cover image: Getty images

First published 2024
by Routledge
4 Park Square, Milton Park, Abingdon, Oxon OX14 4RN

and by Routledge
605 Third Avenue, New York, NY 10158

Routledge is an imprint of the Taylor & Francis Group, an informa business

British Library Cataloguing-in-Publication Data
A catalogue record for this book is available from the British Library

ISBN: 978-1-032-10180-4 (hbk)
ISBN: 978-1-032-10171-2 (pbk)
ISBN: 978-1-003-21402-1 (ebk)

DOI: 10.4324/9781003214021

Typeset in Bembo
by Newgen Publishing UK

CONTENTS

CONTRIBUTORS

A. K. M. Mamunur Rashid has more than 26 years of working experience in different development agencies, including UN, IFIs, NGOs, and academia. He has worked for UNDP in Bangladesh since 2004, focusing mostly on climate change policy, planning, finance, and adaptation. He has also worked in other development areas, including disability and development, promotive and preventive health care, transformative and inclusive food systems, child rights, gender mainstreaming, gender equality, natural resource management, and biodiversity protection. He is currently working as Portfolio Manager for Environment, Energy, and Climate Change for UNDP Iraq. Mamun has completed a postgraduate certificate course on climate change and development from the University of East Anglia and a Master of Social Science in Anthropology from Jahangirnagar University in Bangladesh.

Africa Bauzà Garcia-Arcicollar completed her PhD at the University of Reading, titled 'Towards Just Climate Futures: Exploring Islanders' Narratives of Hope, Movement and Loss'. Her research focuses on the non-material and symbolic dimensions of climate change and life on the islands in order to understand what might be required of justice in the context of climate-related displacement. In particular, her work explores the role of symbolic reparations and memorialization practices in the context of climate change.

Aidan O'Sullivan is a lecturer in the School of Social Sciences at Birmingham City University, UK. He teaches on a range of topics across criminology, sociology, and security studies and is currently developing an MA module on 'Green Criminology in an Age of Climate Collapse'. His research interests are on climate violence and its disproportionate effect on countries in the Majority World. He also writes on the legacy of state violence and counterinsurgency during the Troubles in Northern Ireland. His work has previously appeared in *The Ecologist*, *The Conversation* and the *Journal of Global Faultlines* as well as in edited volumes. Aidan has a PhD from the University of Liverpool (2017) on the policing of anti-austerity movements. He has a BA degree in Sociology and Philosophy and an MA in Sociology from the University College of Cork.

Andrea Lampis is a FAPESP research fellow at the Institute of Energy and Environment (IEE/USP) of the University of São Paulo (2018–2023). At present he is coordinating fieldwork research on energy poverty and vulnerability in São Paulo (Brazil). His research focuses on the governance and justice implications of on-going socio-technical transitions. He holds a PhD in Social Policy from the LSE (UK) and a Post-doctoral degree form CEBRAP (Brazil). He integrates the directive board of RC39 committee (Sociology of Disasters) of the International Sociological Association (ISA) for the 2023–2027 period.

Bento Paulo Rafael has a degree in environmental engineering and disaster management from the Technical University of Mozambique in Maputo. While there, his research on solid waste management focused on the analysis of solid waste management systems of the cities of Maputo and Matola for the Matlemele Sanitary Landfill. In 2019 he monitored the discipline of geology at Saint Thomas University of Mozambique in Maputo and did an internship at Kulima, a non-governmental organization, where he carried out activities related to research, analysis, and elaboration of projects. In 2021, he participated as a field research assistant in the city of Beira within the scope of the climate justice project in the Queen Elizabeth Advanced Scholars (QES) program.

Boaventura Monjane holds a PhD in Postcolonialisms and Global Citizenship (Sociology) from the Faculty of Economics (CES), University of Coimbra, Portugal. He is a postdoctoral researcher at the Institute for Poverty, Land and Agrarian Studies (PLAAS), University of the Western Cape, South Africa; a fellow of the International Research Group on Authoritarianism and Counterstrategies of the Rosa Luxemburg-Stiftung; and an associate fellow at the Centre for African Studies (CEA), Eduardo Mondlane University, Maputo, Mozambique. His areas of research include agrarian movements, rural politics, food sovereignty, and climate change. He has been involved in agrarian social movements, both locally and internationally, working with the National Farmer's Union of Mozambique (UNAC) and the International Secretariat of La Via Campesina. He has contributed to film documentaries *Land of Plenty, Land of but a Few* and *Terra em Suspenso: Ameaças e resistencias en Cabo Delgado*.

Bushra Hameedur Rahman is a professor in the School of Communication Studies, and Chairperson of the Department of Journalism Studies at the

University of the Punjab. Her work takes place within areas of feminism and Muslim feminism, Muslim women portrayals in the western media, Muslim women in religious discourses, West in the religious discourses of Pakistan, peace journalism, and ethical values in communication education and practice. She has authored a number of research articles in reputed journals. Dr Bushra is also the founding member of the Association of Media and Communication Academic Professionals. She has been twice elected Chair of the Islam and Media Group of International Association of Mass Communication Research. She is also an editor at a communication research journal, *Journal of Media Studies*, a publication of the Institute of Communication Studies, University of the Punjab, Lahore-Pakistan.

Caio Penko Teixeira is an independent researcher studying different lines of investigation and thinking around the social movement field. He holds a joint PhD in urban studies at the University of Milano-Bicocca, Italy, and in social psychology at the University of São Paulo, Brazil. He is particularly interested in understanding how collective action and grassroots organizing shape policy regarding climate justice, radical housing, and forced migration.

David Singh has been enrolled in a joint PhD programme at the University of East Anglia (UK) and the University of Copenhagen (Denmark) since 2019. His research interests focus largely on territorialization and land, resource extraction, and green energy infrastructures. David's PhD dissertation analyses the territorial expansion of wind infrastructures in borderland Gujarat and the underlying extraction, dispossession, and destruction dynamics. He also discusses the issue of mediation and caste power in fixing large-scale wind power projects, the (un/re)making of space by identity politics, and the emergence of diverse resistance practices.

Elton Augusto da Amélia Fé has a degree in environmental engineering and disaster management at the Technical University of Mozambique in Maputo. His research focuses on solid waste management at the Technical University of Mozambique and has participated in an internship at the Department of Climate Change of the Ministry of Land and Environment in Mozambique in 2019. In 2021, he participated as a field research assistant in the city of Beira within the scope of the climate justice project in the programme QES.

Giverage do Amaral holds a PhD in environment and society from the Center for Environmental Studies and Research, State University of Campinas (UNICAMP), Brazil; a master's degree in sociology from Fluminense Federal University (UFF), Rio de Janeiro, Brazil; and a degree in sociology from Eduardo Mondlane University, Maputo, Mozambique. He is currently a member of the scientific council, a professor, and a researcher at Wutivi University (UniTiva) in Bairro Belo Horizonte, Maputo, Mozambique, where he is director of the Faculty of Social and Human Sciences. His fields of research and professional experience include community development, management of social intervention projects, social and environmental research (http://lattes.cnpq.br/5669843813535640).

Grace D. O'Donovan is an ESRC-funded doctoral candidate in the School of Social and Political Sciences at the University of Edinburgh. Her research, focused on Southern Africa, contributes to studies in political ecology and political economy, with an emphasis on human and spatial geographies. Her work commits to eco-social and political justice. Grace deploys a critical artsbased research methodology, taking on intersectional, transdisciplinary, and decolonial approaches in interrogating the complex entanglements between climate change, informal economies, plant and social ecologies, (un)sustainable livelihoods, and broader sociopolitical oppressions in the Western Cape, South Africa. She was previously awarded a Global Development Academy studentship for her MSc at the University of Edinburgh, where she focused on challenges around food sovereignty, political economy, and global sustainable development. She also holds an LLB from the University of Stirling and an MA (Hons) in International Relations from the University of St Andrews. Originating from South Africa and Canada, Grace also does documentary environmental photography, capturing visual narratives that speak to ecological, sociopolitical, and economic stresses in the context of climate change and lived socioecological oppressions.

Jessica Omukuti is a research fellow on inclusive net zero for the Oxford Net Zero Initiative and is based at the Institute for Science, Innovation and Society (InSIS) at the University of Oxford. Her research focuses on understanding what net zero climate action means for countries and communities in the Majority World. Jessica's research expertise is on climate justice and equity, climate finance, climate change adaptation, and just transitions in the Global South. She has a PhD in climate justice and climate finance from the University of Reading (2020).

José Maria do Rosário Chilaúle Langa is a PhD in geography from Paulista State University (UNESP), Brazil. He is a postdoctoral researcher at Institute E-Geopolis, France. He was a university professor at the Technical University of Mozambique (UDM), Mozambique, from 2018 to August 2022. He has experience in the areas of planning and development, agrarian geography, history and epistemology of geography, teaching, and curriculum. He also works with the following environmental topics: environmental impacts, climate change, and climate justice. He is a founding member of the Environmental Observatory of Climate Change (ObservA).

Kavya Michael is a researcher at Chalmers University of Technology, Sweden. She holds a PhD in economics from the School of Economics, University of Hyderabad, India. Her expertise lies in analyzing the 'natural environment' through a social science lens. Her research and professional experience broadly lies in examining global environmental change and energy-related issues through a human rights and justice lens. Within this domain, questions of class, caste, and gender have been central to her analysis. She also studies the multiple intersections of climate change/environmental hazards, urban inequality, informality, and inter- and intraregional migration with a special focus on the Majority World. Kavya is a recipient of the IDRC Climate Leadership Fellowship and was hosted at the Political Economy Research Institute, University of Massachusetts, Amherst, USA, as a visiting scholar. She is also an author of the urban background paper of the Global Commission on Adaptation, which underscored the need for urban climate change adaptation measures to be underpinned by justice, equity, and nature for ensuring transformative action. Her current work examines how the various structural elements of gendered injustices and the resultant forms of exclusion in different sectoral contexts interact with the impacts of climate change as well as climate action and energy transitions across the globe. Theoretically, Kavya's research falls within the domains of political ecology, feminist postcolonial theories, and social inclusion/justice.

Leandro Luiz Giatti is an associate professor at the Department of Environmental Health, School of Public Health of the University of São Paulo, Brazil. He has international expertise with sustainability and health, health promotion, participatory research approaches, and the urban water–energy–food nexus. His experience is characterized through interdisciplinary studies and with the main efforts dedicated to socioenvironmental vulnerable groups, also with studies in the Brazilian Amazon Region. Giatti is a member of the Institute of Advanced Studies at USP and an associate editor of the journal *Ambiente & Sociedade* (SCIELO-Brazil). In his academic research projects, he has advised thesis and dissertations of graduate students and also has supervised postdoctoral partners.

Lira Luz Benites Lazaro holds a PhD in Earth System Science from the National Institute for Space Research, Brazil. She also holds a PhD and a master's degree in Latin American Integration from the University of São Paulo, where she has made significant contributions to her field. She is actively involved in research on implementing the Sustainable Development Goals in Brazil, working at the Institute of Advanced Studies of São Paulo University. With a wealth of experience in both research and academia, Lira has demonstrated her expertise in various crucial areas. Her research interests encompass environmental governance; energy transition; policy analysis; energy and climate policy; institutions and management; and the water-energy-food nexus. She is a Working Group member on Energy and Sustainable Development at the Latin America Council of Social Science (CLACSO).

Michael Mikulewicz is an assistant professor of climate justice at the Department of Environmental Studies of the State University of New York College of

Environmental Science and Forestry. As a critical geographer, he studies the intersecting social, economic, and political inequalities caused by the impacts of, and our responses to, climate change and other environmental issues across time and space. His current research is centred on climate and environmental justice, adaptation, post-politics, resilience, urban sustainability, queer theory, and intersectionality. He has published his work in *Annals of the American Association of Geographers, Climate and Development, Geoforum, Area, New Political Economy, Lancet Planetary Health*, among others. He is also the co-editor of the *Routledge Handbook of Climate Justice*, published in 2018. Michael obtained his PhD in human geography from the University of Manchester, UK.

Muhammad Ittefaq is an Assistant Professor in the School of Communication Studies at James Madison University, USA. Ittefaq received his PhD in journalism and mass communications from the University of Kansas; his MA in media and communication science from Technische Universität Ilmenau, Germany; and his MSc in communication studies from University of the Punjab, Pakistan. His research delves into the ways in which people consume and interact with information through new technologies, including how they interpret scientific messages, make decisions related to climate and health, and support policies related to science. He investigates how different segments of society are influenced by the information ecosystems and messages surrounding scientific issues, while also examining how organizations use various persuasive techniques on mainstream and social media to engage with global science-related topics. He has published over 40 peer-reviewed journal articles and book chapters. His work has been published in a variety of academic journals, including the *International Journal of Communication, Journal of Health Communication, Journal of Science Communication, Journalism: Theory, Practice & Criticism, Journalism Practice, Journal of Media Ethics, Media International Australia, Convergence: The International Journal of Research into New Media Technologies, Vaccine, Digital Health, Psychology & Health, Third World Quarterly, Local Environment, Analyses of Social Issues and Public Policy*, among others.

Natacha Bruna is a researcher at the Rural Areas Observatory (Observatório do Meio Rural), an independent research institution in Mozambique. She holds a PhD in development studies from the Political Ecology Research Group at the International Institute of Social Studies at Erasmus University Rotterdam. Her research focuses on rural development, land, agrarian change brought about by resource grabbing as a result of extractivism, and environmental politics. She is the author of *Confronting Agrarian Authoritarianism* (2019, co-authored with Boaventura Monjane), *Land of Plenty, Land of Misery* (2019), *A Climate-Smart World and the Rise of Green Extractivism* (2022), and *Green Extractivism and Financialization in Mozambique* (2022).

Neil J. W. Crawford is a research fellow in climate action in the School of Geography, University of Leeds. Neil's research interests lie in forced migration and displacement, refugee rights, climate and environmental justice, gender and sexuality, urbanism and cities, humanitarian politics, and East Africa. Neil is the author of a number of publications, including *The Urbanization of Forced Displacement: UNHCR, Urban Refugees, and the Dynamics of Policy Change*, published in December 2021 by McGill-Queen's University Press. In 2021, Neil was commissioned as one of two co-editors of *The Climate Connection: Cultural Relations Collection Special Edition*, published by the British Council ahead of the 26th UN Climate Change Conference of the Parties (COP26) in Glasgow, Scotland. In 2023, they were one of the editors of a photobook, *See Change: Visualising the Urban Climate Crisis*, based on a climate change photography masterclass in Kampala, Uganda. They have been a visiting research associate at the Refugee Law Project, Makerere University (Uganda), British Institute in Eastern Africa (Kenya), the University of Nairobi (Kenya), and Hong Kong Polytechnic University (China). Neil holds a PhD in international politics.

Patricia E. Perkins is a professor in the Faculty of Environmental and Urban Change, York University, Toronto. She taught at Eduardo Mondlane University in Maputo, Mozambique, in 1990–1991. She holds a PhD in economics from the University of Toronto and has authored many publications on feminist ecological economics, climate justice, commons, and participatory governance. She is a Lead Author of a chapter ('Demand, Services, and Social Aspects of Mitigation') in the *Intergovernmental Panel on Climate Change Sixth Assessment Report* (2022). She directed international research projects on community-based watershed organizing in Brazil and Canada (2002–2008) and on climate justice and equity in watershed management with partners in Mozambique, South Africa, and Kenya (2010–2012). Her recent international project – with partners in Brazil, Chile, South Africa, Cameroon, Kenya, Mozambique, and Nigeria – worked to build a global network of participatory researchers on climate justice, ecological economics, and commons governance (qesclimatejustice.info.yorku.ca).

Patricia Figueiredo Walker holds a master's degree in environmental studies from the Faculty of Environmental and Urban Change at York University, Toronto. She is the co-author of 'Developing Community-Based Responses to Climate Change' in *Water and Climate Change in Africa* (Routledge, 2013) and 'International Partnerships of Women for Sustainable Watershed Governance in Times of Climate Change' in *A Political Ecology of Women, Water, and Global Environmental Change* (Routledge, 2015). Her current research centres on youth-led climate movements and intergenerational climate justice.

Priya Pillai has worked on environmental and social justice issues for 20 years, with grassroots movements and non-governmental organisations on climate and energy, gender equality, and forest rights. She has worked with Greenpeace, the Right to Food Campaign, Action Aid, and Oxfam. She is associated with the

National Alliance of People's Movement (NAPM) and Mahan Sangharsh Samiti – an indigenous Adivasi community-led-struggle against coalmining in the Mahan forests in Singrauli district, Madhya Pradesh – movements in an independent capacity as a volunteer and solidarity activist. At Mahan, she has specifically worked to implement the Forest Rights Act in the region. In her current role as the Head of Programme – State Climate Action in Asar, she is working with partners in Tamil Nadu and Kerala on climate change and sustainability related issues including Resilience building and adaptation in Agriculture, Energy Transition focussing on Decentralised Renewables, Carbon Neutral Districts and Panchayats, Climate Literacy. She is doing a PhD on the socio-ecological impacts of large- scale renewable energy from the University of Technology Sydney, Australia.

Ruchira Talukdar's research focuses on comparative aspects of climate justice and climate activism between the global North and South. Her PhD thesis compared coal conflicts and protest movements in India and Australia, with an emphasis on the intersections between grassroots and Indigenous movements and mainstream environmental activism. She has worked in environment movement in India and Australia, in Greenpeace, Australian Conservation Foundation and Friends of the Earth, for nearly two decades. She co-founded Sapna South Asian Climate Solidarity, an Australia-based South Asian climate justice network, for effective global North solidarity for just climate futures in the global South.

Sadia Jamil is an assistant professor at the School of International Communications, University of Nottingham, China. She has earned a PhD in journalism (University of Queensland, Australia) and two postgraduate degrees in media management (University of Stirling, Scotland) and mass communication (University of Karachi). Dr Jamil is chair of IAMCR's Journalism Research & Education Section and serving as UAE representative for Asian Media Information & Communication Centre. She is the series editor of IAMCR and Palgrave's book series in global transformations in media and communication research.

Sennan David Mattar is a lecturer at the School of Computing, Engineering and Built Environment in Glasgow Caledonian University. His PhD thesis investigated migration patterns and daily life within the informal settlements in Zambia from a climate justice perspective, and he regularly lectures on the Sustainable Development Goals, climate migration, litigation, and finance in relation to environmental issues. In addition, he organizes workshops and seminars to disseminate knowledge, as well as outreach and knowledge exchange events aimed at building collaborative networks for future research. Dr Mattar has also co-authored a chapter on climate migration in the *Routledge Handbook of Climate Justice* (2019).

Shafiq Ahmad Kamboh is an assistant professor at the School of Communication Studies, University of the Punjab, Lahore, Pakistan. He has completed his PhD in development communication from the Centre for Media, Communication, and Information Research (ZeMKI) at the University of Bremen, Germany. His research focuses on development communication, climate and environmental justice, health communication, science journalism, and the Global South. His research work has been published in reputed research journals including *Journal of Media Ethics, Development Policy Review, Third World Quarterly, International Journal of Communication, Death Studies, Psychology & Health, Analyses of Social Issues and Public Policy, American Journal of Health Education*, among others. He has presented a number of research papers in international conferences organized by global communication research associations like AEJMC, IAMCR, World Forum on Climate Justice, and ISDRS. He has worked on several research projects funded by the local chapters of international development organizations in Pakistan, including UNESCO, WHO, and Micronutrient Initiative.

Siddiqur Rahman has 26 years of teaching and research experience in the field of social/cultural anthropology. His research experience is centered on the broader areas of globalization and Education, Climate Change, Small-Scale Fisheries and Dried Fish, International Migration, and Development – its global challenges and successes. Dr. Rahman holds a permanent teaching position as a professor of Anthropology at Jahangirnagar University (JU), Bangladesh and an adjunct professorship at the University of Manitoba, Canada. Siddiqur holds a PhD in Anthropology from the American University in Washington DC.

Susanne Börner is currently Assistant Professor in Human Geography in the School of Geography, Earth and Environmental Sciences, University of Birmingham. In her previous Marie Curie Global Fellowship in collaboration with the School of Public Health at the University of Sao Paulo, her research explored youth everyday agency to live better with and adapt to resource insecurity and disaster risk in cities of the Global South. She also has several years of experience as a consultant for international development projects on climate change adaptation.

Tanzina Nazia is an Assistant Professor at the Department of Anthropology, Comilla University, Bangladesh. Currently, she is pursuing post graduate studies in Anthropology at the University of Manitoba, Canada. With bachelor and master's degrees in Anthropology from Jahangirnagar University, Bangladesh she has 9 years of experience in academia and research sector. Her research interests include blue economy; small scale fisheries; political ecology; policy and governance; human-environment relations; gender issues; culture and identity politics in Islam.

Thelma I. Vélez (she/her) has been embedded in agroecological and sustainable food systems work for over 15 years. She has worked with grassroots organizations and farmers in various regions of the US, the Caribbean, and India. Her research is at the intersection of social movements, climate justice, and agroecological mobilization. While completing her PhD at Ohio State University, her dissertation centred on the agroecology in Puerto Rico post-Hurricane Maria as a means to create climate resilience and address climate injustices. She is currently the director of research and education at the Organic Farming Research Foundation.

Ukegbu Uwa Kalu is a doctoral researcher at the School of Computing, Engineering and Built Environment, Glasgow Caledonian University, UK. His research interests are in climate change and justice, migration and internal displacement, social protection, and animal rights. He is an occasional lecturer in the Department of Environmental Management, Glasgow Caledonian University, where he lectures on climate justice, society, politics, and sustainability. He holds a bachelor's degree in animal science from the University of Nigeria Nsukka, and a master's degree in climate justice from Glasgow Caledonian University. He received a climate justice scholarship to complete his master's degree and, in 2021, was awarded a Magnus Magnusson Award to undertake the project 'Health Education Service Support Project for Internally Displaced Persons (HESS-IDP)' at New Kuchingoro IDP Camp, Abuja, Nigeria.

Zenaida Luisa Lauda-Rodriguez is a postdoctoral researcher at the Institute of Advanced Studies of the University of São Paulo (Brazil), holds a PhD in Environmental Science at the University of São Paulo, and completed her undergraduate degree in law at National University of Altiplano (Peru). Zenaida is a member of the South American Network for Environmental Migrations (RESAMA) and the Latin American Observatory on Human Mobility, Climate Change, and Disasters (Move-Lam). She is also a member of the executive secretary editorial of the scientific journal Ambiente & Sociedade.

FOREWORD

Some people have access to climate change education and the boardrooms and sites of mass convergence where the climate crisis is discussed. In these spaces you will hear people, especially activists, explain climate justice as a movement that foregrounds the differing social, economic, public health, and other adverse impacts of climate change on underprivileged populations. It is with this understanding that we, as activists, push for a climate agenda that we think everyone should be able to co-shape and respond to.

Other people do not have this access. Let's explore the lives of the farmers from deep down in rural Uganda who, because of the changing weather seasons, have seen their only source of livelihood – mostly agriculture – being dramatically impacted. Or the young people from Mbale in eastern Uganda who, because of heavy rainfall and flooding, have missed out on education because the roads they take to school have been blocked or washed away. Farming and education are two integral parts of society and of people's lives in Uganda, which are being damaged daily by the climate crisis.

Civil society, international non-governmental organizations, and many other stakeholders in Africa, the broader Majority World, and elsewhere continue advancing the principles of climate justice with a focus on gender justice as a key pillar of climate action. However, often the people most impacted are not present for key conversations and are being spoken for by others, if at all. Safe to say that these discussions are happenings far from them – often in the boardrooms of expensive hotels. This raises the question of who is being listened to, what kind of narratives are being put forth, and who is leading efforts to address the climate crisis.

Despite some of these spaces expanding avenues for participation and providing 'inclusion and meaningful engagement', a lot still needs to be done. One vital issue is language. As a young climate advocate, I have had a short but busy time of learning about the multilateral processes on climate change. My first experience was when I participated in the United Nations Environment Assembly (UNEA-3) in 2017. As a first-timer, I was lost in a sea of documents, the need to network with others, and imposing meeting rooms – some I could access and others I wasn't allowed in because of how I was identified. Fast forward five years and I have participated in COP26 in Glasgow, COP27 in Sharm El-Sheikh, and the United Nations General Assembly–convened Stockholm+50 meeting in June 2022. Five years is a long-enough time to grow, and I am now the United Nations Convention to Combat Desertification (UNCCD) Youth Caucus Focal Person. I continue to learn how these spaces work and use my position to advocate for the inclusion of more young people.

My presence in these spaces can be seen as a stepping stone in the right direction. A sign that young people are participating and being included. It has afforded me the opportunity to interact with other young people from different parts of the world, building on my experience in Uganda and East Africa. However, tokenism continues and it can feel as though we are invited to these international climate spaces to check a box, not because what we are saying really matters to leaders. Yet I remain positive; we have also seen young people's voices come out very strongly in recent years. Global actions like the Fridays for Future school strikes have forced leaders to reflect on the climate crisis and its impact on young people. Or the African Youth Initiative on Climate Change, for which I am the East Africa Regional Coordinator, demanding accountability from leaders and that policy rooms, processes, and mechanisms have young people at the centre of these discussions.

Despite the strides being made, I am yet to see adequate representation of the 'poorest and most vulnerable'. In many of the rooms where key climate change conversations take place, and where I now get to be a part of, it is often the already privileged speaking for them. If representation is essential for climate justice, then we are failing on this point. The people who are most impacted should be present in meetings when planning and delivering projects is happening, instead of it happening on their behalf. If these people are the ones being told to 'adapt' and become more 'resilient' to climate shocks, we all should listen, learn, and be corrected by them with regard to what needs to be done.

Some institutions are pushing the 'climate agenda', but are doing so as a key performance indicator and grant-making mechanism, rather than because of shared responsibility or willingness to build true collaboration between the privileged and those hardest hit by recurring climate disasters. The global climate change discourse has a few default faces: a white doomsayer promoting a fairly unorthodox lifestyle and political choices, alongside a fairly moderate policymaker promoting more 'sustainable change'.

While listening to the song 'People of the Land' by Ugandan musician Kenneth Mugabi, I think about how many people need to justify their misuse of natural resources and the Earth — the places they have access to, and which they need for survival, both theirs and their families'. In Luganda, he sings:

Abazzukulu nabagamba ntya?
Abazzukulu nabagamba ntya?
Oluzzi olwali wano lwalaga wa?
Omuzzukulu bwanambuuza ekibira ekyali wano kyalaga wa?
Namugamba ntya?
Munnyonyole ntya sente ezimulwanya zezasaaya ekibira
Munnyonyole ntya kibuga kiramba kyekyaseenza entobazi

What will I tell my grandchildren?
What will I tell my grandchildren?
The well that once was, where did it disappear to?
When my grandchild asks where the forest that once was disappeared to?
What will I say?
How do I explain that my current wealth that they fight to have a share of is from the forest that was cut down
How do I explain to them that in order for us to build a city, a whole wetland was encroached on

The person Kenneth Mugabi sings of what they did in order to survive. I believe this is climate justice with a different face, as climate justice attempts to recognize different social needs, aspirations, and futures.

Behind the big words and even bigger promises of many countries are empty intentions and contradictory actions. All to make a supposed choice between stifling economic growth and preserving the environment and the sustainability of the human race. The choice is not that simple. It is a tough call to make to condemn a sizable amount of your population to hardship or worse, or if you are focused on winning re-election. The belief that these are the only two choices needs to be reconsidered, but this is a different conversation for another day.

The issues I have discussed here create a challenging impasse. We must also focus on the role of resource-rich communities. Many people are only trying to stay afloat. Those in the climate justice movement, myself included, speak of carbon emissions while continuing to take flights to discuss issues of climate change. Impacted farmers are going about their farming while we discuss them. I guess climate justice needs a new face and definition. *Climate Justice in the Majority World* makes an important contribution in this direction by vocalizing silenced knowledges, magnifying important resistance struggles, and engaging in critical discussions on what climate justice means and for whom.

Juliet Grace Luwedde
Global Coordinator of the Youth Constituency, UNCCD Youth Caucus;
East Africa Regional Coordinator, African Youth Initiative on Climate Change (AYICC)

ACKNOWLEDGEMENTS

Climate Justice in the Majority World began in early 2021 during a long lockdown in response to the COVID-19 pandemic. The book has subsequently been the product of many emails, Google Docs, WhatsApp messages, and Zoom calls. What inspired us to embark on this book project was the need to disrupt knowledge and publishing practices that excluded or minimized voices, approaches, and ontologies of climate justice scholars and activists in the Majority World. We hope that it has contributed to fulfilling that mission.

We would like to begin by thanking the over 30 people who have contributed their ideas and insights, through their writing, to this book. At Routledge, we would like to thank Annabelle Harris and Jyotsna Gurung for their help throughout – from the initial book proposal to publication. We would also like to thank them for suggesting the Routledge Advances in Climate Change Research series as the best home for our book. We are incredibly grateful to our endorsers in adding immense credibility to the diverse voices represented in the book.

This book would have been impossible without the invisible labor and unwavering support provided by many people, including our family and friends. A special word of gratitude to Helene and Geo for their support in completing the project, Liana and Liam for happiness and warmth. Thank you, Jon, for always being there. And thank you to Ann and Dennis for your support throughout.

Finally, we want to acknowledge the climate justice scholars and activists from the Majority World who fight tirelessly for a better future for all humans and non-humans. You inspire us.

INTRODUCTION

Climate Justice beyond the Minority World – Towards Decolonial Knowledges

Contributors
Kavya Michael, Michael Mikulewicz, and
Neil J. W. Crawford

The scope and intensity of the impacts of climate change make it one of the defining issues of the twenty-first century. We often hear that humans have brought the planet to the brink of ecological collapse, with scientists sounding the alarm about the Sixth Extinction (Kolbert 2014), melting ice caps (IPCC 2019), water scarcity (Caretta et al. 2022), and other climate-related crises. We hear about global leaders unable to agree on phasing out fossil fuels, despite the grim projections by climate experts, according to which we are running out of time to avoid catastrophic changes to the planet's atmosphere (IPCC 2018). We hear that to save the earth as we know it, everyone must do their share because, after all, 'we are all in this together'. On the face of it, it is difficult to argue with these calls. Climate change will affect, in one way or another, every single place on the planet, with often dire consequences for its human and non-human inhabitants. As of 2022, the world is on track to a warming of 2.8°C by the end of this century compared to pre-industrial levels (UNEP 2022) – far exceeding the 2° C limit agreed on by world leaders at COP21 in Paris in 2015 and approaching double the optimistic objective of 1.5°C. Although these differences may not seem like much, every fraction of a degree signifies millions of lives and livelihoods damaged by or lost to floods, droughts, sea level rise, pests and diseases, and other impacts of climate change. It is hard to argue that humankind cannot continue the current trajectory of overexploiting the earth's finite resources.

Many blame the growing human population for most of our environmental woes including climate change. At the time of writing this introduction in November 2022, the UN, alongside many media outlets, is heralding that this month the world's human population will, for the first time, exceed 8 billion people (UN 2022). But to blame *all* of humanity for our climate predicament would not only be incorrect but also, as we argue, unethical. After all, is the

DOI: 10.4324/9781003214021-1

entirety of humankind equally responsible for causing climate change? Is it true that all global leaders are equally reluctant or unwilling to engage in aggressive climate action? Are we all going to be affected by climate change impacts in the same way? As tempting – and compelling – as the stories of the population bogeyman may be, they also commit a fundamental injustice of portraying humanity as a homogenous global society that has *collectively* caused the climate crisis over the past two centuries or so. We know, however, that this is not the case. Over the last 20 years, a growing body of scholarly work has taken heed from climate activists who for decades have argued that climate change is not just an environmental issue but a social and political issue teeming with various types of inequalities and injustices. Fundamentally, climate change comprises three kinds of injustice. Firstly, it is undeniable that a small *minority* of humankind – aware of it or not – has thrown the earth's systems out of balance. Historically, industrialized countries, predominantly located in Europe and North America, have reaped substantial economic, social, and political benefits from burning fossil fuels – first coal, then oil and gas. Countries in the Majority World have contributed the least to the climate crisis the planet is facing today (World Bank 2013), while colonial extractivist models of development have ensured an inverse relationship between the distribution of climate risks and responsibilities (Sultana 2022). Secondly, compared to more temperate areas, climate impacts are and will continue to be felt most acutely in equatorial, tropical, and subtropical regions where most of the planet's marginalized residents live (IPCC 2022). The Intergovernmental Panel on Climate Change's (IPCC 2022) Sixth Assessment Report (AR6) explicitly demonstrated the differential impacts of climate change in the Majority World, particularly in West, Central, and East Africa, South Asia, Central and South America, Small Island Developing States, and the Arctic. For example, AR6 notes that a global warming level of between 1.5°C and 2°C can lead to increases in the frequency of extreme weather events like droughts, floods, heatwaves, and sea level rise, leading to food insecurity, malnutrition, and micronutrient deficiencies. These consequences would be concentrated in much of Africa, South Asia, Central and South America, and Small Islands. This means extreme and chronic disasters in places where people may be least able to withstand them. Not only that, some countries in the Northern Hemisphere, like Canada, stand to benefit from longer growing seasons in the near to medium term thanks to rising temperatures (Chung 2020) – providing fodder to climate sceptics as opposed to any aggressive emission cuts. Thirdly, most of the mainstream solutions to the climate crisis – renewable energy development, carbon markets, and other technological and market-based mechanisms – originate from, and thus often favour, countries, businesses, and people who have contributed disproportionately to causing climate change in the first place (Mikulewicz and Taylor 2020; Nightingale et al. 2019).

This interlocking set of climate-related injustices shows how important it is, first, to see climate change through the prism of inequality and, second, how urgent it is to find ways to combat this inequality rather than maintaining the status quo or – as sometimes is the case – making it even worse. In essence, this is the main goal of climate justice. Its advocates seek to foreground the various injustices caused by the impacts of climate change and the ways in which we respond to these impacts. At the same time, academics and activists pursuing climate justice are not shy about the term's prefigurative nature – it encompasses within it a vision of a more equitable world where harm is not skewed against those who already bear disproportionate impacts of social and environmental issues. Importantly, it also argues that if we are serious about finding solutions out of the climate conundrum, we need a much more diverse set of knowledges, values, skills, and experiences that would allow humanity to overcome what are quintessentially Western understandings of climate change. These, we argue, favour perceptions of global warming as a techno-environmental issue, a myopic view that elides discussions about the urgently needed political and social transformation towards a more just and sustainable society. Meanwhile, according to climate justice advocates, we must swiftly move beyond the oppressively top-down technological and market-based policies and interventions with a disappointing track record when it comes to reducing emissions and supporting climate adaptation around the world. Despite the lethargic responses to climate change and climate justice concerns by governments at national and international levels, climate justice has become a unifying thread for movements of resistance across the globe (Tokar 2020), which span a diverse spectrum of actors and intersectional identities, including indigenous groups, youth, civil society actors, and gender rights activists, among others. These actors, hailing from different parts of the world, seek to amplify marginalized voices in the climate crisis while also demonstrating examples of transformative and just climate action spaces.

Climate Justice in the Majority World hopes to contribute to this effort through diversifying the body of knowledge on climate change and climate justice. As co-editors, we are all committed to foregrounding academic and activist voices from the Majority World, a commitment that has ultimately resulted in this publication. What we have compiled here are contributions by authors from the Majority World or by those who collaborated closely with colleagues and partners from the Minority World in producing the chapters included in the volume. This final product is a collection of 12 chapters by a total of 32 authors from all over the globe. In this introduction, we provide a brief overview of the state of the art and knowledge gaps in the field of climate justice and explain why we elected to use the term 'Majority World' over various alternatives. We then cede the floor to our contributing authors, whose chapters are summarized below. In the Conclusion, we reflect on the challenges and potential solutions to what we see as a fundamental epistemic injustice in the field of climate studies and climate justice specifically, reflecting our contributors' views on this topic compiled through an online mini survey completed by them.

1 State of the Art and Knowledge Gaps in the Climate Justice Landscape

Climate justice emerged in the 1990s within activist and later academic circles as a concept that challenged the mainstream scientific and technocratic discussions around climate change. The climate justice literature has been highly critical of the focus of adaptation and mitigation policies on the 'naturalness' or 'biophysicality' of the climate crisis. It redefined climate change as an issue of justice and human rights, a question of equity over time and space, one that dangerously intersects class, race, gender, caste, sexuality, and bodily abilities, among others (Sultana 2022; Porter et al. 2021; Schlosberg and Collins 2014; Schlosberg 2012; Mikulewicz et al. 2023). The concept of climate justice originally developed around questions like 'who has benefitted from the colonialism enabled economic development that created the climate crisis?' and 'who bears the cost of climate change?' with emphasis on climate governance at the international level. These conceptions of climate justice were also reflected in the climate policy space until recently, most commonly through approaches including the 'polluter pays model' and 'common but differentiated responsibility'. Ironically, even though the term 'climate justice' appeared in the Paris Agreement in 2015, marking the first time the concept gained official recognition in the climate negotiations space, it was also then that the international climate policy focus shifted from historical responsibilities towards voluntary bottom-up climate solutions at the national level.

Historically, the focus of climate justice literature has largely remained at the global level, focusing on historical responsibilities and differential vulnerabilities between countries, although in recent years there has been an increasing engagement with the term at the local and regional scales (Barrett 2013; Burnham et al. 2013; Fisher 2015). While this shift in focus from international to regional and local levels has given more visibility to stories from the Majority World, we argue that it has also highlighted knowledge gaps within the climate justice literature, which *Climate Justice in the Majority World* attempts to address. The shift in climate justice discourse from the international scale to local and regional levels has further weakened the discussions around historical responsibility with attention shifting to the distributional implications of climate change and climate action. At the local level, the focus largely remains on post-disaster scenarios, including the impact of climate change and climatic vulnerabilities, often missing the bigger picture of the root causes or structural elements. Meanwhile, these structural elements are critical determinants of vulnerabilities as they create the initial conditions on the ground which climate change interacts with (Kashwan and Ribot 2021; Ribot 2022, 2013). Newell et al. (2021) note that a critical gap in local-level analyses of climate (in)justice is the lack of recognition of the intersection of multiple scales: global, regional, and local. For instance, understanding the multiple pathways through which global decision-making processes interact with the national and regional scale is crucial for explaining how power imbalances and the socioeconomic marginalization processes are sustained. This argument is illustrated by the example of climate-smart agriculture in Africa, particularly Ethiopia, Kenya, Rwanda, and Tanzania, where globally driven discussions determine the policy-framing process. At the same time, the primary stakeholders such as smallholder farmers, fisher communities, pastoral and agro-pastoral communities are denied the ability to participate in making decisions that impact their livelihoods and sometimes even reinforce their on-ground vulnerabilities (Newell et al. 2019). The literature has unpacked local realities; however, the focus on proximate causes of vulnerability, rather than on structural factors, can constrain the potential for addressing the root causes of climate change, an issue of crucial importance to climate justice (Michael et al. 2019; Pelling et al. 2015; Tschakert and Machado 2012). The lack of adequate visibility in the climate justice literature of colonial extractivist models of development that generate socioeconomic precarity alongside the climate crisis bears an inherent risk of deepening existing injustices. This can lead to climate policy and action that focus on addressing contemporary vulnerabilities attributed to biophysical climatic changes. However, climate action focusing on the immediate impacts of a disaster event is bound to fail when a system or region is exposed to multiple socioeconomic and ecological stressors. A clear example is the compounding vulnerabilities experienced by communities in the Majority World when climatic disasters occurred alongside the global COVID-19 pandemic, such as Cyclone Amphan in Odisha and West Bengal, India (Boyland and Adelina 2020), or the tidal flooding event in Demak, Indonesia (Seftiani et al. 2021). These cases laid bare the structural nature of socioeconomic marginalizations and precarity, deepening the impacts of both crises. We argue that the current speed and urgency of climate action and energy transitions can deepen pre-existing injustices and vulnerabilities on the ground if the structural elements or root causes are not addressed, and that requires climate justice stories from the Majority World to be told differently.

This brings us to the question of whose knowledge is being valued and whose voices are being heard in the discussions on how to address climate change. The introduction to this book was written during COP27 in Sharm El-Sheikh, where lobbyists from fossil fuel companies outnumbered representatives from at-risk Pacific Island states (Carrington 2022). It is not just the emitters that frequently have a Western or otherwise colonial pedigree. Climate justice itself was pioneered by social movements in the Minority World, and as such, many studies of its use and application remain focused on countries in Europe and North America. From an early focus on racial justice in the United States in the wake of Hurricane Katrina in 2005 to the School Strikes for Climate beginning in Sweden in August 2018, climate justice scholarship has not afforded sufficient attention to the parts of the world most impacted by climate change. There is a widespread recognition and acknowledgement of the unequal distribution of the short- and long-term impacts of climate change which are playing out throughout the Majority World, disproportionately affecting marginalized individuals, communities, and countries that historically have contributed little to the rising global temperatures. However, climate change scholarship has recently been critiqued on playing the 'victimization' narratives in the Majority World (Thomas and Warner 2019; Mikulewicz 2020) while largely disregarding its scholarship and knowledge (including experiential, local, and traditional knowledge). The literature increasingly features case studies from the Majority World; however, this work is often written by scholars in the Minority World. As Sabelo (2022) has argued, the raw data gathered from the Majority World is often developed and utilized for processing into theory in the Minority World, ignoring the lived realities, knowledge, and scholarship that exists in the Majority World. In addition, research and practice in the Majority World has been largely influenced by the development and humanitarian donor landscapes and affected by low institutional capacity, leading to underrepresentation of some locations (Sabelo 2022). The result is that the impacts of climate change on particular regions and spaces *and* the associated solutions generated there remain invisible, compounding the disadvantages experienced by certain areas and groups.

It is widely recognized that climate justice has clear links with environmental justice (Schlosberg and Collins 2014), with some scholars and activists considering the former a branch of the latter, rather than its own movement. A chronological overview of how environmental and climate justice co-evolved over recent decades is beyond the scope of this chapter (see Tokar 2019); it nonetheless should be noted that both traditions are widely considered to have originated in the Minority World, and particularly in the United States. From protests against dumping toxic waste in Black neighbourhoods in Warren County, North Carolina, in 1982, through the First National People of Colour Environmental Leadership Summit in 1991, to the publication of *Greenhouse Gangsters and Climate Justice* by CorpWatch in 1999, and the national and international outcry following the plight of Black New Orleanians during and after Hurricane Katrina, the United States appears to be the locus of climate and environmental justice scholarship and activism. It is therefore no surprise that many climate justice publications emerging from the Majority World still use predominantly Western conceptions of the term. One of the clearest examples of this is the approach that distinguishes between different components of environmental (and climate) justice: procedural, distributive, and recognition. Conceptualized by David Schlosberg (2004), it rests on the work of Western – and more specifically Anglo-Saxon – philosophers, including John Rawls, Iris Young, Nancy Fraser, and David Miller. In fact, this framework is used by a number of chapters in this volume, demonstrating its continued prominence in the field.

However, alternative perspectives on what climate justice is, or should be, started to emerge around the turn of the millennium and were often concentrated around the United Nations Framework Convention on Climate Change (UNFCCC) process. A clear example of this is the 2002 Bali Principles, which make clear reference to the sacredness of Mother Earth, the rights of non-humans, destructive capitalist growth, and other values that are considered to fall outside of the Western 'canon'. A growing number of postcolonial or anticolonial scholars have produced incisive critiques of the global response to climate change, which neglect the historical responsibility of high emitters in the Minority World, the legacy of colonialism in causing climate change and producing vulnerability in the Majority World, and the market-oriented solutions to mitigation and adaptation challenges which are seen as yet another opportunity for economic exploitation of the postcolony (Reo and Parker 2013; Mikulewicz 2020; Nightingale et al. 2019). For example, Wilkens and Datchoua-Tirvaudey (2022) identify three key colonial practices of knowledge production in global climate governance, which in turn create or perpetuate climate injustices. They are the coloniality of knowledge hierarchies (which values certain knowledges over others), the colonization of time in policy-making through the dominant concepts of climate history(ies) and climate future(s) (which homogenizes humanity and conceals social antagonisms in the present), and the coloniality of solutions (referring to the technical fixes to adaptation and mitigation challenges offered by the liberal order). Along the same lines, Kyle Whyte (2020) astutely notes that for Indigenous people in the United States and elsewhere in the world, the climate crisis is but yet another catastrophe caused by (settler) colonialism and rampant capitalist growth. Following these critiques are calls for decolonizing our thinking around, and actions in response to, climate change, which frequently rest on what we often refer to as 'alternative' knowledges and values. These perspectives are highly diverse, yet they more often than not reject the Cartesian rationalism that led to the ontological separation of humans and nature that enabled the former's domination over the latter. Instead, it is argued, we must advocate for more cosmocentric conceptions of climate justice, which in their most distilled form call for the protection of all life, and not just human life (Nunez 2019). Similarly, the short-termism that characterizes capitalist development is often countered with an appreciation for how our choices today will impact future generations of both humans and non-humans, and of the earth in general.

These 'alternative' understandings of climate justice have been relegated by national governments, UN agencies, international financial institutions, international corporations, and other mainstream actors to the margins of the debate on climate change. Approaches rooted in cosmocentrism and global justice are discarded by those in power as irrelevant, unrealistic, or even outrageous, a discursive strategy aimed at asserting the dominant Western liberal order. Nowhere is this more visible than during the annual COP meetings, where Indigenous activists and advocates are actively and repeatedly excluded from key policy discussions and decision-making processes. Often, as in the case of COP26 in Glasgow in 2021, they must resort to disrupting sessions and meetings simply to reassert their presence and voice their opposition to how the world is responding to climate change (Lakhani 2021).

We argue that the global domination of Minority World scholarship and thought leadership on climate justice stems not necessarily from the absence of such scholarship in the Majority World – an argument often raised in more or less informal conversations in academic and policy circles – but rather from the silencing and erasing of alternative epistemologies and ontologies concerned with questions of justice, including in the context of climate change. Much like in the case of the global economic system, the global system of knowledge production – whether it is knowledge on the climate or anything else – is dominated by actors from the Minority World. A recent study on authorship within development research found that only 16% of nearly 25,000 articles published in the top 20 development journals between 1990 and 2019 were authored by Majority World academics, and that these articles had on average fewer citations than those by their Minority World counterparts (Liverpool 2021). Similarly, the study found that only 9% of presenters at international development conferences were based at institutions in the Majority World, depriving academics based there of crucial networking opportunities offered by international gatherings, with those able to attend still having to navigate challenging visa and border regimes. Another recent study found that 89.1% of authors of the most-cited climate science papers from the past five years are affiliated with institutions in North America, Europe, or Oceania, compared to 0.7% who come from Africa (Tandon 2021). Certainly, one of the reasons for this disparity is that academic institutions in the West attract scholars from all over the world due to their prestige, benefits, and opportunities for professional development (this was the case for a number of contributors to this book). However, this is only part of the explanation. Majority World scholars face a number of challenges related to language barriers, access to funding, fewer professional development opportunities, visa restrictions, and outright discrimination (Martin and Dandekar 2022). We come back to these issues in the Conclusion where we discuss these and other challenges in greater detail.

2   Why 'Majority World'?

We support the view that how we describe or talk about different places, spaces, and people matters. This is one of the main reasons we have chosen to utilize the terms 'Majority World' and 'Minority World' in this book and have suggested it as an option to the contributors. We support the use of these terms over the more commonly used 'Global South', 'developing countries', or 'Third World' – where most of the world's people and land can be found – and the 'Global North', 'developed countries' or (less commonly) 'First World' (Dahlburg et al. 1999, 218). While seemingly neutral, these more traditional terms come with significant baggage and have been criticized for perpetuating stereotypes and agency-stripping representations of people who reside in those places. The Global North–Global South distinction is geographically deterministic and misleading – consider the locations of Mongolia and New Zealand on a map compared with how each would be categorized as 'Southern' and 'Northern'. Categorizing countries as 'developed', 'developing', and even 'least developed' implies the existence of a linear process of improvement that concludes with a universally desired state of economic prosperity, which in itself forecloses different visions for what human progress could entail (Shallwani 2015). More importantly, however, these naming conventions depoliticize the way in which we describe global relations. They erase the fundamental inequalities between the (former) colonizers/globalizers and the (formerly) colonized and, in so doing, suggest that humanity's both past and present are devoid of exploitation, violence, and oppression.

Using 'Majority World' and 'Minority World' is thought to address some of these shortcomings. While there is uncertainty about the terms' origins, they are commonly attributed to Shahidul Alam, a Bangladeshi photojournalist and social activist (Punch 2003, 291). Explaining the need for the new terms, Alam notes: 'I didn't want to be third of anything … Certainly we have not chosen you to be the "First World" and us to be the "Third"' (Shafaieh 2022). Focusing on the huge difference in the number of people living in the two 'worlds', as well as on the wealth disparity between them, the Majority vs. Minority distinction also challenges the Western rhetoric of democracy. It exposes the hypocrisy of the fact that a very narrow group of countries that make up less than 20% of the world's population in reality make decisions that affect the entire planet (Alam 2008), not least in the context of climate change. Unlike Global South and Global North, utilizing Majority and Minority World 'invites reflection on the unequal power relations between them' (Punch 2003, 278). As noted by others, 'we need to shift the balance of our worldviews that frequently privilege "western" and "northern" populations and issues', and the terms Majority and Minority World acknowledge where 'the "majority" of populations, poverty, land mass, and lifestyles' are found (Panelli et al. 2007, 13). Finally, Alam (2008, 89) argues that the term Majority World 'also defines the community in terms of what it has, rather than what it lacks'. As such, we should not be talking about countries that are 'third in line' or that 'lack' development, but as sites of formidable diversity and wealth of knowledge that happen not to be considered as equally valid to others.

The use of the Minority vs. Majority language within climate change and environmental debates goes as far back as at least the early 1990s. Writing in the wake of his experiences at the 'abyss of corruption' he found at the United Nations Conference on Environment and Development ('Rio Summit') in 1992, Danny Kennedy (1995) utilizes both terms in his critique of the Minority World's belief that it 'could build Noah's Ark to save us from the deluge of the environment and development crisis facing humanity', 'despite an extensive literature to suggest that the diverse and traditional systems of knowledge in the so-called south or developing world – the Majority – are ecologically superior to modern, developed knowledge systems'. Meanwhile, 'solutions' require an increased investment in the Majority World from the Minority, new technology from the Minority World being shared with the Majority, Minority-educated managers and experts being brought into the Majority World, and a greater push for economic recovery in the Majority World (Kennedy 1995; Hildyard 1993).

As any semantic choice, the use of 'Majority World' and 'Minority World' does not come without issues. The terms have been argued to face the same problems as 'Global South' and 'Global North' by 'unduly homogeniz[ing] both world regions' (Punch 2003, 278) and failing to address that the 'South is in the North and the North is in the South, and privilege and poverty [are] no longer neatly geographically divided' (Pieterse 2000, 131). A similar critique has emerged elsewhere where authors have used Minority and Majority World but have '[ignored] the important phenomenon of Third Worlds in the First and First Worlds in the Third' (Kapoor 2006, 150). Yet, the binary distinction still serves a purpose, and compared to the alternatives, Minority and Majority World is more democratic and acknowledges that they are '*worlds* because they make up complete life-worlds' (Pieterse 2000, 131). We consider the terms to be more democratic in terms of both what they capture and where they originated and – following Alam (2008, 89) – hope that they bring us closer to a time where 'the majority world will reaffirm its place in a world where the earth will again belong to the people who walk on it'.

3 Overview of Chapters

The book spans 12 chapters which examine climate justice issues from across the Majority World. Ruchira Talukdar and Priya Pillai study grassroots resistance to coal mining in the Mahan forests in Central India, where the state government has exacerbated injustices by driving land acquisition for private mining, threatening displacement of communities and attacking community rights over forests. The chapter argues that climate justice concepts that have emerged from within the Minority World should be examined against instances of environmental resistance occurring in the Majority World. In so doing, the chapter shows how 'Southern' climate justice can differ even from marginalized struggles within the Minority World, and that for Southern subaltern communities, climate justice can be conceptualized around what can be described as ecological human rights.

In their chapter, José Maria do Rosário Chilaúle Langa et al. note that Mozambique is one of the countries that is most vulnerable to cyclones and tropical storms with multiple instances during 2019–2021 alone. They argue that there is a serious absence of or disregard for a number of principles of climate justice in the conception, planning, and implementation of national climate policies. What is needed, they argue, is greater participation and involvement of vulnerable Mozambicans in the designing of climate policies and solutions, as well as a rethinking of what climate justice means in the Mozambican context. The chapter depicts the disconnect of climate change governance across scales and how the large political economy of climate change decision-making structures at the national level ultimately exacerbates vulnerabilities on the ground. It further demonstrates a grave violation of procedural justice concerns by invisibilizing the needs and knowledge of the communities most affected by climate change and imposing discriminatory top-down solutions.

Thelma I. Vélez investigates the impact of the 2017 Hurricane Maria on Puerto Rico and the way in which grassroots organizations centred climate justice and agroecology in their recovery efforts. The chapter sheds light on the lived experiences of Puerto Ricans as well as their everyday acts of resistance to oppression and their visions for a more just future. Vélez depicts how climatic hazards intersect with the structural conditions leading to compounding of risks and differential experiences of vulnerabilities. The case of post-Hurricane Maria Puerto Rico shows us the way in which scaling up agroecology has the potential to promote climate-resilient and more sustainable communities while forming part of wider decolonization efforts.

Neil J. W. Crawford et al.'s chapter looks at the consequences of Cyclone Amphan in southwestern Bangladesh. The chapter explores how the impacts of the disaster intersected with other factors (including age, gender, and economic status), compounding existing vulnerabilities, and outlines the different support and aid mechanisms available in the short and longer term following the destruction that occurred in May 2021. The authors argue that it is essential to recognize how people find ways to support each other, including when formal aid mechanisms and structures fail.

Jessica Omukuti and Aidan O'Sullivan draw on zemiology, an emerging paradigm in critical criminology, as a tool for analysing and critiquing legalized social harms as part of state–corporate collusion in the pursuit of neoliberal outcomes. They use this approach in the study of climate finance, and the Green Climate Fund (GCF) specifically. They argue that there are paradoxical harms embedded within GCF, contributing to a climate finance system that fails to address the structural causes of climate change, sustains apolitical climate solutions, and provides an ideological cover for exacerbated climate change that hits the Majority World the hardest.

Lira Luz Benites Lazaro et al. map emerging climate justice debates in Latin America and explore the link between climate issues, demands for social justice, and people's lives and livelihoods. The chapter details the myriad of interconnected challenges and injustices faced across Latin America – from the destruction of forests, floods, and droughts to resource scarcity, desertification, and polluted cities. Amid these wide-ranging issues, young activists across Latin America are using climate justice as a means for fighting environmental destruction. The authors argue that climate and environmental resistance in the region is interlinked with existing conflicts for social justice, as people face environmental threats in their daily lives.

Caio Penko Teixeira discusses the digital advocacy work of social movements focused on claims for socioecological justice by focusing on the work by the Movement of People Affected by Dams (MAB) in Brazil. By drawing on content posted on social media, the chapter shows how grassroots activism, particularly digital activism, can help hold institutions accountable for disasters – in this case those occurring in mines. The chapter visualizes new possibilities for organizing, resistance, and grieving in times of crisis. It centres community-led initiatives and the way in which they can effectively communicate and engage in collective action under the umbrella of climate change activism.

David Singh looks at three wind power projects in western borderland Gujarat in India to understand the extractive, dispossessive, and destructive patterns underlying climate-friendly projects. In India, green energy infrastructures are specifically targeting so-called deserted, empty, and waste lands where residents must face challenges related not only to climate-related uncertainties but also to climate change mitigation efforts. The chapter shows the way that people are pushing back through a range of strategies, from insubordination acts to open resistance.

Siddiqur Rahman and A. K. M. Mamunur Rashid look at climate finance and climate policy and demonstrate how plainland Indigenous people in northern Bangladesh are ignored or even negatively impacted by poor adaptation practices. The authors note that local needs are rarely addressed and that the affected Indigenous people receive insufficient adaptation funds. Rather, they argue, the national government focuses most of its attention on coastal and delta areas, demonstrating what the authors term an 'intracountry climate injustice'.

Shafiq Ahmad Kamboh et al. consider the role of advocacy journalism and its ability to publicize injustices, create dialogue as well as apply pressure for environmentally conscious policies. Through an analysis of 8000 editorials about climate and environmental ethics violations in Pakistani newspapers, they demonstrate how some newspapers give inappropriate coverage of climate change and environmental degradation issues, how editorial priorities can provide misleading takes for readers, and ultimately that a variety of media outlets in Pakistan do not adhere to key parts of UNESCO's Declaration of Ethical Principles in Relation to Climate Change.

Grace D. O'Donovan's chapter focuses on marginalized populations in Cape Town, South Africa, in which social, political, ecological, and economic issues are intertwined with legacies and spatialities of apartheid and settler politics. The author illustrates how the vulnerabilities that are endemic to the political landscape of South Africa are normalized in everyday life, shaping the impacts of climate change in Cape Town. The chapter utilizes political ecology and invasive ecology as frameworks for understanding these issues and the increasing impacts of disasters in the future city.

Africa Bauzà Garcia-Arcicollar considers the assumed finite future of Small Island Developing States (SIDS) through a focused study of the Maldives. This chapter foregrounds resistance against the foreclosure of the island, drawing on the islanders' alternate visions of adaptation and resilience. Exploring visions of the future not rooted in assumed relocation, through drawings and photography by those living in the Maldives, adaptation and resilience become tied to islanders' rights to remain at home.

4   Advancing Climate Justice Scholarship

We identify two key contributions of the book to advance the scholarship on climate justice. The first is foregrounding intersectionality in defining differential vulnerabilities. A number of case studies in this book bring forth the intersection of the social, economic, and ecological factors in determining vulnerabilities on the ground. They depict an understanding of how climate justice issues arise as an outcome of colonial extractivist models, neoliberal development policies, institutional structures, social norms, as well as identities expressive of social relations like class, race, and gender, among others. Moreover, the case studies in this book advance a nuanced understanding of justice issues at multiple levels. We argue that while the vulnerabilities of communities should be highlighted at the local level, it is critical to understand justice issues across scales and how they conglomerate in certain geographical or socioeconomic spaces.

Second, this book stemmed from our collective desire to contribute to the fight for epistemic justice in climate change and climate justice literature. We see foregrounding the excellent work of the authors in this book as a necessary part of this fight and hope that their contributions demonstrate how urgent and necessary it is to open up climate justice scholarship to the perspectives and knowledges from the Majority World. Early-career researchers are also well represented here. Despite offering valuable contributions to climate change and justice scholarship, they can often lack the authorship experience and the professional connections needed to gain access to, and be published by, 'top-tier' journals and press. For several contributors, their chapter represents one of their first academic publications, written during or shortly after their doctoral work. The book is also edited by three non-tenured, early-career researchers.

At the same time, and as the chapters in this book demonstrate, a frequent end result of the complex climate change knowledge production landscape is climate knowledge in a hybridized form. Contributors to this book draw from both Western and non-Western traditions in their work, are based at institutions in both the Majority and the Minority World, and flout the artificial compartmentalization as 'Northern' or 'Southern'. Their work defies what we ultimately see as an arbitrary distinction between alternative and mainstream climate knowledge which, in a typical orientalizing fashion, has served to erase the former in favour of the latter. The cover of *Climate Justice in the Majority World* features an image by photojournalist Katumba Badru Sultan, which is indicative of the book's determination to foreground epistemic justice in the fight for climate justice. The photograph depicts the efforts of students in the city of Mbale to dry their books and papers by laying them out on the roof of their school following heavy rainfall, mudslides, and flash floods in eastern Uganda (Katumba

2022). Through this book there are calls for climate justice scholarship to identify, recognize, and safeguard the diverse knowledge frames that exist in the Majority World in the struggle for sustainable and equitable futures. By presenting alternate, radical visions of local communities, including novel and innovative strategies of mobilization, resistance, and grassroots climate activism, chapters in this book contribute to the much-needed and slowly-but-steadily-growing plurality of climate justice scholarship.

References

Adams, S., Baarsch, F., Bondeau, A., Coumou, D., Donner, R., Frieler, K., Hare, B., Menon, A., Perette, M., Piontek, F., Rehfeld, K., Robinson, A., Rocha, M., Rogelj, J., Runge, J., Schaeffer, M., Schewe, J., Schleussner, C-F., Schwan, S., Serdeczny, O., Svirejeva-Hopkins, A., Vieweg, M., & Warszawski, L. 2013. Turn Down the Heat: Climate Extremes, Regional Impacts, and the Case for Resilience. A report for the World Bank by the Potsdam Institute for Climate Impact Research and Climate Analytics. World Bank, Washington, D.C. http://documents.worldbank.org/cura ted/en/975911468163736818/Turn-down-theheat-climate-extremes-regional-impa cts-and-the-case-for-resilience-full-report

Alam, Shahidul. 2008. 'Majority World: Challenging the West's Rhetoric of Democracy'. *Amerasia Journal* 34 (1): 88–98. https://doi.org/10.17953/amer.34.1.13176027k4q614v5

Barrett, S. 2013. 'Local Level Climate Justice? Adaptation Finance and Vulnerability Reduction'. *Global Environmental Change* 23 (6): 1819–1829. https://doi.org/10.1016/J. GLOENVCHA.2013.07.015

Boyland, Michael, and Charrlotte Adelina. 2020. 'COVID-19, Cyclone Amphan and Building Back Better'. www.sei.org/perspectives/covid-19-cyclone-amphan-and-building-back-better.

Burnham, M., C. Radel, Z. Ma, and A. Laudati. 2013. 'Extending a Geographic Lens towards Climate Justice, Part 2: Climate Action'. *Geography Compass* 7 (3): 228–238. https://doi.org/10.1111/GEC3.12033

Caretta, M. A., A. Mukherji, M. Arfanuzzaman, R. A. Betts, A. Gelfan, Y. Hirabayashi, T. K. Lissner, et al. 2022. 'Water'. In *Climate Change 2022: Impacts, Adaptation and Vulnerability. Contribution of Working Group II to the Sixth Assessment Report of the Intergovernmental Panel on Climate Change* [ed. H.-O. Pörtner, D. C. Roberts, M. Tignor, E. S. Poloczanska, K. Mintenbeck, A. Alegría, M. Craig, S. Langsdorf, S. Löschke, V. Möller, A. Okem, and B. Rama], 1st ed., 551–712. Cambridge: Cambridge University Press. https://doi.org/doi:10.1017/9781009325844.006

Carrington, Damien (2022). 'The 1.5C Climate Goal Died at Cop27 – But Hope Must Not'. *The Guardian*, 20 November. www.theguardian.com/environment/2022/nov/20/cop27-summit-climate-crisis-global-heating-fossil-fuel-industry

Chung, Emily Chung (2020). 'Canada Could Be a Huge Climate Change Winner When It Comes to Farmland'. CBC News, 12 February. www.cbc.ca/news/science/climate-change-farming-1.5461275.

Dahlburg, G., P. Moss, and A. Pence (1999). *Beyond Quality in Early Childhood Education and Care: Postmodern Perspectives*. London: Falmer Press.

Fisher, S. (2015). 'The Emerging Geographies of Climate Justice'. *Geographical Journal* 181 (1): 73–82. https://doi.org/10.1111/GEOJ.12078.

Hildyard, N. (1993). 'Foxes in Charge of Chickens'. In *Global Ecology: A New Arena of Political Conflict* [ed. W. Sachs], 22–35. London: Zed Books

IPCC (2018). *Global Warming of 1.5 °C*. IPCC SR1.5. Geneva: IPCC. www.ipcc.ch/report/sr15/.

IPCC (2019). *IPCC Special Report on the Ocean and Cryosphere in a Changing Climate* [ed. H.-O. Pörtner, D. C. Roberts, V. Masson-Delmotte, P. Zhai, M. Tignor, E. Poloczanska, K. Mintenbeck, A. Alegría, M. Nicolai, A. Okem, J. Petzold, B. Rama, and N. M. Weyer]. Cambridge: Cambridge University Press. https://doi.org/10.1017/9781009157964.

IPCC (2022). *Climate Change 2022: Impacts, Adaptation and Vulnerability. Contribution of Working Group II to the Sixth Assessment Report of the Intergovernmental Panel on Climate Change* [ed. H.-O. Pörtner, D. C. Roberts, M. Tignor, E. S. Poloczanska, K. Mintenbeck, A. Alegría, M. Craig, S. Langsdorf, S. Löschke, V. Möller, A. Okem, and B. Rama]. Cambridge: Cambridge University Press. https://doi.org/10.1017/9781009325844.

Kapoor, I. (2006). 'Reviewed Work(s): Environmental Movements in Majority and Minority Worlds, A Global Perspective by Timothy Doyle'. *International Review of Modern Sociology* 32 (1): 149–151.

Kashwan, Prakash, and Jesse Ribot. 2021. 'Violent Silence: The Erasure of History and Justice in Global Climate Policy'. *Current History* 120 (829): 326–331. https://doi.org/10.1525/CURH.2021.120.829.326.

Katumba, Badru Sultan (2022). 'Students Dry Their Books and Papers after the Mudslide Entered Classroom at Mbale Progressive Secondary School in Namakwekwe on August 1, 2022'. Instagram, 2 August 2. www.instagram.com/p/CgvoPVyouNu/?igshid=MDJmNzVkMjY.

Kennedy, D. (1995). 'Do We Really Want to Get Where We Are Going? An Activist's Appraisal of the Trade and Environment Debate since Rio'. *Australian Geographer* 26 (1): 11–16.

Kolbert, Elizabeth (2014). *The Sixth Extinction: An Unnatural History*. London: Bloomsbury.

Lakhani, Nina (2021). '"A Continuation of Colonialism": Indigenous Activists Say Their Voices Are Missing at Cop26'. *The Guardian*, 3 November. www.theguardian.com/environment/2021/nov/02/cop26-indigenous-activists-climate-crisis.

Liverpool, Layal (2021). 'Researchers from Global South Under-represented in Development Research'. *Nature*, September, d41586-021-02549-9. https://doi.org/10.1038/d41586-021-02549-9.

Martin, Staci B., and Deepra Dandekar, eds. (2022). *Global South Scholars in the Western Academy: Harnessing Unique Experiences, Knowledges, and Positionality in the Third Space*. New York: Routledge.

Michael, Kavya, Tanvi Deshpande, and Gina Ziervogel (2019). 'Examining Vulnerability in a Dynamic Urban Setting: The Case of Bangalore's Interstate Migrant Waste Pickers'. *Climate and Development* 11 (8): 667–678. https://doi.org/10.1080/17565529.2018.1531745.

Mikulewicz, Michael (2020). 'The Discursive Politics of Adaptation to Climate Change'. *Annals of the American Association of Geographers* 110 (6): 1807–1830. https://doi.org/10.1080/24694452.2020.1736981.

Mikulewicz, Michael, and Marcus Taylor (2020). 'Getting the Resilience Right: Climate Change and Development Policy in the "African Age"'. *New Political Economy* 25 (4): 626–641. https://doi.org/10.1080/13563467.2019.1625317 .

Mikulewicz, Michael, Martina Angela Caretta, Farhana Sultana, and Neil J. W. Crawford (2023). 'Intersectionality & Climate Justice: A Call for Synergy in Climate Change Scholarship'. *Environmental Politics*, DOI: 10.1080/09644016.2023.2172869.

Newell, Peter, Shilpi Srivastava, Lars Otto Naess, Gerardo A. Torres Contreras, and Roz Price (2021). 'Toward Transformative Climate Justice: An Emerging Research Agenda'. *Wiley Interdisciplinary Reviews: Climate Change* 12 (6): e733. https://doi.org/10.1002/WCC.733.

Newell, Peter, Olivia Taylor, Lars Otto Naess, John Thompson, Hussein Mahmoud, Patrick Ndaki, Raphael Rurangwa, and Amdissa Teshome. 2019. 'Climate Smart Agriculture? Governing the Sustainable Development Goals in Sub-Saharan Africa'. *Frontiers in Sustainable Food Systems*, 3 August. https://doi.org/10.3389/FSUFS.2019.00055/FULL.

Nightingale, Andrea Joslyn, Siri Eriksen, Marcus Taylor, Timothy Forsyth, Mark Pelling, Andrew Newsham, Emily Boyd, et al. (2019). 'Beyond Technical Fixes: Climate Solutions and the Great Derangement'. *Climate and Development*, July. https://doi.org/10.1080/17565529.2019.1624495.

Nuñez, Alan Jarandilla (2019). 'Mother Earth and Climate Justice: Indigenous Peoples' Perspectives of an Alternative Development Paradigm'. In *Routledge Handbook of Climate Justice* [ed. Tahseen Jafry, Karin Helwig, and Michael Mikulewicz], 420–430. Abingdon: Routledge.

Panelli, R., S. Punch, and E. Robson (2007). 'From Difference to Dialogue: Conceptualizing Global Perspectives on Rural Childhood'. In *Global Perspectives on Rural Childhood and Youth: Young Rural Lives* [ed. R. Panelli, S. Punch, and E. Robson], 1–14. New York: Routledge.

Pelling, Mark, Karen O'Brien, and David Matyas (2015). 'Adaptation and Transformation'. *Climatic Change* 133 (1): 113–127. https://doi.org/10.1007/s10584-014-1303-0.

Pieterse, J. N. (2000). 'Globalization North and South: Representations of Uneven Development and the Interaction of Modernities'. *Theory, Culture & Society* 17 (1): 129–137.

Porter, L., L. Rickards, B. Verlie, K. Bosomworth, S. Moloney, B. Lay, B. Latham, I. Anguelovski, and D. Pellow (2020). 'Climate Justice in a Climate Changed World'. *Planning Theory & Practice* 21 (2): 293–321. www.tandfonline.com/doi/full/10.1080/14649357.2020.1748959.

Punch, S. (2003). 'Childhoods in the Majority World: Miniature Adults or Tribal Children?' *Sociology* 37 (2): 277–295.

Reo, Nicholas James, and Angela K Parker (2013). 'Re-thinking Colonialism to Prepare for the Impacts of Rapid Environmental Change'. *Climatic Change* 120 (3): 671–682.

Ribot, Jesse (2013). 'Vulnerability Does Not Just Fall from the Sky: Toward Multi-Scale Pro-Poor Climate Policy'. In *Handbook on Climate Change and Human Security* [ed. Michael R. Redclift and Marco Grasso], 164–172, Edward Elgar Publishing. https://doi.org/10.4337/9780857939111.00016.

Ribot, Jesse (2022). 'Violent Silence: Framing Out Social Causes of Climate-Related Crises'. *Journal of Peasant Studies* 49 (4): 683–712. https://doi.org/10.1080/03066150.2022.2069016.

Sabello, J. Ndlovu-Gatsheni (2022). 'Epistemic Injustice'. In *Knowledge for the Anthropocene: A Multidisciplinary Approach* [ed. Francisco J. Carrillo and Günter Koch]. 167–177. Northampton: Edward Elgar.

Schlosberg, David (2004). 'Reconceiving Environmental Justice: Global Movements And Political Theories'. *Environmental Politics* 13 (3): 517–540. https://doi.org/10.1080/0964401042000229025.

Schlosberg, David (2012). 'Climate Justice and Capabilities: A Framework for Adaptation Policy'. *Ethics & International Affairs* 26 (4): 445–461.

Schlosberg, David, and Lisette B. Collins (2014). 'From Environmental to Climate Justice: Climate Change and the Discourse of Environmental Justice'. *Wiley Interdisciplinary Reviews: Climate Change* 5 (3): 359–374. https://doi.org/10.1002/wcc.275.

Seftiani, S., I. A.P. Putri, I. Hidayati, L. K. Katherina, V. Ningrum, and D. Vibriyanti (2021). 'Climate Change and COVID-19: The Double Whammy for Coastal Community in Demak, Indonesia'. *IOP Conference Series: Earth and Environmental Science* 824 (1): 012101. https://doi.org/10.1088/1755-1315/824/1/012101.

Shafaieh, Charles (2022). 'Shahidul Alam on the Majority World'. *Harvard Design Magazine.* www.harvarddesignmagazine.org/issues/50/shahidul-alam-on-the-majority-world.

Shallwani, Sadaf (2015). 'Why I Use the Term "Majority World" Instead of "Developing Countries" or "Third World"'. 4 August. https://sadafshallwani.net/2015/08/04/majority-world/.

Sultana, Farhana (2022). 'Critical Climate Justice'. *Geographical Journal* 188 (1): 118–124. https://doi.org/10.1111/GEOJ.12417.

Tandon, Ayesha (2021). 'Analysis: The Lack of Diversity in Climate-Science Research'. 5 October. www.carbonbrief.org/analysis-the-lack-of-diversity-in-climate-science-research.

Thomas, Kimberley Anh, and Benjamin P. Warner (2019). 'Weaponizing Vulnerability to Climate Change'. *Global Environmental Change* 57 (July): 101928. https://doi.org/10.1016/j.gloenvcha.2019.101928 .

Tokar, B., and T. Gilbertson, eds. (2020). *Climate Justice and Community Renewal: Resistance and Grassroots Solutions.* Abingdon: Routledge.

Tokar, Brian (2019). 'On the Evolution and the Continuing Development of the Climate Justice Movement'. In *Routledge Handbook of Climate Justice* [ed. Tahseen Jafry, Karin Helwig, and Michael Mikulewicz], 13–25. Abingdon: Routledge.

Tschakert, Petra, and Mario Machado (2012). 'Gender Justice and Rights in Climate Change Adaptation: Opportunities and Pitfalls'. *Ethics and Social Welfare* 6 (3): 275–289. https://doi.org/10.1080/17496535.2012.704929.

UN (2022). 'World Population to Reach 8 Billion on 15 November 2022'. www.un.org/en/desa/world-population-reach-8-billion-15-november-2022

UNEP (2022). *Emissions Gap Report 2022: The Closing Window – Climate Crisis Calls for Rapid Transformation of Societies.* Nairobi: UNEP. www.unep.org/emissions-gap-report-2022

Whyte, Kyle Powys (2020). 'Too Late for Indigenous Climate Justice: Ecological and Relational Tipping Points'. *Wiley Interdisciplinary Reviews: Climate Change* 11 (1). https://doi.org/10.1002/wcc.603

Wilkens, J., and A. R. Datchoua-Tirvaudey (2022). 'Researching Climate Justice: A Decolonial Approach to Global Climate Governance'. *International Affairs* 98 (1): 125–143.

1

SOUTHERN CLIMATE JUSTICE ACTIVISM IN THE CONTEXT OF AN ENERGY TRANSITION

Forest Rights over Coal in Mahan, Central India

Ruchira Talukdar and Priya Pillai

1 Introduction

The climate crisis requires a drastic reduction of the burning of coal, oil, and gas, collectively called fossil fuels, which make the biggest contribution towards atmospheric carbon dioxide concentration.[1] In response, climate activism has turned its focus towards stopping their extraction. Although community struggles against fossil fuel projects have been an ongoing feature in both the Minority and Majority World or the Global North and the South, such struggles might challenge their extraction not for the damage they are causing to global climate but for their harmful local effects on lands, livelihoods, and cultures (Roy and Schaffartzik 2021). However, with a climate focus now being brought to community-scale anti-fossil fuel fights, especially though not exclusively through interactions with climate activist networks, they have come to signify mobilizations for climate justice (Talukdar 2021). A new global politics of solidarity by climate activist networks towards local community resistances to fossil fuels has emerged as a result. Popularized as Blockadia[2] (see Klein 2014, 295; Martinez-Alier et al. 2016), this new approach aims to break the power of the fossil fuel industry and bring in the energy transition.

Literature on today's fossil fuel resistances describes these solidarities as often constituted of multiple grievances from fossil fuel extraction unified under climate justice as an overarching narrative frame (Goodman 2009). However, a reading of this literature also indicates that the notion of climate justice as an overarching frame for anti-fossil fuel resistances has emerged from movements in the Global North or the Minority World. The climate justice frame encompasses new relations forged between various environmentalisms[3] operating in northern geographies – such as eco-centrism, environmental justice, and Indigenous justice – to collectively resist

DOI: 10.4324/9781003214021-2

fossil fuel extraction. It does not necessarily reflect the sociopolitical context of southern environmentalisms (Talukdar 2021).

In India, for example, climate justice has not become a mobilizing factor for environmental movements of the rural poor, which continue to fight against loss of land and destruction of livelihoods from industrial projects. According to the Environment Justice Atlas (ejatlas.org/country/india), a global database of environmental conflicts, India has the world's highest recorded number of environmental conflicts (at least 271), with coal extraction being the second most common cause after access to water. Rather than climate change, it is the local effects of coal mining and burning that generate discontent around land and livelihoods amongst peasant and Indigenous groups. Reflexively, this phenomenon signifies a highly uneven public sphere in a southern context such as India that prevents the global climate justice narrative from being adopted by movements arising from communities whose livelihoods are local and nature-dependent, whom Dasmann (1988) describes as the 'ecosystem people'.

Williams and Mawdsley (2006) argue that it is not enough to simply see how ideas of justice from the South may tactically align within a 'global' environmental justice movement framed by northern contexts; rather, distinct frames of environmental justice emerging from the South should be treated as such in environmental justice research. Global climate justice activism to stop fossil fuels and bring in the energy transition needs to understand and acknowledge critical differences in the politics and narratives of fossil fuel resistance emerging from the Global South. This chapter draws on the case study of a grassroots resistance to coal mining in the Mahan forests in Central India to delineate critical differences in the politics and narratives of movements in a southern context. It conceptualizes a southern climate justice framework in the context of an energy transition.

Through literature, Section 2 discusses how climate justice in the southern context conceptually differs from that in the North. It proposes a framework for analysing this chapter's case study based on Williams and Mawdsley's (2006) three distinctions between marginalized peoples' movements in the Global North and South. Section 3 gives an overview of the social, political and economic contexts that generate today's environmental conflicts in India. Section 4 delineates the specific context of coal extraction in forested Central India and analyses the Mahan resistance that arose from it. Section 5 discusses the elements of climate justice that emerge from Mahan and their differences from a comparable grassroots anti-coal resistance in the Minority World. Section 6 summarizes the directions that Mahan and India set for theory and praxis in southern climate justice. The chapter draws on the authors' insights from campaign participation and ethnographic research at Mahan. Names of movement leaders from Mahan have been changed to protect their identity.

2   Climate Justice Movements in the Majority World

Marginalized communities' resistances to fossil fuels pose contextually located questions about ecological and social justice at the intersection of climate impacts and transition from coal, oil, and gas (e.g. see Brown and Spiegel 2019). Through their resistance, Indigenous peoples and other marginalized groups in extraction zones who are generally 'framed out' of formal policy discussions are reframing issues of rights, justice, and climate vulnerability on their own terms (Plows 2007).

Grassroots resistance to fossil fuels constitutes a growing field within intersecting literatures in climate justice (see Klein 2014; Whyte 2017) and political ecology (see Brown and Spiegel 2019; Hardt and Negri 2000; Harvey 1996; Martinez-Alier et al. 2016). Political ecology holds threatened livelihoods, linked to altered social and environmental conditions, as central (Bryant and Bailey 1997; Bryant 1998; Guha 1989; Jewitt 1995; Peluso 1992). It assumes that the human impact of environmental change is unevenly distributed, with poor and marginalized groups being disproportionately impacted (Watts 1983).

Climate justice literatures acknowledge the interconnections between colonialism, capitalism, and environmental degradation, and their underlying role in driving climate change (Birch 2016; Bird Rose 2013). For communities that have historically survived environmental injustice under this global system, climate change implies another set of anthropogenic ecological disruptions they had no role in creating and, yet, for which they bear a disproportionate burden (Gottlieb 2005; Quinn-Thibodeau and Wu 2017). Climate change has compounded the significance of the historic resistance of these communities and made their activism global in relevance (Dryzek 2013; Purdy 2016).

Due to historical colonial dispossession and cultural disruption, Indigenous climate justice perspectives enfold a further distinction by being linked to notions of sovereignty and land rights (Esposito and Neale 2016). Indigenous community members working in alliance with environmentalists are not only fighting against the challenge of coal mining and climate change; their historical and continuing struggles are with the structural challenges of colonial and capitalist domination linked to industrialization (Latulippe and Klenk 2020; Whyte 2017)

In the context of the Global South, governments might articulate climate justice as the carbon space to grow vis-à-vis the Global North that is historically responsible for emissions (Goodman 2016), taking a common but differentiated responsibility (CBDR) approach towards climate action (Dubash 2012). Grassroots climate justice networks in the South, however, focus on the disproportionate burden of climate change on the southern poor. They also call for intracountry and intragenerational equity for the poor in a highly unequal southern society[4] (Dubash 2013; Michael and Vakulabharanam 2016; Thaker and Leiserowitz 2014), based on an understanding of what Nixon (2011) calls the slow violence of the poor – that the economic, social, and cultural displacements of vulnerable communities would continue to worsen under climate change.

Williams and Mawdsley (2006) discuss three factors that differentiate sociopolitical contexts for environmental justice movements of marginalized communities in the Global North and South. They argue that this differentiation holds ground for similar patterns of structural marginalization (along race and class in the United States; and caste, ethnicity, class, and gender in India) that cause environmental ills and mobilize movements.

First, even marginalized communities in the North can experience a relatively homogenous public sphere and thus access comparatively effective mechanisms for justice. In postcolonial India, a 'weak–strong state' creates ambitious yet incompletely realized government programmes for public good across various areas, including in participatory environmental management and environmental policy (Rudolph and Rudolph 1987). A lack of enforcement of legislative rights of communities creates a systematic lack of recognition of 'ecosystems people' within governance systems. This inequality of recognition makes their struggles, which fall outside the formal structures of states and institutions, a crucial factor in the process of deliberative democracy for environmental justice in the South (Williams and Mawdsley 2006).

Next, the urban-based middle class that dominates India's public sphere is largely responsible for the visibility of the environmentalism of the poorer sections. In contrast to their northern counterparts who can have a more direct access to platforms for visibility, this mediated representation can create a challenge of accurate representation for southern subaltern peoples. How Western environmental justice research contextualizes southern actors has further bearing on the challenge of their accurate representation.

Williams and Mawdsley (2006) argue that what David Harvey characterizes as a 'sideways looking admiration for those marginalized peoples who have not yet been fully brought within the global political economy of technologically advanced and bureaucratically rationalized capitalism' (1996, 389) can risk using southern actors as symbols of distant 'others' within frames of northern environmental justice without reflecting the complexity of their contextual realities. In particular, transplanting northern environmental discourses to the South has hampered the emergence of a contextualized understanding of the relation between poverty and environmental justice (Lawhorns 2013). But climate justice as an organizing plank can help to globalize the significance of livelihood resistances of the southern poor.

This is especially important given the third factor, that the environment makes a direct contribution to the livelihoods of a large section of the southern population. This chapter analyses the grassroots resistance to coal mining in the Mahan forests in Central India based on Williams and Mawdsley's three distinctions between marginalized peoples' movements in the North and South.

3 How the Indian State Drives Ecological Injustice

India's postcolonial developmental state has played a central role in shaping environmental conflicts and movements of the poor. Consequently, the narratives and politics of movements have also been framed around demands that hold the state as central in redressing environmental injustice. This section outlines changes to the Indian state since independence, and what this transformation implies for today's environmental justice movements and their conceptualization of justice, setting a robust context for the Mahan struggle.

It traces these changes specifically through India's coal sector, which has been synonymous with the 'national interest'. It discusses the extent of violation of communities' human and legal rights to land and forests for coal mining in Indigenous-dominated Central India. This section also outlines how India's climate advocacy, while calling for North–South equity in sharing the global carbon commons, exacerbates injustices towards subsistence communities. Mahan is a case study of climate justice in India in the context of fossil fuel extraction and energy transition, and the points discussed in this section constitute the contextual setting from within which it arose.

3.1 Postcolonial Development and Ecological Injustice

The postcolonial Indian state continued legal regimes and historical patterns from the British colonial era while also adopting its own centralized development pathway since independence in 1947. The state was elevated to the 'commanding heights of the economy', with centralized industrialization being undertaken through public sector enterprises (PSEs) and direct state control of resources (Nayyar 1998).

The Indian Constitution, adopted in 1949, acknowledged historical marginalizations of Indigenous Adivasi people and 'untouchable' Dalits; it contains safeguards for Adivasi lands through geographically demarcated tribal-majority Scheduled Areas with separate legal and administrative frameworks. Scheduled Areas, many of which are in Central India, hold some of India's thickest forests and largest coal and mineral deposits. But the constitutional intent to safeguard Adivasi lands was undermined by the postcolonial state's strong developmental focus. The Indian state drove ecological injustices towards subsistence communities through land evictions and loss of livelihoods for industrialization, with Adivasis being the worst affected: even though they make up only 8.6% of the population, they constitute 40% of all displacements since independence for industrial projects (Negi and Ganguly 2011; Kohli et al. 2018).

People's movements questioned the lack of distributive justice for subsistence communities in the development process, and sometimes succeeded in asserting another vision of development that prioritized sustainability and livelihoods. One of the most notable environmental justice movements was the *Chipko* ('to hug') *Aandolan* (movement), a social movement of peasant communities formed in the Himalayas in the 1970s which succeeded in forcing a community-oriented forest management plan for the region (Guha 1989).

3.2 Neoliberal Development and Ecological Injustice

Since the 1990s, structural changes were made to the Indian economy to align it with private capital and facilitate the entry of foreign direct investments (FDIs).

As compared to an earlier state capitalism, governments now began to acquire increasing proportions of land for private companies and Special Economic Zones (SEZs), raising fundamental questions about whose interests they served and who they disadvantaged (Reddy and Reddy 2007). The reforms specifically reoriented the state's behaviour in favour of private resource extraction and made the state a land broker for private mining (Levien 2011).

The actions of the neoliberal state have been compared to colonial expansion during the late nineteenth and early twentieth centuries that Marx characterized as primitive accumulation. David Harvey's (2003) 'accumulation by dispossession' builds on Marx's 'primitive accumulation' by describing its continuity and extension through privatization and commodification under neoliberalism. Reoriented towards private extraction, the neoliberalizing process in India can therefore be characterized as 'accumulation by dispossession' in which the postcolonial developmental state plays a dominant role as a land broker for private corporations. Opposition to the social and environmental impacts of SEZs formed the first wave of resistances to India's neoliberal growth; these sites became targets for activists' criticism of governments failing to deliver in the public interest (Sharma and Singh 2009).

3.3 Coal and Ecological Injustice

Coal, India's most abundant major fuel, was made central to development. Since electricity use is strongly linked to development in the Human Development Index, state discourse linked coal-led development to the moral imperative of poverty eradication through electricity generation (Ghosh 2016). A national framework of laws and policies gave legal eminence to coal extraction and burning for electricity even while undermining constitutional protections for Adivasi lands (Lahiri-Dutt 2016).

Since the 1990s, policy changes to boost coal and electricity production led to the entry of private players into these sectors. Since 2004, the Indian government has rapidly allocated 194 blocks for coal mining either directly to private companies or to state enterprises who in turn contracted private mine operators, indicating a turn towards private coal economy. Private companies profited through the corporate-owned-and-operated model of mining and power generation due to the constantly rising demand for electricity, producing the phenomenon of 'neoliberal coal' (Lahiri-Dutt 2016). However, government rhetoric continued to equate coal to development, implying that private operations were now indispensable for national energy security (Lahiri-Dutt 2016).

Mining of coal and other minerals rose by 75% between 1994 and 2009, requiring a growing proportion of land to be acquired and forests to be cleared (Shrivastava and Kothari 2012). Mining in general, and coal mining in particular, was granted the highest share of environmental and forest clearances since the 2000s (Nayar 2016). States acquiring an increasing proportion of land for private mining in Central India, including in designated Scheduled Areas, caused 'a ground-clearing' of Adivasis and violated legal provisions for consent and rights over forestlands (Bharadwaj 2018). The state's policies and practices to facilitate increased and privatized coal production and electricity generation exacerbated environmental injustices towards communities.

3.4 Paradoxical Climate Justice

India is highly vulnerable to a changing climate due to global warming's disproportionate impact on the poor.[5] Yet, while emphasizing its disenfranchisement under colonization and right to grow as a postcolonial nation, India paid insufficient attention to its own vulnerabilities during the initial phase of climate dialogues (Raghunandan 2019). Successive governments articulated climate justice in terms of India needing the carbon space to grow (Goodman 2016).

India is now the world's third highest greenhouse gas (GHG) emitter, but its per capita emissions remain one of the world's lowest due to the negligible carbon footprint of a large rural and poor population. Environmental actors like Greenpeace India redirected the gaze towards this internal disparity (and consequently the divide in the emissions of the rich 1% versus the majority poor) and called for domestic accountability through climate action to protect the vulnerable poor (Anathapadmanaban et al. 2007). Advocates of intracountry responsibility for climate change in a post-economic liberalization India have called for the emissions of elites to be targeted (Michael and Vakulabharanam 2016).

By now India has made a substantial commitment to source 40% electricity from renewables by 2030 in its Nationally Determined Contribution (NDC) submission for the Paris Climate Summit in 2015. However, the Narendra Modi government's auctioning of 40 new coal mines in 2020 undermines India's efforts to transition towards clean energy and exposes the contradiction in the government's approach of 'talking renewables and walking coal' (Roy and Schaffartzik 2021). Another contradiction between India's climate plan of increasing renewable energy and how these policies are driving injustice through land dispossession, livelihood disruption, and the inability of local communities to access electricity from renewable projects indicates that clean energy is replicating some fundamental issues of coal generation in India (Roy and Schaffartzik 2021). It signifies that justice for India's subsistence people remains tied to lands and livelihoods both before and after the advent of the climate crisis.

3.5 Undermining Democratic Rights over Forests: Legal Injustice

India's forests have served as sites of conflict since colonial times through repressive laws such as the Forest Act 1878 that restricted people's access to the forest commons. The British colonial government also brought in the Land Acquisition Act 1892 (LAA) that vested arbitrary powers in the state for land acquisitions through the doctrine of the eminent domain.

These colonial laws continued unchanged for decades after independence, while, in several respects, independent India continued aggregating resources and land for industrialization in the same vein as the erstwhile colonial state (Padel 2016). Democratic reforms such as the Forest Rights Act 2006 (FRA) and the Right to Fair Compensation and Transparency in Land Acquisition, Rehabilitation, and Resettlement Act 2013 (LARR) were eventually passed to redress historic injustices towards Adivasis and their access to forestlands, and to overhaul land acquisition's colonial legacy of dispossession. For the first time since independence, communities were given a say about how the state should deal with their land and forests[6] (Ramesh and Khan 2015).

The FRA recognizes both community-based and individual rights over forestlands. It could substantially alter living conditions of forest-dependent communities if properly implemented. But its implementation was beset with challenges due to the strong state–corporate nexus. State governments eager for mining revenues forged consent for mining on Adivasi lands (Chowdhury 2016) and withheld forest rights through slow and flawed implantation and high rejection rates (as high as 50% in Central India) of claims (Bharadwaj 2018). Consequently, a mere 3% of community rights were recognized across India after 10 years of the FRA (CFR-LA 2016). Legal dilutions, particularly unprecedented weakening of democratic provisions and environmental protections by the Narendra Modi government since 2014, further risked forest rights for communities (Chowdhury 2016). A new focus for community resistances came to both fight for the implementation of the FRA and prevent its dilution (Sundar 2011).

Together, these five dimensions and imperatives of the developmental Indian state drive ecological injustices towards communities and constitute the political context from which their resistances arise in the present times.

4   A Forest Rights Movement in India's Energy Capital

In 2015, a grassroots mobilization in the Singrauli district in the Central Indian state of Madhya Pradesh, a region known as the energy capital on account of producing 10% of India's thermal power, saved their forests from coal mining. The four-year resistance in the Mahan forests that fringe one of India's largest coal basins ended when the Indian government, acting on a Supreme Court order, cancelled the coal mine.

The incident that shaped Mahan's resolve to fight occurred in 2013, when local government and company agents, or *dalals*, colluded to forge villagers' signatures on a referendum on mining. The movement was motivated by the realization that they had legal rights over the forests they had depended on for generations. Activists from the Indian arm of the international environmental non-governmental organization (ENGO) Greenpeace had been building awareness about forest rights in Mahan since 2010. An alliance eventually formed between the globally oriented Greenpeace activists and the local forest community, and together they ran a movement that ended with the cancellation of the mine. The three subsections delineate the background and parameters by which climate justice can be understood at Mahan.

4.1 Displacement and Deforestation

Singrauli is South Asia's biggest industrial area containing some of India's oldest state-owned thermal plants and coal mines. Its saga of displacements began in the 1960s with the development of large dams followed by the coal industry from the 1980s. The region stands out in terms of the intensity of land acquisitions and displacements, with Adivasis comprising nearly half the number of those displaced (Pillai et al. 2011).

People's mass movements and subsequently laws such as the LARR created rights for displaced people that did not exist before. But in Singrauli, such efforts were undercut by a pervasive favouritism towards industry and treating sustainability with disdain. One of the critical fallouts from the destruction of Singrauli's environment was through fly ash contamination of water sources from coal plants; a high concentration of toxic acidity turned water bodies into what locals called 'death water' that could corrode the flesh (Dokuzović 2012).

The second decade of economic liberalization brought a drastic increase in India's power generation, and Singrauli became a centrepiece in this scheme. Some of India's largest private energy companies such as Essar and Reliance and public private partnerships (PPPs) led by state governments entered Singrauli from 2005 (Pillai et al. 2011). A third wave of development-related displacements began by 2008 when five new designated super thermal power projects threatened the future of over 4000 families (Sharma and Singh 2009).

Coal mines began to be allocated in Singrauli's last remaining intact forests. This portends a new complexity in Singrauli's displacement saga since Adivasis in particular had lived in the forests for generations without property titles (Sharma and Singh 2009). Between the beginning of industrialization in the 1960s and 2011 when the Mahan resistance started, the region had lost one-third of its forest cover (Pillai et al. 2011). A fact-finding report concluded that violation of communities' rights had created chronic unemployment and expressed concerns about lack of awareness on the part of communities about forest rights[7] (Pillai et al. 2011). The social fallout of Singrauli's ecological collapse from extensive coal mining and burning is compounded by the fact that communities are directly and primarily dependent on the land and forests for their survival, a reality in the Global South, turning them into what Gadgil and Guha (1995) term as 'ecological refugees'.

4.2 How the State Drove Ecological Injustice at Mahan

Both central and state governments demonstrated a strong corporate bias and played a critical role in risking the livelihoods of 50,000 people across 54 villages surrounding the Mahan forests through facilitating the coal mine (Padel 2016).

The forests are a catchment for major rivers; mining also risked destabilizing their watershed. Mahan is one of South Asia's oldest and most intact Sal tree forests, a prominent tiger habitat, and elephant corridor (Chakravartty 2011).

4.2.1 Central Government's Conflict between Coal and Forests

In 2006, the Indian government allocated Mahan to a private venture between Essar Power and the aluminium manufacturing Hindalco Ltd. The 1000-hectare open-cut Mahan mine would produce 8.5 million tonnes of coal per year over 14 years to supply Essar and Hindalco's power plants (Fernandes 2012). But the allocation became a point of conflict between the coal and environment ministries under the United Progressive Alliance government (2004–2014), with the former arguing for protecting these high-quality forests (Ramesh and Khan 2015). The environment ministry was subjected to pressures by CEOs from Essar and Hindalco who claimed that not proceeding with the mine would harm the national interest. The government eventually sidestepped the environment ministry to approve mining through a high-level ministerial group under the finance minister in 2011. Mahan was then granted a Stage 1 Forest Clearance[8] based on 36 conditions, including the completion of various studies and compliance with the FRA (Greenpeace India 2014).

4.2.2 State Government's Violation of the FRA

The government of Madhya Pradesh was required to conduct village council meetings, or *Gram Sabhas*, across the 54 potentially impacted villages around Mahan to determine whether the community consented to mining under the FRA. Having now become aware of their rights, the community demanded that forest clearance not be granted till the process of recognizing their individual forest rights (IFRs) and community forest rights (CFRs) was completed as required under the FRA (Kohli et al. 2012).

But the local administration and company *dalals* acted in collusion to disrupt the Gram Sabhas.[9] Another Gram Sabha in Amelia to determine consent for mining was followed by a large-scale forgery of signatures. And the state government altogether avoided the community consent process in the other 53 villages, indicating the extent of FRA violations at the local level (Greenpeace India 2014). In 2013, the central government issued a final forest clearance for the Mahan coal mine on the basis of a fraudulent village council resolution in 2013 (Press Trust of India 2014).

The high-level intervention that stopped the construction of the coal mine and the destruction of forests came through a Supreme Court order in 2014, which called for the de-allocation of Mahan and 213 other coal mines that had been 'illegally and arbitrarily' allocated by the Indian government to private companies between 1993 and 2011. The 'coal scam' case demonstrated the corruption and corporate favouritism in resource governance[10] (Rajagopal 2014). Paradoxically, the next government under Narendra Modi that de-allocated Mahan proceeded to crack down on the movement, and particularly Greenpeace, for opposing coal, revealing a consistency in the state's corporate bias (Talukdar 2018a). The state's double movement – through the centre's biased approval process and the state's violations of the FRA – served to drive ecological injustice at Mahan.

The state's double movement and the disruption of legal forest rights for communities are demonstrative of the paradox of ambitious yet incompletely realized government programmes for public good in the southern context (Rudolph and Rudolph 1987) and indicative of a systematic lack of recognition of 'ecosystems people' within governance systems. The case of Mahan is demonstrative of how this phenomenon is deepened under neoliberalism where states clearly prefer corporate interests over the rights of communities.

4.3   Forest Rights and Democracy: The Movement

As a first step, Greenpeace activists sought to understand the community's level of awareness about forest ownership. A Greenpeace survey showed that the forests were indispensable for the locals, but they felt disempowered due to a lack of awareness of their rights:

> Every year, I stay in the forests for a month to collect mahua flowers and sell in the market. We also collect tendu leaves, chironji, harra, bamboo, mushroom. But if the government gives away these forests to the company we will have no other means to live. We will not get compensation because we have no rights.
> *(Respondent from Amelia, in Kohli et al. 2012, 14)*

Mahan's transformation into an organized resistance was steered by various imperatives. Sunita, aged 25, from Budher village, one of the very few women leaders in the movement, first learnt about the effects of coal mining in Central India at a youth awareness camp organized by Greenpeace in 2010. But she joined the movement only after *dalals* started visiting Budher to convince people about the mine, and threats and violence from the state–corporate complex became a daily affair. Another movement leader Narayan had seen big NGOs come and go and wondered if Greenpeace would last in the region. The willingness to trust Greenpeace grew over time and in direct response to the threat of mining. Eventually, the forgery of signatures made it clear that the state was acting against people; it served as a catalyst for him to join with Greenpeace and fight for rights.

4.3.1   Mobilization

Eleven villages formally came together under the banner of Mahan Sangharsh Samiti (MSS – Mahan Resistance Front) in 2012. With the understanding that the law acknowledged their historic connection with the forests, the word *adhikaar* (right) entered the movement lexicon. The MSS logo depicted a ring of dancing people around a mahua tree (that serves as an economic and cultural lifeline for Central India's forest-dwellers), signifying people's custodianship of the forest.

Although women risked losing more from mining – grazing grounds for cattle and collecting mahua flowers and tendu leaves – grinding daily work and hesitancy to speak in front of men initially kept them from joining the MSS. When the Greenpeace activist leading the mobilization, a woman, organized separate gatherings, they gained confidence and raised the issue of domestic violence that subsequently became a regular part of MSS meetings. On caste, although a divide persisted in the Mahan society at large, Greenpeace activists observed upper-caste MSS members' acceptance grow over time. Harassment of lower-caste members by Amelia's upper-caste village head who supported mining, alongside threats of murder by *dalals*, strengthened solidarities across social divisions through a shared discontent against the state–corporate complex.

4.3.2   Tactics and Countertactics

The police intimidated MSS members to prevent them from organizing meetings and slapped false charges. Disruptions and arbitrary arrests increased at the peak of the conflict in 2014 after final forest clearance had been granted based on the fraudulent Gram Sabha, and MSS members, mostly women, went to the forest to prevent company contractors from felling trees. Sunita had witnessed the curious incident that led to an arrest. When company workers who were marking trees for felling left the forest without taking their equipment, MSS deposited them with the police. But that night the police arrested two MSS members and two Greenpeace activists on charges of stealing the equipment. Despite such disruptions, the community persisted with the forest blockades for five months. Despite numerous threats, three villages in Mahan succeeded in organizing Gram Sabhas and submitting claims for CFRs by 2016 (Talukdar 2018b).

Like several of India's peoples' movements, women put themselves at the forefront at Mahan at the cost of significant abuse. Forest officials and company agents harassed women stopping trees from being marked and felled. A local member of the legislative assembly threatened MSS women with rape. The police refused their complaints and even planted false cases (Pillai 2019).

Through the association with Greenpeace, MSS participated in activities that could be considered unusual from a grassroots perspective. Twenty-seven MSS members participated in a Greenpeace banner-drop action outside the Essar headquarters in Mumbai in January 2014. None of them had travelled outside their state of Madhya Pradesh before. They were arrested and detained overnight along with Greenpeace activists. The trial continued for four years and required them to travel at regular intervals. These actions demonstrated the leap of faith MSS took in working with Greenpeace. On one of these multi-day journeys in a crowded public bus from Amelia to Mumbai, Narayan saw windmills for the first time; on learning that they generated electricity from wind, he wondered why, instead of coal, the government could not consider installing windmills at Mahan.

4.3.3 Victory or Respite?

Celebrations continued through the night of 30 March 2015 after the Modi government announced its decision to cancel the coal mine and were followed by offerings to the forest god *Dih Baba* in the morning. Although the movement claimed victory over mining, given the state–corporate nexus, they had not won security for the forests, their rights, and a sustainable future. Even though the government acknowledged the high quality of Mahan forests while cancelling the coal mine, it allocated the adjoining coal block (approximately 15 kilometres from Amelia) in the same forests to a public sector corporation, failing yet again to apply a common rationale in making decisions that are critical for the environment and communities (Pillai 2017).

The corporate bias exacerbated under the Modi government, as demonstrated through crackdown on livelihoods and human rights movements and civil society organizations, in which Greenpeace became a primary target. Crackdowns and freezing of its funds from 2014 forced Greenpeace to reduce campaign and staffing and reorient its climate activism away from anti-coal mining mobilizations towards less risky approaches of exposing the economic implausibility (and ecological impacts) of new coal investments.[11] A final funding attack by the government in 2018 forced Greenpeace to shed most campaigns; it now operates skeletally out of India (Talukdar 2019). Members of the Greenpeace Mahan team continued with the movement in their individual capacities. This state of play left many questions unanswered about the future of Mahan.

4.3.4 Forest Rights as Democracy and Climate Justice

The victory brought a sense of empowerment. At the second anniversary celebrations of their victory in 2017 at Amelia village, bright yellow banners with slogans celebrating people's forest rights lined the walls of the function tent.

These slogans – such as *Jangal hamara apka hai, nahi kisi ka baap ka hai* (the forests belong to you and me, not to the government or company), *Jan Jan ka naara hai, van adhikar hamara hai* (a people's chorus for forest rights), and *gaon gaon ki yahii pukaar, le ke rahenge van adhikar* (village after village will claim forest rights) – had served as inspirational chants for the movement during the forest blockades, while the assertion *Purkho ka naata nahi todenge, jangal zameen nahi chodenge* (we will respect our ancestral land, we will not give up our forests) had grown to become an anthem for the movement.

Greenpeace connected this celebration to the day of global climate action by the 'Break Free from Fossil Fuels' network. The largest banner in the tent said *Loktantra Zindabad* (Long Live Democracy); this became a definitive slogan after the 2014 crackdowns by the Modi government, and Greenpeace won a legal case against the government where the judge upheld the ENGO's democratic right to dissent. MSS saw their fight to stop the coal mine as a struggle to democratize India's development. *Democracy Zindabad* served as a meeting ground for the demands of an international ENGO's climate change campaign in India and the demand for forest rights and livelihood security by a forest community.

The extent of violation of people's legal rights in India's mining regions, daily interference by the state–corporate apparatus in communities' lives including acts of violence towards movements, shrinking democratic freedoms including the ability to protest, and the compounded threat from coal mining to communities entirely dependent on the forests for survival, are outstanding themes of climate justice mobilizations in a Global South context. Greenpeace's role in the Mahan mobilization, including generating awareness about the community's forest rights, is indicative of an 'uneven public sphere' in a southern context where urban elite allies play a crucial role in connecting local subaltern movements at national and international scales. This too comprises an outstanding theme of climate justice mobilizations in the southern context. The central role of women speaks to their core dependency on the forests and its resources. These dimensions shape the politics and narratives of community activism against fossil fuels that need to be considered in their own right as dimensions of climate justice activism emerging from a southern geography.

5   Analysis: Forest Rights and Democracy as Climate Justice for Central Indian Communities

The disparities in the political contexts of the Majority and Minority World, even in the case of democracies (e.g. see O'Neill 2012), and consequently the different political challenges for grassroots activism and framings of environmental conflicts in the South (Haynes 1999), stand acknowledged in the field of comparative environmentalism. This disparity can be attributed as a reason for the relative lack of perspectives and conceptualizations of climate justice in anti-fossil fuel activism from the Global South. Comparing global grassroots research that faces similar challenges (see Peluso and Watts 2001; Taylor 1995) is understood as one way of bridging the disparity in North–South comparative environmentalism research.

In the dissertation 'Cutting carbon from the ground up: A comparative ethnography of anti-coal activism in India and Australia', Talukdar (2021) compared contexts, politics, and narratives of anti-fossil fuel activism in the Majority and Minority Worlds, focussing on the countries mentioned as cases. It compared the Mahan–Greenpeace resistance with a multipronged resistance by an Indigenous community, local farmers, and a national climate movement against the Carmichael coal mine in central Queensland. Unlike in Mahan, in the latter case, claims of various actors came together to create a collective climate justice political narrative. Drawing on the main dimensions of the Mahan resistance from

Section 3, this section establishes a critical differentiation of southern climate justice activism in the context of an energy transition from the Indigenous climate justice struggle of the Wangan and Jagalingou traditional owners in central Queensland against the Carmichael coal mine. This section discusses a southern climate justice framework based on these differences.

5.1 Difference in Indigenous Climate Justice: Climate Justice vs. Forest Rights

Against the history of Indigenous dispossession in Australia's settler colonial society, the land rights campaign of the Wangan and Jagalingou traditional owners against the Carmichael coal mine signified an assertion of sovereignty and articulated a form of Indigenous climate justice based on the demand to redress historic injustices. Their international links with Indigenous fossil fuel struggles in North America strengthened a Global North Indigenous claim for sovereignty over traditional lands in fossil fuel resistances; reciprocally, it collectivized sovereignty and land rights as a Global North Indigenous climate justice demand in the context of an energy transition. As (relatively) empowered civil society actors with access to platforms as compared to Mahan, they linked their grievances to climate change of their own accord. Also, against the context of a pro-coal and climate-denialist national politics in Australia,[12] their articulation of climate injustice as an extension of their grievances against coal mining reflects a dialectical process of meaning-making.

However, the same cannot be said about Mahan, where the Indian arm of the global ENGO Greenpeace connected the local struggle with the issue of climate change. Greenpeace's urban activists added new dimensions to a global conceptualization of climate justice in the context of an energy transition; they 'southernized' a northern campaign frame of (End)ing Coal as climate justice by connecting it to India's destruction of forests for coal and violation of forest rights (Fernandes 2012) and connecting Mahan's perspectives of eco-social justice to the global framework. Conversely, they globalized the Mahan struggle by interpreting the movement's significance as a quest for climate justice (Talukdar 2019).

In contrast to the (relative) independence of the Wangan and Jagalingou, Mahan's association with Greenpeace needs to be regarded in terms of broadening the significance of their struggle: beyond access (raising the issue of forged consent with the central government), platforms (Radio Sangharsh), and support (through being made aware of their rights), Mahan also benefitted from a critical reinterpretation of their struggle as an essential act of dissent in the Indian democracy by urban civil society when Greenpeace won the legal challenge against the state's crackdown. Greenpeace's intervention signifies the crucial role urban environmental actors can play in the South, as translators between Global North and South narratives and movement frames. Such a role is not given sufficient consideration by Western climate justice scholars, but deserves a critical reflection within climate justice conceptualization and praxis (see Joshi 2021).

The lack of connection with climate change issues amongst India's livelihood-focused people's movements is on account of socioeconomic and political differences. First, unlike the Wangan and Jagalingou, the people of Mahan have a lived presence at the site of the project. This reality adds further criticality to the human rights issue of the communities' loss of lands from coal mining and other developments in the South. Being overwhelmed with the daily struggle for survival, as also seen in Singrauli, environmentalism of the poor's assertions remains grounded in immediate injustices. This must not be mistaken as the environmental parochialism of the southern poor. Instead, it signifies how industrialization poses imminent and persistent risks to their survival and security as noted in Sections 2 and 3.

Next, in comparison to well-educated urban activists, environmentalism of the poor movements is composed of largely rural populations that can often lack a scientific understanding of the issue of climate change, even though they are attuned to changing weather patterns. The South Asian People's Action on Climate Crisis (SAPACC), formed in 2019, is a very recent and unique collaboration between livelihood-focused people's movements, Indigenous groups, trade unions, and farmers across South Asia. It marks an emergent space in mass activism in South Asia, making climate change central to people's movements and attempting to link the existent malcontents of ecosystems-dependent subsistence communities with the broader problem of climate change (Adve 2020).

Finally, unlike in Australia, the Indian government carefully positions itself as a supporter of global climate action. While India's 'moral' position on climate justice as needing the carbon space to grow generates contradictions in terms of climate's worsening effects on its vulnerable, ecosystems-dependent communities, the position does not generate the specific political imperative such as in Australia to mobilize against the government on the issue. While interpreting southern anti-coal activisms as an assertion of climate justice, Western environmental justice researchers need to 'tie' climate justice to various risks that communities vulnerable to destructive industrialization in the South encounter.

5.2 Framing a Southern Climate Justice Activism in the Context of Energy Transition

The impact of the political economy of coal in southern geographies such as India on 'ecosystems people' is creating a human rights challenge that deserves attention in the conceptualization of climate justice activism in the context of an energy transition.

India's ambition of granting sovereign rights to Adivasi and other forest communities was undermined by its own plans to increase coal-driven economic growth. A state–corporate nexus that results in the state denying people's rights, often through violence, has added new dimensions to environmental injustice towards communities in the neoliberal era. This is evident in Singrauli and the Mahan forests in the Central Indian landscape that serves as the mineral backdrop for economic growth. Set against Singrauli's development and eco-social rupture, the Mahan resistance highlights how previous patterns of structural disenfranchisement of subsistence peoples are exacerbated under neoliberalism across scales, from national resource governance to local administrative actions towards people's livelihood rights.

Narratives of resistance emerging from this context have a critical bearing on the framing of southern climate justice. The Mahan movement turned on the demand for legal forest rights; at a broader level, the movement demanded democracy and a say in decision-making on coal development that directly affects them. By giving decision-making powers to communities, the FRA allows for a direct assertion of democracy (Kothari 2016); however, the struggles communities face to secure their forest rights effectively make their assertion an act of 'radical democracy' (Sundar 2011).

This precariousness of forest rights for communities poses a challenge for theorizing climate justice in a transitional southern context (O'Neill 2012). A multidimensional global understanding of climate justice needs to acknowledge this challenge; it needs to recognize both the importance of these newfound rights for communities and their vulnerability in the context of government violations. The documentation of the violation of forest rights at Mahan, the risks and intimidations to the local MSS, and the government crackdown on Greenpeace contribute to environmental justice research on human rights violations in environmental conflicts.

As discussed earlier, transplanting northern environmental discourses to the South has hampered a contextualized understanding of the relation between poverty and environmental justice (Lawhorns 2013). Like many subsistence communities, Mahan residents mostly still live without electricity, with the central grid not having arrived there due to issues in last-mile connectivity (Talukdar 2017). Being doubly disadvantaged through a lack of energy access and environmental injustice, the challenges of southern subsistence communities need to be considered within a postcolonial development paradigm of the need to democratize development. However, Mahan also aspired towards decolonization, through aspirations for a forest economy-based future that became central to discussions in the years following the struggle. Decolonization is an emerging but necessary area for consideration by communities with recently acquired forest rights.

Climate justice activism to stop fossil fuels and bring in the energy transition can acknowledge that multiple and urgent human justice issues in the South result in climate justice often being used as a proxy issue by southern campaigns to link with the global movement. Climate justice research needs to consider mechanisms through which the land and forest rights of southern communities can be strengthened as necessary actions towards their climate justice.

6 Conclusion

In the context of an energy transition that is essential to avoid dangerous climate change, mobilizations to 'leave fossil fuels in the ground' are being understood as climate justice activism. An indirect political association with the issue of climate change, as seen at Mahan, can be considered a southern characteristic of climate justice activism. It raises the need to reflexively include the local ecological justice demands of southern communities in the politics and narratives of global climate justice activism. Global climate change activism and research need to understand critical differences in the contexts and imperatives of southern activisms and contextualize southern environmental actors.

Dimensions of ecological justice emerging from Mahan further qualify the three distinctions in contextual realities between structurally marginalized groups and their environmental protests in the North and South that Williams and Mawdsley (2006) identify. First, the ecological resistances emerge within a context of double turn by the Indian state, which favours corporations and consequently denies legally granted rights to subsistence groups. Next, given the persistence of an unequal public sphere, the role played by urban middle-class activists, in linking global climate justice campaigns (framed in the North) with subaltern environmental actors in the South, assumes greater significance. Due to the persistence of an unequal public sphere, it is the actions and advocacy of urban environmental actors that can draw attention to these multiple crises of democracy and link them to global framings of ecological justice through critical reinterpretations. Finally, the immediacy of environmental challenges in the lives of subsistence communities makes the need for a serious consideration of their security and sustainability as a central concept in a global climate justice framework. A reflexive understanding of emancipatory and non-hegemonic collaborations between them and global climate justice networks is essential for defining, supporting, and centring their justice and their visions for a sustainable future in global climate justice activism and research.

Notes

1 For a 50:50 chance to keep global warming within 2 degrees, 88% of the world's coal reserves, 52% gas reserves and 35% oil reserves need to be considered 'unburnable fuel' and left in the ground (Steffen 2015). It is well understood that meeting the goals of the 2015 Paris Agreement requires a rapid phasing out of fossil fuels and for those fossil fuels not yet extracted to be left in the ground.

2 Blockadia is however not a new concept; the idea originated from a peaceful uprising in the Niger Delta against the oil corporation Shell, after oil spills destroyed the lands of the Ogoni and Ijaw peoples (Environment Justice Atlas 2014).

3 Environmentalism, a term that broadly stands for a collection of ideologies, politics, and actions towards the environment, has essentially been variously realized by movements emerging from different socioeconomic and socioecological contexts.

4 In *Ecology and Equity*, Gadgill and Guha (1995) propose an ecological and sociological classification of a highly unequal Indian society into three broad categories – omnivores, ecosystems people, and ecological refugees – based on the size of their respective

ecological footprint, their ability to influence policy, and the social power they wield as a demographic. A narrow set of elite omnivores have the social, economic, and political power to command resources, which comes at a cost to masses of ecosystems people, creating a further problem of ecological refugees, displaced and dispossessed communities who are forced to migrate to cities.

5 Predicted impacts include displacements driven by sea level rise and coastal erosion (Hazra et al. 2002), increasing frequency and duration of heat stress (Somanathan et al. 2017), impacts of monsoon variability on agriculture on which 65% of the population relies (Pai et al. 2017), and risks to water supplies (Adve 2019).

6 The LARR made it mandatory to seek approvals from affected communities through consent and social impact assessment (SIA) clauses, as well as to resettle and rehabilitate title-holders and livelihood losers. It set compensation formulas at four times the value of rural and twice that of urban land and also contained provisions related to return of unused lands and food security (Kohli et al. 2018).

7 The report states that, according to the district collector of Singrauli, although 4000 individual rights have been issued under the FRA between 2008 and 2010, with 7000–8000 other applications in the pipeline, only 64 community rights applications had been made, and hardly any granted, during the same period.

8 Clearance to fell trees in the forest needs to follow a separate process from the environmental approval process and requires the consent of the community under the Forest Rights Act 2006.

9 Company-hired goons disrupted council proceedings in Amelia, Mahan's largest village, to prevent people from registering their CFRs (*Economic Times* 2013).

10 The coal scam caused an estimated loss of over $150 billion in government revenue and gains by large private companies including Essar and Hindalco (Rajagopal 2014).

11 A Greenpeace analysis argued that the super thermal power plant attached to the next proposed coal mine in the Mahan was financially risky, unnecessary, and posed a great health risk through worsening air pollution in North India (Greenpeace India 2018). An analysis by the Institute of Energy Economics and Financial Analysis (IEEFA) showed that India's existing supply glut of electricity and the rapidly declining cost of renewable energy made the current project economically illogical (Buckley et al. 2018).

12 India's national political context on coal significantly differs from that of Australia where governments have ideologically championed coal exports, discouraged renewables and denying climate change.

References

Adve, N. 2019. 'Impacts of global warming in India: Narratives from below', in N. Dubash (ed.), *India in a Warming World*, Oxford University Press, pp. 65–78.

Adve, N. 2020. 'South Asian coalition links climate with social struggles', *Ecologist*, 26 February, viewed 20 September 2020, https://theecologist.org/2020/feb/26/south-asian-coalition-links-climate-social-struggles.

Ananthapadmanabhan, G., Srinivas, K., & Gopal, V. 20077. 'Hiding behind the poor', *Greenpeace India*, viewed 20 March 2020, www.greenpeace.org/india/Global/india/report/2007/11/hiding-behind-the-poor.pdf.

Bharadwaj, S. 2018. *The Legal Face of the Corporate Land Grab in Chhattisgarh*, Janhit People's Legal Resource Centre.

Birch, T. 2016. 'Climate change, mining, and traditional Indigenous knowledge in Australia', *Multidisciplinary Studies in Social Inclusion*, vol. 4, no. 1, pp. 92–101.

Bird Rose, D. 2013. 'Slowly-writing into the anthropocene', *TEXT*, vol. 20, pp. 1–14.

Brown, B. & Spiegel, S. J. 2019. 'Coal, climate justice and the cultural politics of energy transition', *Global Environmental Politics*, vol. 19, no. 2, pp. 149–168.

Bryant, R. L. 1998. 'Power, knowledge and political ecology in the third world: A review', *Progress in Physical Geography*, vol. 22, no. 1, pp. 79–94.

Bryant, R. L. & Bailey, S. 1997. *Third World Political Ecology*, Psychology Press.

Buckley, T., Shah, K., & Garg, V. 2018. 'The Khurja Power Project: A recipe for an Indian stranded asset', *Institute of Energy Economics and Financial Analysis*, viewed 20 September 2020, http://ieefa.org/wp-content/uploads/2018/10/Khurja-Thermal-Power-Project_10.2018.pdf.

CFR-LA. 2016. *Promise and Performance, Ten Years of the Forest Rights Act in India*, Citizen's Report on the Promise and Performance of Scheduled Tribes and Other Traditional Forest Dwellers (Recognition of Forest Rights) Act 2006. https://rightsandresources.org/publication/promise-performance-forest-rights-act-2006-tenth-anniversary-report/.

Chakravartty, A. 2011. 'Mahan at all costs', *Down to Earth*, 31 October, viewed 14 August 2019, www.downtoearth.org.in/news/mahan-at-all-costs-34230.

Chowdhury, C. 2016. 'Making a hollow in the Forest Rights Act', *The Hindu*, 7 April, viewed 20 September 2019, www.thehindu.com/opinion/columns/Making-a-hollow-in-the-Forest-Rights-Act/article14226592.ece.

Dasmann, R. 1988, 'Towards a biosphere consciousness', in D. Woster (ed.), *The Ends of the Earth: perspectives on modern environmental history*, Cambridge University Press, England.

Dokuzović, L. 2012. 'Bhikharipore Singrauli: A case for just development', 30 September, viewed 27 September 2019, http://sanhati.com/excerpted/5621/.

Dryzek, J. S. 2013. *The Politics of the Earth: Environmental Discourses*, Oxford University Press.

Dubash, N. 2012. 'Climate politics in India: Three narratives', in N. K. Dubash (ed.), *Handbook of Climate Change in India: Development, Politics and Governance*, Oxford University Press, pp. 197–207.

Dubash, N. 2013. 'The politics of climate change in India: Narratives of equity and co-benefits', *Wiley Interdisciplinary Reviews: Climate Change*, vol. 4, no. 3, pp. 191–201.

Economic Times. 2013, 'Tribal Affairs Ministry orders probe into Mahan Coal Block allocation', 22 July, viewed 30 September 2019, https://economictimes.indiatimes.com/industry/indl-goods/svs/metals-mining/tribals-affairs-ministry-orders-probe-into-mahan-coal-block-allocation/articleshow/21229133.cms?from=mdr.

Environment Justice Atlas. 2014, 'Oil Extraction Forces Ogoni to Consume Benzene Water for Survival, Nigeria', *Atlas of Environment Justice*, viewed 20 March 2020, <https://ejatlas.org/conflict/oil-extraction-forces-ogoni-to-consume-benzene-water-for-survival-nigeria>.

Esposito, A. & Neale, T. 2016. 'Never squib the rights issue in favour of conservation', in E. Vincent & T. Neale (eds), *Unstable Relations: Indigenous People and Environmentalism in Contemporary Australia*, UWA Publishing. pp. 336–355.

Fernandes, A. 2012. 'How coal mining is trashing tigerland', *Greenpeace India Society*, viewed 14 August 2019, www.greenpeace.org/india/en/publication/984/how-coal-mining-is-trashing-tigerland/.

Gadgil, M. & Guha, R. 1995. *Ecology and Equity: The Use and Abuse of Nature in Contemporary India*, Routledge.

Ghosh, D. 2016. 'We don't want to eat coal: Development and its discontents in a Chhattisgarh district in India', *Energy Policy*, vol. 99, pp. 252–260.

Goodman, J. 2009. 'From global justice to climate justice? Justice ecologism in an era of global warming', *New Political Science*, vol. 31, no. 4, pp. 499–514.

Goodman, J. 2016. 'The "climate dialectic" in energy policy: Germany and India compared', *Energy Policy*, vol. 99, no. C, pp. 184–193.

Gottlieb, R. 2005. *Forcing the Spring: The Transformation of the American Environmental Movement*, Island Press.

Government of India. 2014. 'India's Intended Nationally Determined Contribution: Working towards climate justice', viewed 20 March 2020, www4.unfccc.int/sites/ndcstaging/PublishedDocuments/India%20First/INDIA%20INDC%20TO%20UNFCCC.pdf.

Greenpeace India. 2014. 'NGT renders forest clearance to Mahan Coal Ltd. invalid after SC verdict; Greenpeace, MSS celebrate but vow to oppose any attempt to re-allocate Mahan', 26 September, viewed 20 September 2019, www.greenpeace.org/india/en/press/2481/ngt-renders-forest-clearance-to-mahan-coal-ltd-invalid-after-sc-verdict-greenpeace-mss-celebrate-but-vow-to-oppose-any-attempt-to-re-allocate-mahan/.

Greenpeace India. 2018. 'Khurja coal plant unviable: Solar project would benefit region more', October 2018, viewed 20 March 2019, www.greenpeace.org/india/en/story/3255/analysis-on-khurja-supercritical-thermal-power-plant-proves-solar-will-be-a-better-investment/.

Guha, R. 1989. *The Unquiet Woods: Ecological Change and Peasant Resistance in the Himalaya*, Oxford University Press.

Hardt, M. & Negri, A. 2000. *Empire*, Harvard University Press.

Harvey, D. 1996. *Justice, Nature and the Geography of Difference*, Wiley-Blackwell.

Harvey, D. 2003. *The New Imperialism*, Oxford University Press.

Hazra, S., Ghosh, T., DasGupta, R., & Sen, G. 2002. 'Sea level and associated changes in the Sunderbans', *Science and Culture*, vol. 68, no. 9–12, pp. 309–321.

Haynes, J. 1999. 'Power, politics and environmental movements in the Third World', in C. Rootes (ed.), *Environmental Movements: Local, National and Global*, Frank Cass, pp. 222–242.

Jewitt, S. 1995. 'Europe's "others"? Forestry policy and practices in colonial and postcolonial India', *Environment and Planning D: Society and Space*, vol. 13, pp. 67–90.

Joshi, S. 2021. *Climate Change Justice and Global Resource Commons: Local and Global Postcolonial Political Ecologies*, Routledge.

Klein, N. 2014. *This Changes Everything: Capitalism vs. the Climate*, Simon and Schuster.

Kohli, K., Kapoor, M., Menon, M., & Vishwanathan, V. 2018. *Midcourse Manoeuvres: Community Strategy and Remedies for Natural Resource Conflicts in India*, CPR–Namati Environmental Justice Program.

Kohli, K., Kothari, A., & Pillai, P. 2012. 'Countering coal?', *Kalpavriksh and Greenpeace India*, viewed 15 September 2019, www.greenpeace.org/india/en/publication/989/countering-coal-community-forest-rights-and-coal-mining-regions-of-india/.

Kothari, A. 2016. 'Decisions of the people, by the people, for the people', *The Hindu*, 18 May, viewed 20 September 2019, www.thehindu.com/opinion/op-ed/Decisions-of-the-people-by-the-people-for-the-people/article14324692.ece.

Lahiri-Dutt, K. 2016. 'The diverse worlds of coal in India: Energising the nation, energising livelihoods', *Energy Policy*, vol. 99, pp. 203–213.

Latulippe, N. & Klenk, N. 2020. 'Making room and moving over: Knowledge co-production, Indigenous knowledge sovereignty and the politics of global environmental change decision-making', *Current Opinion in Environmental Sustainability*, vol. 42, pp. 7–14.

Lawhorn, M. 2013. 'Situated, network environmentalism: A case for environmental theory from the south', *Geography Compass*, vol. 7, no. 2, pp. 128–138.

Levien, M. 2011. 'Special economic sones and accumulation by dispossession in India', *Journal of Agrarian Change*, vol. 11, no. 4, pp. 454–483.

Martinez-Alier, J., Temper, L., Del Bene, D., & Scheidel, A. 2016. 'Is there a global environmental justice movement?' *Journal of Peasant Studies*, vol. 43, no. 3, pp. 731–755.

Michael, K. & Vakulabharanam, V. 2016. 'Class and climate change in post-reform India', *Climate and Development*, vol. 8, no. 3, pp. 224–233.

Nayar, L. 2016. 'Unlock Mantri: Javadekar's key to success', *Outlook India*, 5 July, viewed 20 September 2019, www.outlookindia.com/website/story/unlock-mantri-javadek ars-key-to-success/297036.

Nayyar, D. 1998. 'Economic development and political democracy: Interaction of economics and politics in Independent India', *Economic and Political Weekly,* vol. 33, no. 49, pp. 3121–3131.

Negi, N. S. & Ganguly, S. 2011. *Development Projects vs. Internally Displaced Populations in India: A Literature Based Appraisal*, Centre on Migration, Citizenship and Development.

Nixon, R. 2011. *Slow Violence and the Environmentalism of the Poor*, Harvard University Press.

O'Neill, K. 2012. 'The comparative study of environmental movements', in P. F. Steinberg & S. D. VanDeever (eds), *Comparative Environmental Politics: Theory, Practice and Prospects*, MIT Press, pp. 115–142.

Padel, F. 2016. 'Investment induced displacement and the ecological basis of India's economy', in S. Venkateshwar and S. Bandyopadhyay (eds), *Globalisation and the Challenges of Development in Contemporary India,* Springer, pp. 147–169.

Pai, D. S., Guhathakurta, P., Kulkarni, A., & Rajeevan, M. N. 2017 'Variability of meteorological droughts over India', in M. N. Rajeevan & S. Nayak (eds), *Observed Climate Variability and Change over the Indian Region*, Springer Geology, pp. 73–87.

Peluso, N. L. 1992. *Rich Forests, Poor People: Resource Control and Resistance in Java*, University of California Press.

Peluso, N. L. & Watts, M., eds. 2001. *Violent Environments*, Cornell University Press.

Pillai, P. 2017. 'An ongoing battle', *Frontline*, 15 September, viewed 27 September 2019, https://frontline.thehindu.com/social-issues/an-ongoing-battle/article9831547.ece.

Pillai, P. 2019. 'Coal mining and ecological fragility', in P. Ray (ed.), *Women Speak Nation: Gender, Culture, Politics*, Routledge.

Pillai, P., Gopal, V., & Kohli, K. 2011. 'Singrauli: The coal curse – A fact finding report on the impact of coalmining on the people and environment of Singrauli', *Greenpeace India*, viewed 27 September 2019, www.greenpeace.org/india/en/publication/1006/ singrauli-the-coal-curse/.

Plows, A. 2007. 'You've been framed: Why publics mistrust the policy process', *Genomics Network Newsletter*, vol. 6, pp. 22–33.

Press Trust of India. 2014. 'Moily planning to give oil field to Essar cheaply, alleges AAP', *Economic Times*, 17 April, viewed 27 September 2019, https://economictimes.indiati mes.com/news/politics-and-nation/moily-planning-to-give-oil-field-to-essar-chea ply-alleges-aap/articleshow/33828233.cms.

Purdy, J. B. 2016. 'Environmentalism was once a social justice movement: It can be again', *The Atlantic*, 8 December, viewed 20 March 2020, www.theatlantic.com/science/arch ive/2016/12/how-the-environmental-movement-can-recover-its-soul/509831/.

Quinn-Thibodeau, T. & Wu, B. 2017. 'NGOs and the climate justice movement in the age of Trumpism', *Development*, vol. 59, pp. 251–256.

Raghunandan, D. 2019. 'India in international climate negotiations', in N. Dubash (ed.), *India in a Warming World*, Oxford University Press, pp. 187–204.

Rajagopal, K. 2014. 'Supreme Court quashes allocation of 2014 coal blocks', *The Hindu*, 24 September, viewed 15 September 2019, www.thehindu.com/news/national/supr eme-court-quashes-allocation-of-all-but-four-of-218-coal-blocks/article6441855.ece.

Ramesh, J. & Khan, M. A. 2015. *Legislating for Justice: The Making of the 2013 Land Acquisition Law*, Oxford University Press.

Reddy, V. R. & Reddy, B. S. 2007. 'Land alienation and local communities: Case studies in Hyderabad-Secunderabad', *Economic and Political Weekly*, vol. 42, no. 31, pp. 3233–3240.

Roy, B. & Schaffartzik, A. 2021. 'Talk renewables, walk coal: The paradox of India's energy transition', *Ecological Economics*, vol. 180, no. 106871, pp. 1–12.

Rudolph, L. I. & Rudolph, S. H. 1987. *In Pursuit of Lakshmi: The Political Economy of the Indian State*, Chicago Indian Press.

Sharma, R. N. & Singh, S. R. 2009. 'Displacement in Singrauli region: Entitlements and rehabilitation', *Economic and Political Weekly*, vol. 44, no. 51, pp. 62–69.

Shrivastava, A. & Kothari, A. 2012. *Churning the Earth: The Making of Global India*, Penguin.

Somanathan, E., Somanathan. R., Sudarshan, A., & Tewari M. 2017. 'The impact of temperature on productivity and labor supply: Evidence from Indian manufacturing, working paper, EPIC-India.

Steffen, W. 2015. 'Unburnable carbon: Why we need to leave fossil fuels in the ground', *Climate Council of Australia*, viewed 20 March 2020, www.climatecouncil.org.au/uplo ads/a904b54ce67740c4b4ee2753134154b0.pdf.

Sundar, N. 2011. 'The rule of law and citizenship in Central India: Postcolonial dilemmas', *Citizenship Studies*, vol. 15, no. 3–4, pp. 419–432.

Talukdar, R. 2017. 'Hiding neoliberal coal behind the Indian poor', *Journal of Australian Political Economy*, no. 78, pp. 132–158.

Talukdar, R. 2018a. 'Sparking a debate on coal: Case study on the Indian government's crackdown on Greenpeace', *Cosmopolitan Civil Societies: An Interdisciplinary Journal*, vol. 10, no. 1, pp. 47–62.

Talukdar, R. 2018b. 'Democracy zindabad! A day in the life of an anti-coal resistance in India's energy capital', *New Matilda*, 10 February, viewed 20 March 2020, https:// newmatilda.com/2018/02/10/democracy-zindabad-day-life-anti-coal-resistance-ind ias-energy-capital/.

Talukdar, R. 2019. 'Profit before people: Why India has silenced Greenpeace', *New Matilda*, 29 March, viewed 20 June 2020, https://newmatilda.com/2019/03/29/pro fit-before-people-why-india-has-silenced-greenpeace/?fbclid=IwAR2NwczXun5q 1Ug-g3kSxvaOJLInTfwNlnS6Up3Xxts7A7XZF_wbvOsHKYc.

Talukdar, R. 2021. 'Cutting carbon from the ground up: An ethnography of anti-coal movements in India and Australia', unpublished thesis, University of Technology Sydney.

Taylor, B., ed. 1995. *Ecological Resistance Movements: The Global Emergence of Radical and Popular Environmentalism*, SUNY Press.

Thaker, J. & Leiserowitz, A. 2014. 'Shifting discourses of climate change in India', *Climate Change*, vol. 123, pp. 107–109.

Watts, M. 1983. *'Silent Violence': Food, Famine and Peasantry in Northern Nigeria*. University of Georgia Press, Athens and London.

Whyte, K. 2017. 'Way beyond the lifeboat: An Indigenous allegory of climate justice', in D. Munshi, K. Bhavnani, J. Foran, & P. Kurian (eds), *Reimagining Global Climate Justice*, University of California Press.

Williams, G. & Mawdsley, E. 2006. 'Postcolonial environmental justice: Government and governance in India', *Geoforum*, vol. 37, pp. 660–670.

Appendix: Map of the Proposed (and Suspended) Mahan Coal Mine Site

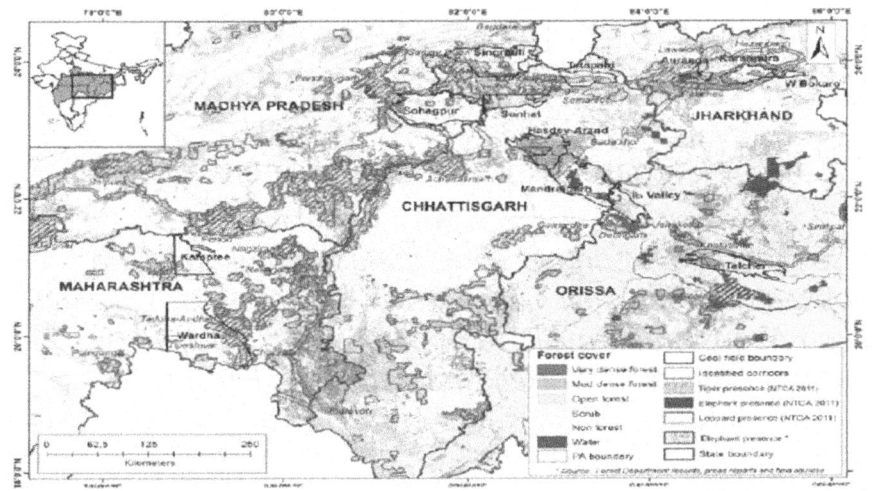

Range and Quality of Forest Cover in Central India. Sourced from A. Fernandes, 2012, 'How coal mining is trashing tigerland', *Greenpeace India Society*, viewed 14 August 2019, www.greenpeace.org/india/en/publication/984/how-coal-mining-is-trashing-tigerland/

2

EXTREME CLIMATIC EVENTS AND CLIMATE CHANGE POLICIES

A Call for Climate Justice Action in Mozambique

*José Maria do Rosário Chilaúle Langa, Natacha Bruna,
Boaventura Monjane, Giverage do Amaral,
Elton Augusto da Amélia Fé, Bento Paulo Rafael,
Patricia Figueiredo Walker, and Patricia E. Perkins*

1 Introduction

Climate change is the greatest global challenge of the twenty-first century, but there is a notable difference among countries' ability to respond to its impacts. Those in the Global South have faced numerous investments by polluting multinationals from countries in the Global North, particularly firms that extract fossil fuels. However, these countries often do not have the capacity to impose compliance with the requirements for globally accepted standards of socioenvironmental sustainability, and they face great challenges in monitoring and remediating environmental impacts in their territories. All of this takes place in a context characterized by lack of documentation or in-depth scientific information. In their quest for resources, many countries in the Global North promote or cause displacement, wars, unscrupulous devastation of biodiversity, disregard of the precautionary principle, and huge population movements in areas of great social and environmental vulnerability. This highlights the need, especially in poorer countries, to include everyone's voices and perspectives in climate-related and economic decision-making, since centralized, top-down policies are likely to ignore important local realities, knowledge, and lived experience that is crucial for effective and efficient climate action.

Mozambique's low-lying central agricultural areas, its long coastline facing the southern Indian Ocean, its location downriver from large dams managed for hydropower, and its colonial history emphasizing settlement and infrastructure along the coast expose millions of Mozambicans to huge Indian Ocean storms and resulting floods as well as periodic severe droughts.[1] Since it is an impoverished country, with a GDP per capita of about US$500 – one of the poorest in the world – the probability that natural disasters will evolve into calamaties is very

DOI: 10.4324/9781003214021-3

high (INGC 2009). People living with greater socioeconomic vulnerability are among the most at risk from extreme weather events such as heavy rains, floods, prolonged droughts, water insecurity, and cyclones (Amaral 2018) due to their lack of adaptive capacity or ability to cope with and respond to the adverse effects of climate change. This is the essence of climate injustice: those most affected by the climate crisis are also least responsible for it, and least likely to share in the benefits of high-emission energy consumption.

As some of the world's biggest carbon emitters fail to adopt stringent plans for reducing their GHG emissions, climate injustices are unfolding everywhere, and especially in the Majority World. Yet, these major emitters suffer minimal consequences of environmental devastation, due to their wealth and abundance of technical and financial means for adaptation. This worsening situation invites us to (re)think the concept of climate justice in dynamic terms, as capitalism evolves and both climate change and income distribution worsen, inflicting climate injustices on more and more people.

The term climate justice derives from the concept of environmental justice and is used to refer to disparities in terms of negative impacts suffered and responsibility for the causes of climate change (Roberts and Parks 2009). It is partly the result of unsound policies for ecological protection in societies marked by economic and social inequities and of marginalized people's lack of space, voice and agency, both locally and globally. In this chapter we report on our research which has used a range of methods to explore how the concept of climate justice is being understood and adopted by different groups and individuals who are directly involved in environmental management in Mozambique.

Our team includes researchers from the Maputo-based Environmental Observatory for Climate Change (ObservA), professors and students from the Technical University of Mozambique, researchers linked to the Centre for African Studies at Eduardo Mondlane University, researchers from the international QES-Climate Justice Network, and community members from the city of Beira (including universities, the Beira City Council, the Provincial Government of Sofala, and civil society organizations). We have come together to (1) create a space to debate, update, and (re)think the concept of climate justice and (2) bring our research results back to local and national communities, enabling their application. Our work is ongoing, and this chapter reports on some preliminary results.

Fieldwork and collaboration with local people have provided invaluable opportunities for the team to rethink the concept of climate justice in the Mozambican contest. Our participatory research has also allowed community groups to become directly involved in the process of creating climate risk maps. In order to add accurate information to maps that can both increase and reduce the climate risks of a given community, we have found it necessary to *actively listen* to community members.

Additionally, the project team has conducted a literature review and survey of the climate change and climate justice discourse in Mozambique. We surveyed mentions of the terms 'climate change' and 'climate justice'[2] in newspapers and government legislation and policy documents in order to understand whether and how the concept of climate justice is being used and to what extent this is influencing the development of public policy on climate change. In addition to newspaper and document analyses, we are reviewing the process of policy implementation as well as donor/funder guidelines for the design and implementation of the national Climate Change Strategy for Mitigation and Adaptation and other international mechanisms, such as the Reducing Emissions from Deforestation and Forest Degradation (REDD+) programme. In accordance with the research priorities of various team members, this analysis combines research carried out in Maputo, the capital of Mozambique; in the city of Beira and the village of Nhambita in Sofala province; and in Zambézia province (see Map 2.1). Team members conducted several fieldwork trips between 2019 and 2021 in Sofala and Zambézia provinces. In Sofala province we visited the Nhambita Community, near a REDD+ reforestation project, interviewed peasants affected by the project, and visited their farms and the tree-planting areas. In Zambézia province, we visited communities residing in the buffer zone of the Gilé National Reserve, a re-established conservation area under REDD+. The goal was to understand how the reestablishment of the Reserve impacted rural people's lives and how effectively the community development projects set up under REDD+ compensated for local people's loss of livelihood strategies. In Beira, we interviewed local officials and community groups about storms, flooding, infrastructure damage, and needed policy support. Finally, in Maputo, where most of us are based, we have explored climate justice concepts and their applicability with various academic, NGO, and government groups in a range of settings. Throughout our meetings, interviews, and focus groups, we emphasize two-way communication, listening more than talking, with mutual information-sharing to raise the awareness of all involved about the climate situation our country faces, what needs to be done, and what we can do. Our collaborative research is ongoing, and future stages will explore these topics further and pursue specific proposals related to climate justice in Mozambique.

This chapter is organized as follows. Section 2 overviews the context for understanding climate justice in Mozambique, outlining interactions among civil society organizations, government climate policies, and legal frameworks. Section 3 discusses national climate-related policies and also includes a summary of how 'climate justice' is treated in Mozambican news reports and public documents. Section 4 considers Mozambique's geographic vulnerability to extreme weather events and its recent history of climate-related disasters, with a special focus on the city of Beira, and the climate resilience led by community and civil society groups in the face of extreme weather events. Section 5 pulls together our conclusions on how people around the country are advancing a dynamic form of climate justice in Mozambique.

2   Framing Climate Justice in Mozambique

The term *justice* seems well known and mostly understood, but time and time again, we see that there may be many understandings of this concept. According to Ngoenha (2019, 5), justice is 'the stitching together of consensus', and consensus is the most effective political means to eliminate conflicts. In a society where there is no consensus regarding the laws that govern communities and resources, or the institutions which make and implement those laws, this lack of consensus produces disappointments, misunderstandings, and sometimes conflict.

In Mozambique, the environmental movement has a long history, with NGOs playing a notable role in increasing the prevalence and quality of environmental debates. The term *climate justice* was not often used until quite recently. The mainstream use of this term in Mozambique can be largely attributed to the work of the NGO Justiça Ambiental (JA!). For example, since 2019, JA! has organized events to recognize International Climate Justice Day (29 November), and it publicizes global climate reports and other current events, highlighting climate justice impacts (Lemos 2022; Cabo Delgado 2020). Through its collaborations with international environmental organizations such as Friends of the Earth International, JA! helps provide access to global climate justice debates.

JA! is a volunteer-based Mozambican NGO, founded in 2004 by a group of people who had grown concerned about the way Mozambique was developing in the unregulated global economy. Thanks to the work of JA!, Mozambicans started to hear more about environmental justice issues. But the fight for environmental rights in Mozambique precedes this NGO.

The 1997 creation of Mozambique's Environment Law (Lei do Ambiente, No. 20/97) opened a consolidated space to defend the environment. Since then, this has happened in a more structured and systematic way, and we have gradually witnessed the emergence of the concept of climate justice in Mozambique. Eduardo Mondlane University introduced an environmental education diploma programme in 2010 (Cossa 2013), and university courses are increasingly being offered on environmental issues, which suggests greater interest in actions to defend the environment. Social movements demanding environmental justice, by adopting a 'justice' frame, open the way for considering 'justice' in a climate change context as well. In our view, it is crucial to rethink this concept and the basic structures for more inclusive public policies that respond to the issues faced by those most affected by climate change. Although the concept of climate justice seems to be a fundamental priority for social movements fighting for environmental justice, the use of the term climate justice is inconsistent.

In recent years, we have seen civil society organizations redefine themselves within the climate change context; for example, in 2011 the NGOs Livaningo, Kuwuka JDA, Kulima, and six others formed the National Civil Society Organizations Platform on Climate Change in Mozambique (PNOSCMC) to strengthen civil society advocacy on climate adaptation (Danida 2019, 7). The Mozambican

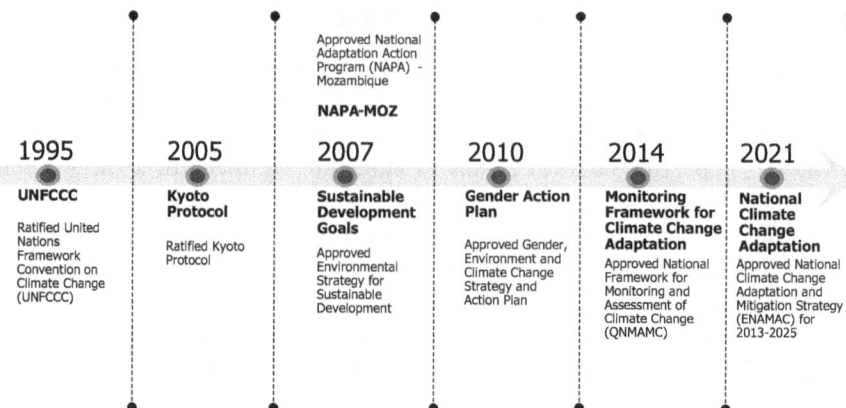

FIGURE 2.1 Timeline of Mozambique's approval/ratification of climate change agreements.

Source: Authors.

government has also refocused on climate change – the greatest example of this being the closing in 2020 of the National Institute for Disaster Management and the creation of the National Institute of Disaster Risk Management and Reduction. Additionally, the Disaster Risk Management and Reduction Law (No. 10/2020) was approved. 'This Law establishes the Disaster Risk Management and Reduction Legal Regime, which comprises risk reduction, disaster management, sustainable recovery for the construction of human, infrastructural and ecosystem resilience, as well as adaptation to climate change' (Mozambique 2020, 1163). Climate and environmental justice goals are implicitly present in this law. These recent actions have accompanied the approval of a body of laws and strategic documents that seek to address climate change, as shown in Figure 2.1.

These legal instruments create spaces and opportunities for Mozambicans to think about and manage the effects of climate change. Furthermore, they open space for practical actions to operationalize the principles of climate justice – something that unfortunately does not currently often happen.

3 Climate Policy in Mozambique: Does Climate Justice Matter?

Mozambique has been actively creating mechanisms to respond to its extreme climate vulnerability, from ratifying various international agreements on climate change to restructuring its administrative mechanisms to manage the issue. With external funds (loans and grants), the Mozambican government followed donor and funder guidelines towards the conception and implementation of a Climate Change National Strategy for Mitigation and Adaptation. As a result, in November 2012, the national strategy was approved by the Council of Ministers. Mainly based on the United Nations Framework Convention on Climate Change (UNFCCC), this document 'establishes action guidelines to build resilience, including the reduction of climate risks, in communities and in the national economy and promote low-carbon development and the green economy, through its integration in the sectoral and local planning process' (MICOA 2012; MITADER 2017). The strategy has three main objectives: (1) to make Mozambique resilient to the impacts of climate change by reducing climate risks, restoring and ensuring the sensible use and protection of natural capital; (2) to reduce GHG emissions by sustainably using natural resources, promoting access to financial and technological resources, and reducing degradation by promoting low-carbon development; and (3) to build institutional and human capacity to implement this strategy. However, specific climate justice provisions are missing.

The creation of the Department of Climate Change (DMC) in the new Ministry of Land, Environment and Rural Development (MITADER), which replaced the Ministry of Coordination of Environmental Affairs (MICOA) in 2015, is particularly noteworthy (World Bank 2013). Similarly, there is a strong movement by other ministries to address the country's extreme vulnerability.

While the government develops climate policies and strategies, researchers and scholars produce knowledge to inform and influence decision-making, as does civil society. The latter also ensures and monitors the implementation of policies and strategies. The media also plays an important role as it helps to inform the public. According to Castells (2000), the social and power relations of contemporary societies are mediated by various media modalities.[3] Thus, the 'public' space is, to a large extent, influenced by the media, which outlines what is or is not legitimate, and what should or should not be prioritized. This helps to explain the use or absence of certain concepts, and how these are adopted by communities and public authorities who use them to develop public policy.

To investigate the use and adoption of the concept of climate justice, we examined the largest and oldest newspaper in Mozambique, *Notícias* (founded in 1926). Published in Portuguese in Maputo, the capital city, and with a national online edition including sections focusing on news from the provinces, *Notícias* receive support from the Mozambican government (Bussotti 2021; see also Vieira 2011). We looked at whether the term climate justice appeared, and if so in what context(s). To track possible changes over time in recent years, this was done for the months of March and April (the end of the rainy season and cyclone season) in 2019, 2020, and 2021 (see Figure 2.2).

As illustrated, the term *climate justice* was not used in the country's largest newspaper at all during the prime cyclone season in 2019–2021. Although climate change was often mentioned in the newspaper, it was seldom the main focus of discussion, which in a way minimizes its importance. A few articles discussed vulnerability or other topics that implicitly relate to climate justice, without using the term explicitly.

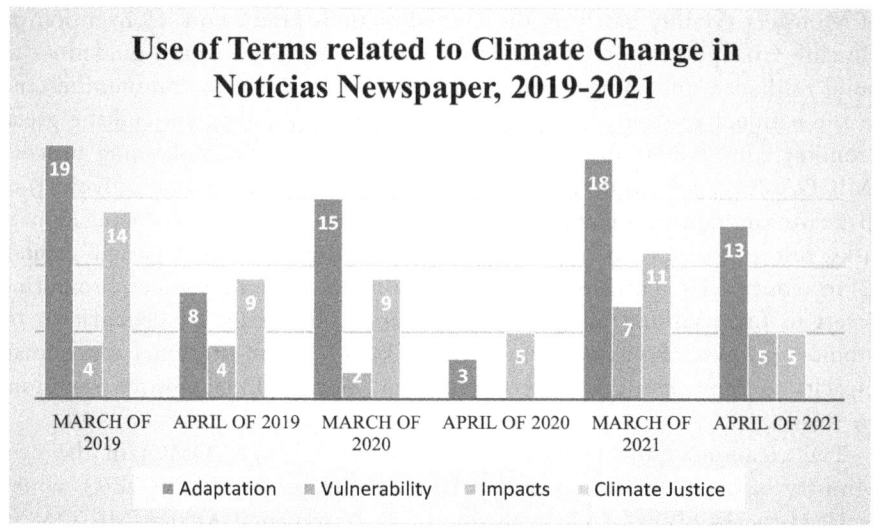

FIGURE 2.2 Mentions of climate-related terms in *Notícias* news articles, 2019–2021.
Source: Authors.

An in-depth analysis of the excerpts reveals repetition of certain news. Although this repetition reinforces certain facts about climate change, it also undermines the debate, as attentive readers may perceive this repetition as lack of information or news, or simply an editing strategy to fill the newspaper. Furthermore, this indicates a lack of intention/purpose when reporting on a certain climate change-related issue.

When *climate change* was centrally addressed, this was often done by representatives of UN organizations. As an advocate for the centrality of this concept, the UN in general seeks to mobilize resources to finance mitigation and especially adaptation actions, which has sometimes resulted in funding related to REDD+, for example. When climate change is discussed as a central topic, it is possible to see how multisectoral it is and how it affects various areas. To our disappointment, the term *climate justice* was not used even once in March 2019, even after Cyclone Idai (see Pereira 2019 – the government's post-Idai 'Needs Assessment', which also does not mention the term). A close analysis of Mozambique's national strategy for climate change mitigation and adaptation (outlined above) reveals a near absence of climate justice considerations. The strategy's objectives are intended to be realized through the implementation of multiple land-based policies such as REDD+ and climate-smart agriculture (CSA). This strategy is the country's main policy response to global environmental concerns. The policy focuses on several strategies and mechanisms to mitigate and adapt to climate change: REDD+, which aims at reducing emissions from deforestation and forest degradation, and CSA and other strategies to promote 'green' investments, such as tree planting and biofuel production. REDD+ interventions target 'indigenous' forests for conservation and community-based reforestation through community development projects. Many international studies have shown that REDD+ and other similar schemes negatively impact rural populations and their livelihoods through dispossession and resource-grabbing (Bruna 2019; Temper et al. 2020; Veltmeyer and Petras 2014; Fairhead et al. 2012).

CSA has also been criticized by small-scale farmers' movements and food justice activists throughout the Majority World. Its opponents see this mechanism as a continuation of a project first begun with the green revolution by the World Bank's poverty reduction projects and the corporate interests involved. This is the same industrial agriculture approach that increases GHG emissions, using the 'climate-smart' label to promote corporate practices as solutions to climate change (Anderson 2014; LVC 2014).

Government climate change policies and discourses thus turn a blind eye to the social, economic, environmental, and climate change vulnerabilities of the country, as well as the negative implications of extreme weather events and the need to adapt and mitigate. Moreover, the design, conception, and implementation of existing mitigation and adaptation policies do not reflect the historical causes and trajectory of the current environmental crisis. Although Mozambican climate policy emphasizes the need to integrate climate and development goals, it is disconnected from the wider development strategy of the country (Naess et al. 2015). Pilot climate mitigation and adaptation projects in Mozambique claim to be enhancing community development initiatives. For instance, the Gilé National Reserve under REDD+ in Zambézia province includes social development and income-generating projects for rural households and the implementation of CSA at the level of subsistence agriculture. The failed Nhambita Community Carbon Project in Sofala province sought to capture carbon through agroforestry and sell carbon credits on the voluntary markets. Its implementer, a UK-based company, Envirotrade, claimed that the project would alleviate poverty in the region. However, a study by Via Campesina showed that 'in addition to using their land to plant trees …, communities are also expected to protect and patrol a defined area of just over 10,000 hectares, from which Envirotrade also sells carbon credits through the REDD+ mechanism' (Via Campesina Africa 2012). Although most of these policies claim to benefit local communities through integrated development projects, evidence from Mozambique and our ongoing research on the Gilé National Reserve and Nhambita community carbon project show that REDD+ projects are highly problematic (Bruna 2022; Via Campesina Africa 2012). Some of the problems include: (1) lack of community participation and marginalization of local needs and aspirations; (2) inefficacy in the implementation of community development projects; (3) adverse implications for rural livelihoods regarding subsistence, food production, and access to key forest resources; (4) expropriation and poverty intensification; and (5) land and resource conflicts and fragmentation within communities.

An article in *Notícias* (April 2, 2019; p. A35) wryly highlights:

> Climate change has caused populations in the northern part of Gaza to face food shortages; this is due to the semi-arid climate in the districts of Chigubo, Massangena, Mapai, Guijá and Chicualacuala. The fields do not produce [yields], because it does not rain for a long period of the year, especially at times when there is no access to the park for hunting, which serves as an alternative for the subsistence / survival of the populations. I remember a local citizen who suggested that 'the elephants were going to vote in the elections, as they are more protected than the people'. It is curious, however, that no one openly says that they intend to hunt animals in the park or that they need the forest for the production and sale of charcoal.

Overall, we see an unfair and maladjusted policy situation under the guise of finding solutions to the climate and environmental crises. For instance, huge volumes of financial resources are directed to *mitigation* policies in rural Mozambique, despite the fact that Mozambique's CO_2 emissions per capita are among the lowest in sub-Saharan Africa and the world – around 0.1 metric tonnes compared to the sub-Saharan African average of 0.7–0.8 metric tonnes for low-income countries, and 3.5 metric tonnes for middle-income countries (World Bank 2007). Multiple policies and projects are designed in a top-down manner, disregarding local and national priorities and aspirations to meet global environmental goals, while undermining rural development in Mozambique. REDD+ projects elsewhere have been criticized for having poor engagement with local communities, and studies also point to the potential risks of land-grabbing by outsiders and loss of local user rights to forests and forest land (Larson 2013; Duchelle et al. 2018).

In sum, we identify an absence or disregard of the principles of climate justice in the conception, planning, and implementation of public policies to address climate change in the country, not only within Mozambique but also by international partner organizations. This clearly constitutes a call for more participation by vulnerable social groups in designing climate policies and in climate justice debates. In Section 4, we go back to basics: what are Mozambique's major climate vulnerabilities from a climate justice perspective, and what are local people's options for addressing them in the absence of government support or justice-oriented policy interventions?

4 Vulnerability and Resilience: Grassroots Climate Justice

As noted, cyclones and resulting floods are the most prominent and catastrophic extreme weather events affecting the country. According to the Mozambican Ministry for the Coordination of Environmental Affairs, the cyclones that often threaten Mozambique form either in the Indian Ocean or the Mozambique Channel which separates the mainland from Madagascar (MICOA 2005). Most of the cyclones that devastate Mozambique reach the coast between Pemba (population 175,000) and Angoche, or near the city of Beira, the second largest city in the country (population nearly 600,000), which is located along the Pungue river estuary (see Map 2.1). The port of Beira, a gateway for the central interior of the country as well as the landlocked countries of Zimbabwe, Zambia, and Malawi, is located 'in a marshy region and on coastal dunes' (Manuel 2015, 7).

The cyclone season generally runs from November to April, peaking in January. Although on average there is one cyclone per season, they have been increasing in frequency and intensity, with concomitant impacts on local populations (see Table 3.1). Over the past 20 years, in the period between 1999 and 2000, there were three, while between 2019 and 2021, Mozambique was affected by five cyclones (Idai, Kenneth, Chalane, Eloise, and Guambe), two of which reached category 4 intensity, causing catastrophic flooding with wind gusts of around 230 km/hour. This change in the annual frequency of cyclones or severe depressions may be attributable to climate change (Hope 2019).

The data presented in Table 3.1 make it clear that a new cyclone scenario is developing in Mozambique. We are witnessing (1) a change in the time of year

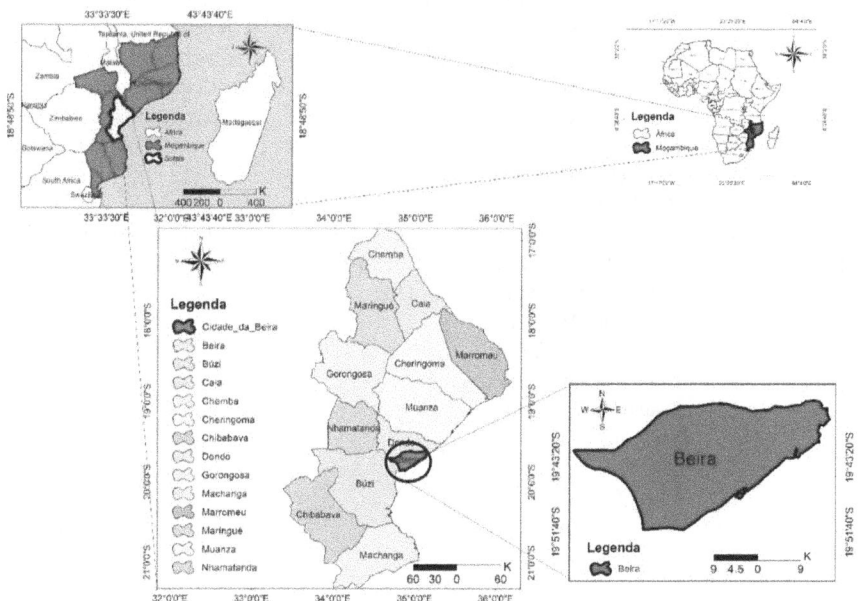

MAP 2.1   The city of Beira in its national context.

Source: Assembled by authors Bento Paulo Rafael and Elton Augusto da Amélia Fé using maps from DNA (2008), mapping database of territorial units in Mozambique.

TABLE 3.1 2019–2021 cyclone events, characteristics, and impacts in Mozambique

Cyclone/ depression name and date	Entry point and affected areas (provinces and countries)	Event characteristics	Social impact
Idai –1–19 March 2019	Entry point: City of Beira, Mozambique; affected areas: Provinces of the central region of the country, Manica, Sofala, Zambézia, Malawi and Zimbabwe	Category 3; wind speed 195–205 km/h; one of the worst tropical cyclones on record in the southern hemisphere	518 casualties, more than 2.5 million people experienced the direct effects of the cyclone, with hundreds of thousands in need of help; 400,000 displaced people needed urgent help, valued at more than 40 million dollars
Kenneth –17– 28 April 2019	Entry point: City of Pemba, Mozambique; affected areas: Cabo Delgado Province, Madagascar, and Tanzania	Category 4; wind speed 230–305 km/h; the strongest tropical cyclone to make landfall in Mozambique on record	52 casualties and damage estimated at (at least) $100 million
Chalane –19 December 2020–3 January 2021	Entry point: City of Beira, Mozambique; affected areas: Sofala, Zambézia and Manica Provinces, Madagascar and Zimbabwe	Category 2; wind speed 110 km/h	73,500 people (14,421 families) affected, 7 deaths and 13 injuries; 475 classrooms were destroyed, affecting around 57,000 students
Eloise –16–25 January 2021	Entry point: City of Beira, Mozambique; affected areas: Sofala Province, Manica, Tete Zambézia, Malawi, Zimbabwe, South Africa, and Swatine	Category 2; wind speed 150–165 km/h	19 fatalities and 4 missing

TABLE 3.1 (Continued)

Cyclone/ depression name and date	Entry point and affected areas (provinces and countries)	Event characteristics	Social impact
Guambe–10–23 February 2021	Entry point: Vilanculos, Mozambique; affected areas: Inhambane Province, Gaza, Maputo, South Africa, and Eswutine	Category 2; wind speed 120–155 km/h	4938 families had their tents and shelters destroyed/partially destroyed. Of these, 4393 are located in Sofala, and 545 in Manica province. The most affected districts are Dondo (1793 households), Buzi (1629), Nhamatanda (802), and Sussundenga (471). Two schools near the sites were damaged. Both are located in Sofala, in the resettlement sites of Mutua (Dondo district), and Ndedja 1 (Nhamatanda district). Five health centres near sites were also damaged, in the localities of Maxiquiri alto/Maxiquiri 1 (Buzi), Macarate and Mutindiri sede (Chibabava), Mutua (Dondo), and Ndedja 1 (Nhamatanda).

Source: Authors.

when cyclones take place, (2) an increase in the number of cyclone events in Mozambique, and (3) more adverse or catastrophic impacts. This combination makes it difficult if not impossible for the country to properly respond to these events, given its existing financial, infrastructure, and organizational challenges. Additionally, cumulative impacts become more evident. For example, Cyclones Chalane and Eloise, which hit the city of Beira, affected the same people who had been previously affected by Cyclone Idai. Although Cyclones Chalane and Eloise happened two years after Idai and Kenneth, Beira was not able to respond any more effectively to the negative impacts of these later events. These events highlighted the deep social inequality and precarious conditions faced by the poor, opening space for a deeper debate on the effective participation of vulnerable groups in climate change discussions in Mozambique.

To explore whether such negative impacts accumulate over time, and how they disproportionately affect those population groups who are at greatest risk, it makes sense to think about climate risk through a social and climate justice lens. It is necessary to create possibilities of inclusion and participatory opportunities for citizens for an effective and quick response to each event, so that human exposure and vulnerability can be reduced in the coming years. This may involve creative new solutions and risk mitigation approaches to meet the needs of local populations who are not well served by existing policies, international aid, or market-based strategies.

For example, Acselrad (2002) comments that people who highlight the importance of a logical relationship between social injustice and environmental degradation are those who do not necessarily trust the market as an instrument to overcome environmental inequality and promote the principles of environmental justice. Acselrad (2002, 51) states:

> There is clear social inequality in the exposure to environmental risks, resulting from a logic that goes beyond the simple abstract rationality of technologies. Combating environmental degradation offers us the opportunity to obtain democratization gains and not just gains in efficiency and market expansion. This is because there is a logical connection between the exercise of democracy and the capacity of a society to defend itself against environmental injustice.

Our ongoing research in Beira involves interviewing people affected or displaced by the repeated cyclones, along with government officials, community leaders, and civil society organizations, to map local needs and views on the best ways to advance climate justice. We are drawing on principles of citizen science, or the idea that breaking down barriers between academic researchers and community members recognizes that all have a role in the production of scientific knowledge. These concepts have also guided our practical actions as researchers, brought light to the exclusion of the most vulnerable in climate change debates, and shown how the use and adoption of climate justice-oriented processes of education, consultation, and political participation, besides ideally influencing the development of climate policy, can empower community members and organizations to take action themselves.

Socioeconomic resilience is important in addressing climate change, especially from a climate justice perspective. Mozambican president Filipe Jacinto Nyusi spoke during the 2021 high-level virtual dialogue on COVID-19 Emergencies and Climate Change in Africa, which included discussions on strategies to combine

COVID-19 emergency responses, resilience/capacity building, and adaptation to climate change. Speaking of Mozambique's experience, Nyusi (ADBG 2021) said:

> We suffer from prolonged drought, intense heat, floods, cyclones and other associated factors, such as rising sea levels, saline intrusion, and wind resulting in damage, including the loss of thousands of human lives, public infrastructure and private facilities such as health units, schools, roads, bridges, energy transport networks and residences ... We are in a conundrum ... On the one hand, there is a decrease in national and international funding for resilience-building and climate change adaptation programs, [while] other social and economic determinants further aggravate[e] the situation, reversing previous gains in financing for development.

This highlights the need for policies to establish a framework for managing and responding to the impacts of extreme weather events as well as reducing those impacts on vulnerable communities. An international tool for increasing socioeconomic resilience is the Sendai Framework for Disaster Risk Reduction 2015–2030, known as the Sendai Protocol (UNSIDR 2015), considered to be the most important instrument for disaster risk reduction to date. The Sendai Protocol, which replaced the Hyogo Framework for Action 2005–2015, establishes four priority actions to increase the resilience of vulnerable communities to disasters in the context of sustainable development: (a) identify and understand disaster risks; (b) strengthen governance on disaster risks and capacity to manage these risks; (c) invest in resilience for disaster risk reduction; and (d) be prepared for an effective response and improve recovery, rehabilitation, and reconstruction.

One way to apply this framework of priorities in Mozambique is to consider the issue of socioeconomic resilience in relation to the expenses the country incurs with each extreme weather event (see Figure 2.3). From this analysis, it becomes clear not only that disaster-related costs have tended to increase, but that the country is unable to respond and/or cover the financial costs that arise following each sequential event. According to the Ministry of Economy and Finance (2020), the options for financing natural disasters require a special strategy, given the difficulty in predicting the magnitude of the shock. The state budget allocates 0.1% or more of state revenue to the Contingency Plan (CP), which is aimed at the prevention and management of natural disasters in Mozambique. CP funds are being used to cover a growing proportion of the damage costs due to natural disasters, and even so, recent disaster costs (estimated at 0.05% of Mozambican GDP in 2019, and rising) exceed what can be paid from the national budget including contingency funds.[4]

This relationship between the cost of events and what is allocated to respond to those events is fundamental in thinking about vulnerability. For Ojima and Marandola Jr. (2012), vulnerability is understood as the reverse of sustainability, and risks are structurally produced; in this sense, the promotion of specific

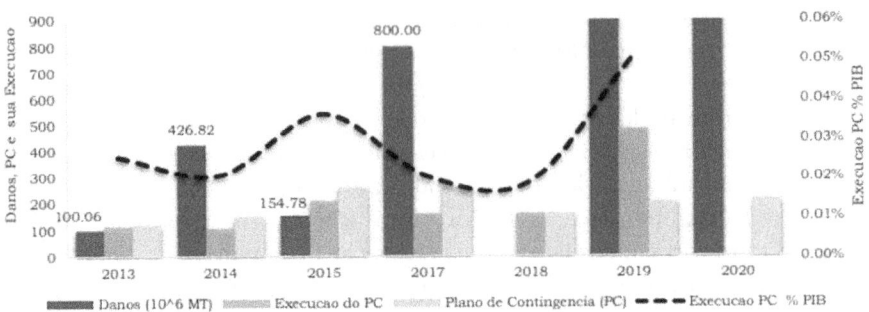

FIGURE 2.3 Mozambique's natural disaster financing, 2013–2020.

Source: Ministry of Economy and Finance (2020) citing calculations based on data from MEF/CGE and INGC (2013–2019). Amounts are given in million meticais.

management actions aimed at reducing environmental vulnerability also creates new opportunities for sustainable development. According to these authors, because of social inequalities and omnipresent risks of many kinds, no person (or group, city, region, country, etc.) can be totally vulnerable or completely protected (Ojima and Marandola Jr. 2011). Magis (2010, 401) similarly argues that 'communities can develop resilience by actively building and engaging the capacity to thrive in an environment characterized by change … community resilience is an important indicator of social sustainability'.

Thus, it would be impossible to create a completely sustainable city, for example, since urban sustainability implies much more than a city that recycles its garbage, consumes clean energy, and reduces its GHG emissions; unaddressed structural social inequities make sustainability a 'distant utopia' (Ojima and Marandola Jr. 2011, 23). This makes perfect sense in the context of Mozambique, mainly because people and groups already bring with them vulnerabilities created by other sectors such as health, education, poverty, etc. Although these arguments were originally made in reference to Brazilian cities, they clearly apply to Mozambique as well.

Unfortunately, Mozambique is far from being socioeconomically resilient, and existing public policies aimed at fostering socioeconomic resilience are few and not effective. Going far beyond individual and group vulnerabilities, this is because the country itself depends on foreign aid to supplement its budget. Despite these challenges, Mozambicans have been exploring other mechanisms to advance sustainability. Section 5 discusses the demonstrated, and potential, contributions of climate justice perspectives in these initiatives.

5 Conclusion: Dynamic Climate Justice from the Ground Up in Mozambique

More than almost anywhere else, people in Mozambique are subject to grave climate injustices. Due to its degree of exposure and vulnerability to climate change, the country is forced to deal with severe, interrelated, and escalating climate challenges. Economic crises, also far outside Mozambique's influence or control, heighten the impacts and lessen the country's ability to respond. A climate justice framing of the country's climate change mitigation/adaptation policies and strategies is useful for setting priorities: the risks and needs of the poorest should be addressed first. In a procedural sense, this means that spaces and opportunities must be created for the most vulnerable to articulate their needs and advocate for locally appropriate solutions – the basic principle of procedural justice.

Our analysis of political discourse in the national newspaper *Notícias* indicated, however, that climate change itself is rarely the focus of news articles, and we found little indication of climate justice being explicitly addressed. Our analysis also indicated that national institutions seldom address the issue or concept of climate change as a focus, nor do they always present the concept correctly. We did find UN statements directly addressing climate change (though not climate justice), which suggests some clarity and cohesion between UN organizations at least. Hopefully, this cohesion will influence the development of more equitable climate policies in Mozambique as well as other countries.

There is a growing focus on climate justice at the grassroots, supported by academics and civil society organizations, who are organizing to reduce the impacts of climate change on the poor. For example, ObservA, along with the environmental NGOs Livaningo and Centro Terra Viva, organized a symposium in Maputo about climate action on 7–8 November 2019. Government officials and JA! participated, and climate justice issues received prime focus. But the Mozambican government's implementation of some strategies clearly demonstrates a disconnect between theory and practice. REDD+ projects, highlighted in World Bank-funded government programmes, are far from promoting climate justice and supporting equitable development. Instead, they lead to lack of participation and marginalization of local needs and aspirations; ineffective community development projects; adverse implications for rural livelihoods, food production, and access to forest resources; expropriation and intensification of poverty and land and resource conflicts; and fragmentation within communities. In short, mechanisms like REDD+ are exacerbating and perpetuating injustices under the guise of finding solutions to climate and environmental crises.

By articulating and focusing on these challenges, denouncing climate injustices, linking researchers and students with community groups, local NGOs, and activists, and collaborating with progressive international organizations to share Mozambican perspectives with others, Mozambican civil society organizations, in partnership with academics and other activists, are creating spaces to support, safeguard, and defend those Mozambicans who are most wronged by climate chaos.

Notes

1 The Republic of Mozambique is located in the southern hemisphere. Its climate is tropical, with days and nights about the same length year-round, which explains its

relatively small temperature variations (Muchangos 1999, 13). Due to Mozambique's geographic location, the types of weather are determined by the location of the low equatorial pressure zone, tropical anti-cyclonic cells, and Antarctic polar fronts. The hot, rainy season starts in October and ends in March. Cyclones are more frequent in January–February and cause thunderstorms and heavy rains, with strong winds that at times reach more than 100 km/h (Muacuveia 2019).

2   In Portuguese, these terms are 'mudanças climáticas' and 'justiça climática' – with definitions as broad and wide-ranging as in English, closely following the academic literature. In African languages which are the mother tongues of most Mozambicans, these terms are harder to translate and much less often used.

3   Along with power grid limitations, Mozambique's limited internet coverage (with only about 5% of the population having internet access) means that radio is the main source of information throughout the country. Using radio and communications via mobile phones for disaster preparedness is an important way to build climate justice-focused grassroots risk management capabilities (Nhamo and Eloff 2021).

4   Mozambique is forced to turn to the World Bank and other international agencies to assist with disaster costs, which were estimated at $3.2 billion in 2019 (Pereira 2019, 5).

References

Acselrad, H. (2002). Justiça ambiental e construção social do risco. *Desenvolvimento e Meio ambiente*, 5.

African Development Bank Group (ADBG). (2021, 7 April). GCA Leaders' Dialogue: H. E. Filipe Nyusi's Remarks. Accessed 18 June 2022. www.afdb.org/en/events/leaders-dialogue-africa-covid-climate-emergency

Amaral, G. A. D. (2018). *Mudanças ambientais, percepções de risco e estratégia de adaptação aos eventos extremos em Moçambique: estudo de caso em Machanga*. Tese de Doutoramento, Universidade Estadual de Campinas.

ANAC. (2015). National Strategy and Action Plan of Biological Diversity of Mozambique (2015–2035). República de Moçambique, Ministério da Terra, Ambiente e Desenvolvimento Rural. www.cbd.int/doc/world/mz/mz-nbsap-v3-en.pdf.

Anderson, T. (2014). Why 'Climate-Smart Agriculture' Isn't All It's Cracked Up To Be. *The Guardian*, 17.

Bruna, Natacha (2019). Land of Plenty, Land of Misery: Synergetic Resource Grabbing in Mozambique. *Land*, 8(8): 113.

Bruna, Natacha (2022). A Climate-Smart World and the Rise of Green Extractivism. *Journal of Peasant Studies*. https://doi.org/10.1080/03066150.2022.2070482.

Bussotti, Luca (2021). The Coverage of Non-communicable Diseases in Mozambique: The Case of the Newspaper *Notícias* (2006–2018). *Saúde Social*, 30(2). https://doi.org/10.1590/S0104-12902021190308.

Cabo Delgado (2020, 25 June). Stop the Ravaging of Cabo Delgado. Justiça Ambiental. www.foei.org/wp-content/uploads/2020/06/EN-Moz-letter-with-all-signator ies-final-2.pdf.

Cossa, Eugenia (2013). Bringing Together Community and Academic Knowledge: The Eduardo Mondlane University Faculty of Education Environmental Education Club. In P. E. Perkins, ed., *Water and Climate Change in Africa* (pp. 51–56). Routledge.

Danida (2019). Natural Resources Management Committees in Sofala: Governance, Rights and Climate Change. www.sustainableenergy.dk/wp-content/uploads/2020/01/6.2-Mozambique-Component-description.pdf.

Duchelle, A. E., G. Simonet, W. D. Sunderlin, and S. Wunder (2018). What Is REDD+ Achieving on the Ground? *Current Opinion in Environmental Sustainability*, 32: 134–140.

Eckstein, D., V. Kunzel, and L. Schäfer (2021). *Global Climate Risk Index 2021*. Briefing Paper. Germanwatch. www.germanwatch.org/sites/default/files/Global%20Clim ate%20Risk%20Index%202021_2.pdf.

Fairhead, J., M. Leach, and I. Scoones (2012). Green Grabbing: A New Appropriation of Nature? *Journal of Peasant Studies*, 39(2): 237–261.

Hope, M. (2019). Cyclones in Mozambique May Reveal Humanitarian Challenges of Responding to a New Climate Reality. *Lancet Planetary Health*, 3(8): e338–e339.

Instituto Nacional de Gestão de Calamidades (INGC) (2009). Study on the Impact of Climate Change on Disaster Risk in Mozambique: Synthesis Report. www.biofund. org.mz/biblioteca_virtual/main-report-ingc-climate-change-report-study-on-the-impact-of-climate-change-on-disaster-risk-in-mozambique.

La Via Campesina (LVC) (2014). Climate: Real Problem, False Solutions. No. 4: Climate-Smart Agriculture. https://viacampesina.org/en/climate-real-problem-false-soluti ons-no-4-climate-smart-agriculture/.

Larson, A. M., M. Brockhaus, W. D. Sunderlin, A. Duchelle, A. Babon, T. Dokken, … & T. B. Huynh (2013). Land Tenure and REDD+: The Good, the Bad and the Ugly. *Global Environmental Change*, 23(3): 678–689.

Lemos, Anabela (2022, 26 March). Pelo menos 132 mil milhões de dólares em financiamento para combustíveis fósseis estão a bloquear Africa de uma transição justa, mostra novo relatório: países africanos mantidos em estrangulamento de combustíveis fósseis por dinheiro do exterior. https://justica-ambiental.org/2022/03/26/pelo-menos-132-mil-milh oes-de dolares-em-financiamento-para-combustiveis-fosseis-estao-a-bloquear-africa-de-uma-transicao-justa-mostra-novo-relatorio/

Magis, K. (2010). Community Resilience: An Indicator of Social Sustainability. *Society and Natural Resources*, 23(5): 401–416.

Manuel, A. F. (2015). Análise da perigosidade de tsunami para a cidade da Beira, Moçambique. Doctoral dissertation. Repositório da Universidade de Lisboa, Communities and Collections, Faculdade de Ciências (FC), FC - Dissertações de Mestrado

Ministry for Coordination of Environmental Affairs (MICOA). (2005). *Avaliação da vulnerabilidade as mudanças climáticas e estratégias de adaptação*. MICOA.

Ministry for Coordination of Environmental Affairs (MICOA). (2012). Estratégia Nacional de Adaptação e Mitigação de Mudanças Climáticas: 2013–2025. https://cgcmc.gov. mz/attachments/article/194/Estrategia%20Nacional%20de%20Adaptacao%20e%20Mi tigacao%20das%20Mudancas%20%20Climaticas%20versao%20final.pdf.

Ministério de Economia e Finanças. (2020). *Relatório de Riscos Fiscais 2021*. Edição: Direcção de Gestão do Risco. Maputo. https://www.mef.gov.mz/index.php/todas-publicacoes/ instrumentos-de-gestao-economica-e-social/relatorios-de-riscos-ficais/930-relato rio-de-riscos-fiscais-2021/file?force-download=1

MITADER. (2017). *Environmental and Social Management Framework*. Ministry of Land, Environment, and Rural Development.

Moçambique. (2020). Lei n.º 10/2020: Lei de Gestão e Redução do Risco de Desastres.

Muacuveia, R. R. M. (2019). Urbanização contemporânea em Moçambique: papel dos instrumentos de planejamento urbano na ocupação do espaço. Doctoral dissertation. Universidade Federal de Uberlândia.

Muchangos, A. D. (1999). *Moçambique, paisagens e regiões naturais*.

Næess, Lars Otto, P. Newell, A. Newsham, J. Phillips, J. Quan, and T. Tanner (2015). *Climate Policy Meets National Development Contexts: Insights from Kenya and Mozambique*. Global Environmental Change. https://doi.org/10.1016/j.gloenv cha.2015.08.015.

Ngoenha, S. E. (2019). *(In)justiça: terceiro grande consenso moçambicano.* Real Design Editora.

Nhamo, Godwell, and Mariki M. Eloff (2021). ICT Readiness for Disaster Risk Reduction: Lessons from Tropical Cyclone Idai. In G. Nhamo and K. Dube, eds., *Cyclones in Southern Africa, Vol. 2: Foundational and Fundamental Topics.* Springer, pp. 87–103. https://doi-org.ezproxy.library.yorku.ca/10.1007/978-3-030-74262-1.

Ojima, R., and Marandola Jr., E. (2010). 'Indicadores e políticas públicas de mudanças climáticas: vulnerabilidade, população urbanização.' *Revista Brasileira de Ciências Ambientais,* vol. 18. www.researchgate.net/publication/52012448_Indicadores_e_politicas_publicas_de_adaptacao_as_mudancas_climaticas_vulnerabilidade_populacao_e_urbanizacao/citations#fullTextFileContent

Ojima, R., and E. Marandola Jr. (2012). O desenvolvimento sustentável como desafios para as cidades brasileiras. *Cadernos Adenauer,* 1: 23–36.

Pereira, Francisco (2019). Mozambique Cyclone Idai Post Disaster Needs Assessment. www.ilo.org/wcmsp5/groups/public/---ed_emp/documents/publication/wcms_704473.pdf

Roberts, J. T., and B. C. Parks (2009). Ecologically Unequal Exchange, Ecological Debt, and Climate Justice: The History and Implications of Three Related Ideas for a New Social Movement. *International Journal of Comparative Sociology,* 50(3–4): 385–409.

Temper, Leah, S. Avila, D. Del Bene, J. Gobby, N. Kosoy, P. Le Billon, J. Martinez-Alier, P. Perkins, B. Roy, and A. Scheidel (2020). Movements Shaping Climate Futures: A Systematic Mapping of Protests against Fossil Fuel and Low-Carbon Energy Projects. *Environmental Research Letters,* 15(2): 123004.

UNISDR. (2015). *Sendai Framework for Disaster Risk Reduction 2015–2030.* www.preventionweb.net/files/43291_sendaiframeworkfordrren.pdf.

Veltmeyer, H., and J. F. Petras (2014). *The New Extractivism: A Post-Neoliberal Development Model or Imperialism of the Twenty-First Century?* Zed Books.

Via Campesina Africa. (2012). Carbon Trading and REDD+ in Mozambique: Farmers 'Grow' Carbon for the Benefit of Polluters. https://grain.org/es/bulletin_board/entries/4531-carbon-tradingand-redd-in-mozambique-farmers-grow-carbon-for-the-benefit-of-polluters.

Vieira Mário, Tomás (2011). *Assessment of Media Development in Mozambique Based on UNESCO's Media Development Indicators.* UNESCO. https://unesdoc.unesco.org/ark:/48223/pf0000216942.

World Bank. (2013). Implementation Completion Report Review. https://documents1.worldbank.org/curated/pt/346681520728179052/text/Mozambique-MZ-Climate-Change-DPO.txt.

World Bank. (2007). *World Development Report 2008: Agriculture for Development.* World Bank.

3

THE INTERSECTION OF CLIMATE JUSTICE AND AGROECOLOGY IN PUERTO RICO POST-HURRICANE MARIA

Voices from the Ground

Thelma I. Vélez

1   Introduction

> Hurricane Maria landed on a legacy of austerity, neglect, and colonialism in Puerto Rico and opened the floodgates to those who prosper on the pain and loss of people of color, those responsible for climate change. Climate Justice is the resistance to a history of extraction of land and labor in the Global South. We know this is a fight for our survival and we are ready.
>
> *(Elizabeth Yeampierre J.D., co-chair of Climate Justice Alliance;*
> *Executive Director, UPROSE)*

When Hurricane Maria made landfall in Puerto Rico on 20 September 2017, the archipelago was decimated and the damage was unprecedented. In the days following the storm, roughly 3.4 million people lacked access to basic necessities, such as food, shelter, water, and medical care (Sutter and Ewan 2017; Severino et al. 2018). The US federal disaster response lacked gumption, and the weaknesses in the response aggravated the crisis, increasing vulnerabilities across Puerto Rico. Further exacerbating the situation were US policies that prohibited other nations from providing supplementary disaster relief for Puerto Ricans (Willison et al. 2019; Cortés 2018). The case of Puerto Rico post-Hurricane Maria exemplifies a critical phenomenon stemming from disasters, specifically the potential for socioecological instabilities to galvanize uprisings when state deficiencies become salient (Drury and Olsen 1998; Xu et al. 2016).

In the wake of Hurricane Maria, grassroots organizing championed a recovery centred on climate justice. The Our Power Puerto Rico #JustRecovery campaign (herein referred to as Just Recovery) took root within a week of the storm. The focus

DOI: 10.4324/9781003214021-4

of Just Recovery is, and has been, largely to support recovery and transformation in Puerto Rico by stimulating a shift away from ecologically and socially destructive systems towards alternatives empowering Puerto Ricans via agroecology and climate, social, and environmental justice initiatives (Rowsome 2017).

The work presented in this chapter is rooted in critical qualitative inquiry and postcolonial critical realism (Cannella et al. 2016; Tinsley 2021). Critical qualitative inquiry has emerged over the past several decades as a leading form of research that resists the positivist ontological struggles and redirects research towards a reflexive and decolonizing process. Postcolonial critical realism is an ontological framework that examines the structures and relationships between racialized and colonial discourses as embedded in the social world (Tinsley 2021). Scholars engaged in this avenue of work do not position themselves as objective or politically neutral observers who stand outside of the social world. Rather, they are committed to exposing and critiquing the forms of inequity and discrimination that are embedded in society, as well as capturing the lived experience, meaning, and interpretations of the persons most directly affected by the cases they are involved in (Cannella et al. 2016). Critical scholars also recognize that research is a power-oriented activity, and thus are intentional in finding ways to eliminate or reduce the unequal power dynamics (Rodriguez 2018).

While this chapter provides a scholarly contribution exploring Puerto Rico's ongoing struggles, it is critical to note that grassroots organizations and frontline communities in no way require scholarly validation of their lived experiences. This chapter is geared towards advancing discourse among the general public seeking to comprehend how climate justice mobilization and everyday acts of resistance provide a means to resist oppression and envision a just future. Thus, my intention here is not to validate the claims made by frontline communities; rather, this work is intended to amplify discourse on the struggles and efforts of communities and grassroots organizations experiencing the effects of climate change and also to highlight the solutions put forth by this specific frontline community.

The findings presented here are the result of a mixed-methods qualitative case study with data triangulated from participant observation, semi-structured interviews, and content and discourse analysis. Participant observation took place at public forums, *mercados* (markets), *charlas* (talks), *talleres* (workshops), and *brigadas* (brigades) held between January 2018 and June 2019. Agroecological promoters and producers across Puerto Rico have been hosting *brigadas*, *talleres*, and *charlas*, both on farms and at markets where growers congregate. The workshops and talks ranged in scope and topic. While some focus on farming strategies, such as climate-specific growing tactics, composting, and seed-saving techniques, other topics were enterprise-related, such as how to reach markets and form cooperatives and how to incorporate beekeeping to diversify income.

For the content analysis, I used secondary data from extant interviews with organizational leaders, organizational reports, and social media posts. Additionally, I conducted 15 in-depth interviews in Puerto Rico in the summer of 2019, with

Field Sites in Puerto Rico

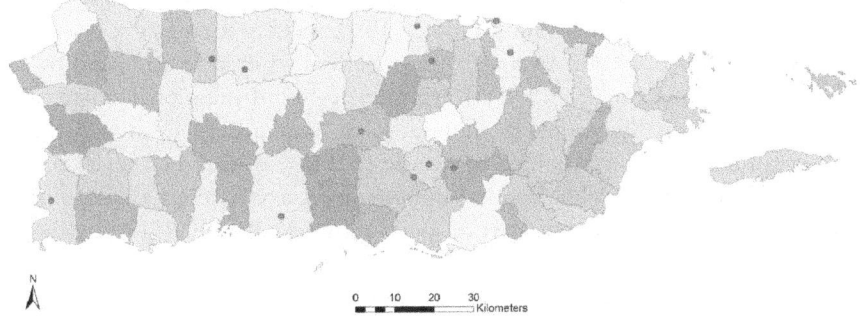

FIGURE 3.1   Map of field sites visited in Puerto Rico, June–July 2019.

Productores y Promotores de agroecología (Producers and Promoters of Agroecology), including some affiliated with Organización Boricuá and the Climate Justice Alliance. The interviewees were selected using a combination of convenience and snowball sampling and ranged in age from 23 to 68 years, including eight women and seven men. The data was hybrid-coded manually using methods described by Saldaña (2021) and Creswell (1994), with an initial round of in vivo and emergent coding and subsequent rounds of thematic coding. Interviews and coding occurred in Spanish; thus all quotes are presented in the speaker's native language, followed by the English translation.

Complementing this analysis is personal experience as a brigade volunteer on agroecological farms in Puerto Rico throughout June of 2019. I participated in three *brigadas* and volunteered alone on two other farms. I engaged in manual labour on these farms as a form of sweat equity (Sbicca 2015). Most of the help I provided entailed weeding, seeding, planting, and preparing soil beds under the hot tropical sun. My time on these farms remains one of my most prized aspects of this work, as it afforded me the opportunity to connect more deeply with the movement and community. See Figure 3.1 for the field sites in Puerto Rico.

2   Collective Agency and Community Resilience Framework

In this chapter I address several questions related to agroecological expansion and climate justice and resilience in Puerto Rico from the perspective of the people on the ground. I ask: (1) How can people ensure their communities are sustainable and resilient in the face of future climate disasters: what does the process entail? and (2) Why is agroecology so important to agroecological promoters and producers in Puerto Rico: what does it offer?

To address the first question, I draw on White's (2018) theoretical framework of Collective Agency and Community Resilience (CACR). The CACR framework builds upon social movement literature and theories centred on everyday strategies of resistance. It draws on the day-to-day activities that community members engage in to become more self-sufficient and self-reliant and embeds it within collective and political dimensions (White 2018). Everyday strategies of resistance refer to 'form(s) of resistance that are often overlooked or overshadowed by ... organized social movements ...[it is] typically less confrontational, incurs less repression, and is usually enacted by individuals or small groups' (White 2018, 6).

The term *collective agency*, coined by White (2018, 7), 'involves a social actor's ability to create and enact behavioural options necessary to affect their political future. As such, it is an intrinsic part of social activism'. White (2018) highlights that much of the scholarship on agency has historically focused on psychological origins and the impact agency has on the individual. She proposed the concept *collective agency* as a means to expand beyond the psychological and individual realm and apply it to the collective dimension of communities (White 2018). White (2018) centres her analysis on community dynamics, as she explores the intersection of agency, resilience, and adversity. The CACR framework focuses on the process of developing systems, structures, knowledge, skills, and relationships that help build alternatives to the existing political and economic structures that are oppressive and non-inclusive (White 2018). Thus, *community resilience* entails paying careful 'attention to community-based forms of social organization that plan for and respond to political violence, environmental/climate change, and other forms of natural and human-induced disasters' (White 2018, 8).

The construct of resilience is often described as the ability to rebound to a previous state post-shock (Chang and Shinozuka 2004; Kelman et al. 2016). While the term *resilience* is used by grassroots organizations and agroecologists in Puerto Rico, it is often coupled, or even replaced, with the term *resistance*. In Puerto Rico, the idea of returning to 'how things were' pre-hurricane has been outright rejected. Grassroots organizers explain that these communities do not wish to return to the unequal, extractive, and oppressive dynamics that have long prevailed (Yeampierre 2018). Instead, they wish to disrupt the status quo and reimagine a future where they have more autonomy and self-sufficiency. Semantics aside, in this chapter, I use these concepts resilience and resistance interchangeably noting that resilience in Puerto Rico is not intended as a means to return to old ways; rather, it is a path to resist outdated systems and prepare for future shocks, natural or sociopolitical. The use of the CACR framework aligns well with the study of agroecology in Puerto Rico, as it overlaps with agriculture and food practices as a form of resistance to oppressive political systems.

3   Hurricane Maria and Puerto Rico's Colonial Legacy

In the wake of Hurricane Maria, power was out on nearly 90% of the island, and the majority of the 3.4 million inhabitants lacked basic necessities (Sutter and Ewan 2017). Flooding covered most of the island, and downed trees and the debris of destroyed homes were scattered across roads. The island lost over 80% of its agricultural production, estimated at over $780 million (Robles and Sadurni 2017). The devastation received ample news coverage in the United States and across much of the world, with headlines such as 'About 1 million Americans without running water, 3 million without power. This is life one month after Hurricane Maria' (Sutter and Ewan 2017). Three months post-disaster, only 69% of power had been restored, making it the largest blackout in US history (StatusPR 2018).

Hurricane season in the Atlantic Basin, which includes the Atlantic Ocean, Caribbean Sea, and Gulf of Mexico, generally runs from 1st June through 30th November every year. Geographically, Puerto Rico and the broader Caribbean region are particularly vulnerable to hurricane activity. However, the 2017 Atlantic hurricane season, which Hurricane Maria spawned from, was one of the most intense and destructive seasons in the past 50 years (Vaccaro 2017; Johnson 2017). The National Oceanic and Atmospheric Administration (NOAA) recorded 17 storms in the 2017 season, with ten reaching hurricane status, six of which were major category storms (Vaccaro 2017).

Hurricane Harvey produced heavy rainfall and flooding in Texas, shattering previous storm-related rainfall records by over one foot (Belles 2017). Hurricane Irma, which made landfall in Florida, set a world record in terms of sustained intensity at 185-miles-per-hour winds for 37 hours (Belles 2017). Hurricane Maria fluctuated between category 4 and 5 as it tracked across the Caribbean. At the time Hurricane Maria hit, both the US Virgin Islands and Puerto Rico were recovering from the damage of Hurricane Irma just two weeks prior. The sheer size of Hurricane Maria was more than five times the size of the entire archipelago of Puerto Rico, see Figure 3.2.

In Puerto Rico, immediate shortages of critical resources, as well as damaged infrastructure and communication pathways, led to social instability and increased vulnerability for frontline communities. While US states and territories are eligible to receive federal disaster assistance, aid for Puerto Rico was poorly handled (FEMA 2018). Puerto Rico's residents needed food, shelter, water, and medical care immediately following the storm, but the US federal response to the disaster was weak. Further, a 100-year-old piece of protectionist legislation – the 1917 Marine Merchant Act (also referred to as the Jones Act) – exacerbated the issue by prohibiting humanitarian aid and disaster support from nations seeking to provide relief to the people of Puerto Rico in the aftermath of the storm. The Jones Act is just one of many policies in place that stem from the extractive legacy of colonialism that still remains in effect to this day.

Still under US sovereign control, the devastation left by Hurricane Maria brought international attention to the unequal power relations between the island and the US federal government. Further, the issue of Puerto Rico's food system became a prominent talking point in the media and within organizations mobilizing to support disaster recovery. Puerto Rico, a land capable of producing a vast wealth of agricultural commodities, relies on US imports to meet over

FIGURE 3.2 Hurricane Maria off the eastern coast of Puerto Rico on 19 September 2017, hours before landfall.

85% of food and nutritional needs. The negative impact US imperialism has had on Puerto Rico's food system is documented in Dietz (1986) and Nelson (2015). Puerto Rico's relationship with the federal government is complicated and tenuous, first as a colony, then an unincorporated territory, and now as the only commonwealth of its kind. Though Puerto Ricans are US citizens, they are not granted all constitutional freedoms or full voting rights. For instance, Puerto Rican US citizenship does not guarantee equal protection, due process, or all civil rights identified in the 14th Amendment (Dick 2016).

The documented treatment of Puerto Rico by the US administration after Hurricane Maria is indicative of the Trump administration's inadequacies in addressing environmental and economic devastation. Less than three weeks after the storm made landfall, the then US president threatened to withdraw federal disaster assistance in reaction to the mayor of San Juan's critiques of the slow federal response. FEMA has since recognized deficiencies in handling the disaster relief post-Hurricane Maria in their internal after-action report (FEMA 2017). Two years after the storm, investigative reporters discovered several warehouses with disaster supplies in Puerto Rico that were never distributed (Perret 2020).

The US federal response to Hurricane Maria epitomizes a legacy of neglect and inequity in Puerto Rico. Willison et al. (2019) quantified the disparities in the federal response comparing aid to Texas, Florida, and Puerto Rico, showing significantly more resources and money were expedited for the mainland states of Texas and Florida than for the territory of Puerto Rico. The inconsistencies in disaster response are striking given the severity of damage caused by Hurricane Maria far outweighed the other two storms, specifically in terms of infrastructure damage and loss of human lives (Willison et al. 2019). In August 2018, almost one year after the storm, the Puerto Rican government was pressured into releasing an official death toll. The official tally of excess lives lost post-Hurricane Maria is now 2975, making Hurricane Maria the deadliest natural disaster recorded in US history (Severino et al. 2018).

4   Agroecology and #JustRecovery Puerto Rico

While the federal government was scrambling to provide aid after the disaster, a coalition of grassroots organizations quickly mobilized to support the people of Puerto Rico. In the wake of the storm, the priority of grassroots organizations and frontline communities was to sustain lives while rebuilding a Puerto Rico rooted in justice using the vision of locals (Klein 2018). Within a week of the storm, the Climate Justice Alliance (CJA) alongside UPROSE (formally the United Puerto Rican Association of Sunset Park) spearheaded the Our Power Puerto Rico #JustRecovery campaign and coalition. In the case of Puerto Rico, climate justice is deeply entwined with agroecology. Though the Just Recovery campaign was initiated by the CJA, the coalition explicitly sought to support the actions of local organizations on the ground in Puerto Rico. The CJA emphasized their role was to bring together environmental, social, and economic justice groups to support Puerto Rican leadership on the island to create a recovery centred on a green economy, agroecology, food sovereignty, renewable energy, and self-sufficiency (Dutta et al. 2019). This represents a core value of the Just Recovery movement, such that those impacted by the disaster have a say in, and control over, the kinds of aid and resources they feel are most needed to move forward.

The Just Recovery campaign decided early on to commit resources and capital to assist Organización Boricuá de Agricultura Ecológica (Organización Boricuá), a well-established non-profit organization based in Puerto Rico. For over 30 years, Organizacion Boricuá has been diligently pursuing agroecological transformation throughout the island. Their mission is to 'promover la soberanía alimentaria y la conservación del ambiente, siendo un ente facilitador para la comunidad a través de la educación y capacitación en agricultura ecológica' ('promote food sovereignty and conservation of the environment by being a facilitating entity for the community through education and training in ecological agriculture') (Organización Boricuá 2020). Organización Boricuá has been integral to promoting organic, sustainable, and local agriculture in Puerto Rico and has built alliances with local and international organizations such as La Via Campesina (the largest and most recognized food sovereignty movement on the planet).

Agroecology is considered a practice and science incorporating sustainable and ecological considerations in agricultural settings (Altieri 1999). As the name implies, agroecology embeds the design and management of agriculture in sustainable food systems and treats it as nested within the regional ecosystem. Thus, agroecological farming systems are inherently more resilient to extreme climate events such as hurricanes, droughts, and floods (Altieri et al. 2015). While practices vary by region, climate, topography, and social factors, they centre on

> diversifying farms and farming landscapes, replacing chemical inputs with ecologically-based materials and processes, reducing waste by closing material cycles, reducing fossil-fuel energy use by maximising biomass accumulation and internalising energy flows, optimising biodiversity, and stimulating interactions between different species, as part of holistic strategies to build long-term fertility, healthy agroecosystems, and secure and just livelihoods.
>
> *(Gliessman et al. 2019, 92)*

However, while agroecology is a science and practice, the heart of agroecology lies in resistance to deep-rooted injustices of a globalized food system and cannot be divorced from a social justice or political-economic lens (Holt-Gimenez and Shattuck 2011; Rosset and Martínez-Torres 2012). Peasants and activists have long advocated the need to incorporate agroecological methods to create sustainable food systems and resilient communities (La Via Campesina 2018). Indeed, the global food sovereignty movement known as La Via Campesina has been working towards such transitions for over three decades. Gleissman et al. (2019, 91) suggest, 'agroecology provides a framework for understanding "levels" for the transition to sustainable food systems', and go even further to say, 'efforts to achieve food sovereignty and diversified agroecological systems have the potential to dismantle concentrations of power'. Ian Pagan-Roig, founder of El Josco Bravo, a farm and agroecology field school in Puerto Rico, shared this same sentiment after Hurricane Maria. Ian said, 'debido a la emergencia climática, realmente entendemos que hay urgencia. El problema representa una seria amenaza para la humanidad y la agroecología ofrece alternativas' ('due to the climate emergency, we really understand that there is urgency. This problem represents a serious threat to humanity and agroecology offers alternatives') (Jake Price 2019).

5 Injustice in Puerto Rico: Colonial Roots

The focus of Just Recovery is, and has been, largely to support recovery and transformation in Puerto Rico by stimulating a shift away from ecologically and socially destructive systems towards alternatives empowering Puerto Ricans via agroecology and climate, social, and environmental justice initiatives (Dutta et al. 2019; Rowsome 2017). Though the campaign seeks to champion climate justice and strengthen agroecological presence in Puerto Rico, at the heart of this mobilization is a clear requisite to resolve unjust and unequal power dynamics in the US–Puerto Rico relationship. Thus, Just Recovery is underpinned by decolonial and counterhegemonic principles evidenced by calls for actions explicitly demanding an end to colonialism and US imperialism in Puerto Rico (Yeampierre 2018).

Demonstrations and calls for action focus on attaining a sustainable transformation on the island; others are contesting Puerto Rico's colonial status, lack of voting rights, and representation in congress, including demands to end the Jones Act (Figure 3.3).

In addition to protests, demonstrations, and calls for action demanding lasting political change, the faculty from the University of Puerto Rico, Rio Piedras, collaborated with Organización Boricuá and CJA just three months after Hurricane Maria to host a forum titled 'De los Desastres del Capitalismo al Capitalismo del Desastre: Resistencias y Alternativas' ('From the Disasters of Capitalism to Disaster Capitalism: Resistance & Alternatives') (Figure 3.4). This forum brought a large gathering of activists, farmers, scholars, and legal consultants involved in the Just Recovery movement to Puerto Rico. Key speakers included Elizabeth Yeampierre, a lawyer and leader in the CJA, as well as Naomi Klein, a leading journalist and author of *The Shock Doctrine: The Rise of Disaster Capitalism* (2007) and *The Battle for Paradise: Puerto Rico Takes on the Disaster Capitalists* (2018).

FIGURE 3.3   Poster of the national call-to-action for Just Recovery organized by UPROSE and CJA. Image source: Climate Justice Alliance (2017).

FIGURE 3.4 Poster advertising a forum on disaster capitalism held at the University of Puerto Rico, Rio Piedras, in January 2018. The invitation to attend the live stream of this event was shared with me by one of the organizers.

Understanding colonialism in Puerto Rico requires operationalization of the term. Colonialism relates to the process of subjugating a nation and its peoples through cultural, political, economic, religious, racial, and even philosophical dominance. This often occurs through conquering and exploiting people and nations. Embedded in colonialism is the perpetual reproduction of hegemonic exploitation through the spread of Eurocentric/Western-centric notions of 'progress' that have come to dominate global sociopolitical and economic affairs. Coloniality refers specifically to the structure of power and knowledge relations that support the process of development, modernization, and globalization and also determine 'permissible forms of economic exchange, sexuality, gender, subjectivity, knowledge and education' (Quijano 2008).

The modern world system has taken an extractive and capitalist form, exacerbated by a hierarchical division between the Global North (Minority World) and Global South (Majority World) (Wallerstein 1974; Chase-Dunn 1998). Wallerstein's (1974) world systems theory speaks specifically to unequal exchange and describes the process by which core nations, or those with the most 'advanced' economic, industrial, and labour sectors, come to dominate the world's economic system and exert control over the 'lesser developed' semi-periphery and periphery nations. It should come as no surprise then that colonialism and subsequent resource extraction enacted by core nations has led to an exhaustive model, whereby stratification, inequities, and injustices are perpetuated as periphery nations strive to emulate, and attain, the unsustainable standards established by the Global North.

The common use of core or periphery 'nations' does not neatly apply to Puerto Rico in that Puerto Rico is technically a part of the United States, a core nation; however, as a territory, it is still subject to the struggles and challenges shared by periphery and semi-periphery regions. Velez (2021) provides a more thorough analysis of Puerto Rico's political-economic history elucidating how core nations, such as the United States and Spain, capitalized on the opportunity to exploit land and agricultural resources in Puerto Rico for their own economic gain while also exposing the lasting influence of colonialism on the modern-day food system. In Section 6, I devote attention to collective action and how agroecology is seen as a tool for decolonizing and building resilience.

6 Agroecology on the Ground: Capacity Building for Community Resilience

Earlier in this chapter I posed two questions: How can people ensure their communities are sustainable and resilient in the face of future climate disasters, and what does that process entail? And, why is agroecology so important to agroecological promoters and producers in Puerto Rico, and what does it offer? I address the first question by drawing on the CACR framework to highlight the collective actions occurring on the ground in Puerto Rico in an effort to build more sustainable and resilient communities. To answer the second question, I contextualize the most prominent themes that arose from my interviews with *Productores y Promotores de agroecología*.

7 How Can People Ensure Their Communities Are Sustainable and Resilient in the Face of Future Climate Disasters?

Collective action for climate justice and community resilience in Puerto Rico looks different at the national scale (macro level) than what is occurring on the ground in Puerto Rico (micro level). Though there have been a variety of demonstrations and marches in Puerto Rico that were organized and attended by members of Organización Boricuá, the day-to-day work of creating sustainable and resilient communities occurs through less contentious, ordinary acts of resistance. Mobilization at the micro level takes the form of day-to-day practice and knowledge sharing that is needed to support the growth of agroecology across the island. Organización Boricuá prioritizes capacity building through supporting education and reinforcing farmer-to-farmer knowledge networks (Organización

Boricuá 2020). On the ground, the acts of resistance entail disciplined and steady efforts to build cooperatives and community-supported agriculture, attend local farmers markets, participate in brigades, host and attend educational workshops and talks. The community-supported agricultural knowledge networks in Puerto Rico are critical to ensuring resilience to climate change, and that people have the ability to feed themselves when faced with future disasters (natural or sociopolitical).

In summer of 2019, a paramount meeting, 'El Primer Encuentro Nacional de Promotores y Productores Agroecológicos' (the First National Encounter of Agroecological Promoters and Producers), was held in Manati, Puerto Rico. The meeting had over 130 attendees representing agroecologists from across the island who are working diligently to scale up agroecology in Puerto Rico (Figure 3.5). Many of those in attendance were people who have passed through Finca El Josco Bravo's agroecology field school. The educational programming at Finca El Josco Bravo was initiated in 2015 by Ian Pagan-Roig, a trained agronomist who seeks to educate residents of Puerto Rico in agroecological practice and science as well as in the sociopolitical aspects of resistance and food sovereignty. Now in its eighth cohort, this programme is a cornerstone to the steadily growing population of *Promotores y Productores agroecológicos*.

In 2019 I was invited to observe the final examination process of El Josco Bravo's sixth cohort just a few weeks after the *Primer Encuentro Nacional*. One of the graduating students, a 68-year-old man, made the following statement, 'En los 70's yo estaba mucho en los movimientos sociales y siempre me sentía

FIGURE 3.5 Mosaic of agroecological promoters, producers, and educators across Puerto Rico that were present at the First National Agroecology Meeting in Puerto Rico. Photo of the mosaic map was taken by Paco Cardona.

como un idealista, pero la agroecología me hacer sentir como un realista' ('in the 70's, I was very involved in social movements and I always felt like an idealist, but agroecology makes me feel like a realist'). He spoke of the legacy of colonialism and how agroecology was a tangible solution to build a better future. This sentiment was reflected amongst the group as several respondents spoke of their everyday actions as a means to build sustainable communities, prepare for hurricanes, and for when the government fails them once more. The shared vision of an agroecological Puerto Rico informs their vision for community resilience and is evidence of the group's collective agency. Agroecological expansion is collectively conceptualized as a strategy to resist colonialism through the creation of self-sustaining communities that are resilient to a changing climate and political uncertainty.

8 Why Is Agroecology So Important to Agroecological Promoters and Producers in Puerto Rico?

In this section I turn attention to the second question I posed: Why is agroecology so important to agroecological promoters and producers in Puerto Rico, and what does it offer? To answer this question, I pull together a narrative using the recurring themes from the interviews I conducted in Puerto Rico in summer of 2019. The most prominent themes to arise from the data indicated that all of the individuals included in this work see 'agroecology as a tool' for (1) decolonizing, (2) sovereignty, (3) resisting capitalism. All of the participants were well aware of the negative effects of 'capitalism' and 'colonialism' in Puerto Rico, and these themes closely intersected the desire for 'auto-suficiencia' (self-sufficiency), 'sovereignty', and the need to transition to a 'circular economy'. Other prominent themes included anti-'individualism', engagement in 'community building', and 'ecological sustainability'. Further, throughout the interviews, content analysis, and participant observation came a cross-cutting theme centred on 'inclusivity' within the movement, with special attention to agroecology as a 'woman-led' movement.

Agroecological revitalization in Puerto Rico emphasizes moving beyond extractive, corporate solutions. Puerto Rico has a long history of resource and labour extraction that served to advantage the US government and externalize profits and growth. Interestingly, in-depth discourse related to this subject was revealed during a public *taller* (workshop) on composting that was held at a farmer's market in San Juan at the Mercado de la Cooperativa Madre Tierra. In the workshop the speaker emphasized that while composting is an ecologically sustainable practice used to return nutrients to the soil, it closely parallels the principles of a circular economy. In a 30-minute workshop dedicated to the topic of composting, the speaker focused more than half of his time explaining the need to end the linear extractive economy. He specifically highlighted that under US rule, Puerto Rico operates in a linear fashion that mostly benefits those external to the island, where resources are extracted and their wealth is exported to benefit external investors with little returned to the island. He explained that composting resembles the closed-loop system of a circular economy which is a means to retain wealth in Puerto Rico. Three important aspects to a circular economy in Puerto Rico are the creation of cooperatives, the use of local resources, and support for local enterprises.

Closely aligned with the notion of a circular economy were two cross-cutting themes that arose from interviews: first, the need to move beyond individualism and, second, community building and societal good as central components of agroecology. During a public forum, one individual asked, 'Cómo vamos a sanar ese problema del individualismo?' ('How are we going to heal the problem of individualism?'). Another person broached the subject in an interview by saying, 'Hay que trabajar en colectivo, ¿por qué ser individualista?' ('We have to work as a collective. Why be individualistic?'). Several of the agroecological producers emphasized that they valued collectivism and explained that they deliberately established their enterprises as cooperatives and/or non-profit organizations.

At the heart of agroecology and climate justice is the goal of empowering communities to realize their vision for a sustainable future. Referring to their farm, one person said 'Nuestro modelo es un LLC sin fines de lucro, con fin social y enlace comunitario … estamos trabajando para que este espacio sea de la gente' ('our model is a non-profit LLC, with social purpose and community ties … we are working so that this space can be for the people'). After Hurricane Maria, the community surrounding this particular agroecological farm was in dire need of assistance. Rather than prioritizing rebuilding their farm after the storm, the group dedicated themselves to feeding the surrounding community and rebuilding nearby homes. They established a community kitchen on their farm and collectively prepared meals for the surrounding communities. Even when everything was a disaster, there was no electricity, limited food, and water, they found a way to sustain themselves. It is worth noting that none of those involved in this agroecological farm even live in that community, yet their farm became a space for convening.

The devastation left by Hurricane Maria, and recognition that climate change will continue to bring hurricanes and storms to the archipelago, intensified the need to focus on becoming self-sufficient. When interviewees spoke of *auto-suficiencia* (self-sufficiency), it was not at the individual level. Rather, people spoke of self-sufficiency for all of Puerto Rico, meaning *sovereignty*. As one interviewee expressed, 'Agroecología se trata de auto-suficiencia, el poder de subsistir sin tener que depender de los Estados Unidos' ('agroecology is about self-sufficiency, the power to subsist without having to depend on the United States'). Another person aptly stated, 'el gobierno de los Estados Unidos no le importa lo que le pasa a le gente aqui … asi que nosotros tenemos que levantarnos y regresar a la tierra y producir nuestros alimentos' ('the government of the United States does not care what happens to the people here … so we have to lift ourselves up and return to the land and produce our food'). Similar sentiments were shared by others, including one person who said, 'en cualquier momento Estados Unidos dice, bueno no, ya no me conviene pa' mis intereses, pues mira, los dejo' ('at any moment, the United States can say, well, it's no longer convenient for my interests, let's leave them'). These quotes highlight how the people view US dependence as an integral problem that leaves Puerto Rico's residents vulnerable in the face of severe natural disasters and climate change.

Political themes related to resisting colonialism and capitalism were also quite prominent. These two themes were deeply entwined, with one person saying, 'El capitalismo es colonialismo' ('capitalism is colonialism'). Colonialism and capitalism are positioned as big hurdles to overcome, with agroecology being a tool with which to do so. One person emphasized, '[Agroecología] es una herramienta para decolonizar' ('agroecology is a tool for decolonizing'), and another stating, 'No podemos negociar nuestra sistema alimentaria en términos desiguales' ('We can't negotiate our food system on unequal terms'). There is obvious recognition that Puerto Rico's food system is extremely vulnerable to climate disasters, and that the forced reliance on US imports for food is a legacy of colonial rule. One interviewee highlighted how agroecology differs from the historical agricultural labour that Puerto Rican peasants were subjected to. He explained that agroecology bears no resemblance to the enslavement of Puerto Ricans that occurred on sugar plantations run by US foreign investors. He said, 'Ya eso pasó … que nos esclavizaron con la agricultura … ya no mas' ('that already happened … that they enslaved us with agriculture … enough, no more').

Many interviewees were also concerned with the influx of 'los ricos' ('rich people') after the storm and 'el capitalismo del desatre' ('disaster capitalism'). The term disaster capitalism, coined by Klein (2007), refers to the process of government and wealthy elites capitalizing on the vulnerabilities in regions recovering from huge shocks. Klein describes a post-disaster period as one vulnerable to strategic moves that promote neoliberal development and for-profit corporate 'solutions' to recovery that undermine the work of more democratic, community-centred initiatives. In Klein's (2018) book, *The Battle for Paradise: Puerto Rico Takes on the Disaster Capitalists*, she explains that the process is already underway post-Hurricane Maria as wealthy elites have begun moving in and buying up real estate, and government and private entities have been promoting neoliberal development projects such as the privatization of public utilities and education.

Doing away with capitalism is perhaps one of the deeper themes that was shared by the more radical people I interviewed. In 2019, one of the individuals I interviewed shared a graphic that was used to promote an agroecology *taller* and *brigada* on social media. It had the phrase 'The autonomy of the village is the greatest fear of the state and capitalism'; in the graphic was an image of a skull and government uniform buried under a planted garden bed with the caption 'Abono para el huerto' ('compost for the garden'). While the graphic may seem extreme, many people brought up the need to fight against a variety of topics linked to capitalism, including large corporate interest, 'agroindustriales' or agroindustrial enterprises, and 'Monsanto'. There is a great deal of disdain for Monsanto (now Bayer) and other agrochemical and seed companies in Puerto Rico. Seed corporations reportedly dominate over 9700 acres of public and private land in the island (Mercado 2017). This land is considered prime agricultural land that the government has leased to corporations for transgenic testing. The workers and surrounding communities have exposed the negative health and environmental impacts of these companies, with little action taken to stop or prevent future harms (Mercado 2017). Several of the interviewees even mentioned that they helped organize or participated in a variety of marches or political actions to push back against the rise of GMO testing on the island.

In addition to the aforementioned themes, 'inclusivity' arose as a key aspect that highlights the intersectional nature of climate justice and agroecology. The term 'inclusivity' is commonly used within justice-organizing spaces to ensure inclusion of everyone: women and men; Black, Indigenous, and persons of colour (BIPOC); as well as lesbian, gay, bisexual, transsexual, queer, intersex, asexual (LGBTQIA+) people of all ages. Systemic and structural racism was brought up in several interviews and forums, including a reference to BIPOC leadership as a strength. One person specifically emphasized that they respected the Afro-Indigenous and Puerto Rican roots of Elizabeth Yeampierre, a leader within the CJA.

The term 'inclusivity' was not used explicitly in public forums, *talleres*, *brigadas*, and interviews; instead, this theme arose through linguistic analysis and observations on the ground. It is worth noting that Spanish is a gendered language. In traditional Spanish, certain words are gendered masculine by default, and in other cases the suffix 'o' or 'os' is used for masculine form, and 'a' or 'as' is used to reflect the feminine form. The altering of morphemes, or the smallest grammatical unit of speech, can be seen in the words 'campesinos' (peasants – masculine form) or 'campesinas' (peasants – feminine form). In agroecological spaces, inclusive language takes the form of disrupting traditional gendered morphemes and suffixes, replacing the 'o' or 'a' with the letter 'e' (considered gender-neutral), or even with the letter 'x' or symbols, such as'@'. Examples include *campesines*, *campesinxs*, or *campesin@s*. Across the island, there was consistent use of altered morphemes to disrupt the gender aspects of the Spanish language and create gender-inclusive spaces at 'brigadas' and 'talleres'. Some examples of gender-neutral messaging include 'productor@s' (producers), 'muchxs' (many), 'campesinxs' (peasants), and 'jibares' (Puerto Rican slang used to refer to a traditional rural farmer). Further, the word 'elle', a proposed equivalent for the gender-neutral pronoun 'they', was shared at various brigades. The fact that introductions in many of these spaces began with the deliberate act of pronoun sharing indicates that inclusivity is deeply embedded within the agroecological movement in Puerto Rico.

Another key takeaway from this study of climate justice and agroecology in Puerto Rico relates to the 'woman-centred' or 'woman-led' nature of these movements. While this concept is sometimes linked to 'feminism' in scholarly spaces, criticisms of Eurocentrism and the Western origins of the term make it asynchronous and even rejected by some agroecologist women on the ground. Regardless of the semantics used to describe the phenomenon, the reality is that agroecology in Puerto Rico is a women-centred/led movement. This became most apparent when the majority of men interviewed indicated that women were at the forefront of the movement and pegged women as true leaders in these spaces. These comments usually arose unprompted or after asking for recommendations of people to interview. Several women were identified as being pivotal to the success of agroecological expansion across Puerto Rico. One interviewee described an amazing woman in Puerto Rico's agroecology movement and said, 'Ella rompió el mold … la idea, el stereotype de que la agricultura es pa'l hombre' ('she broke the mold … the idea, the stereotype that agriculture is [just] for men'). Someone else provided the names of four women whom he thought were ideal to interview and then followed his list with, 'Ahí tú va viendo … un giro de la mujer siendo el papel principal de la agroecología, en educación, en producción. Eso es bien importante' ('there you start seeing … a pivot with women playing the principle role in agroecology, in education, in production. That is very important').

This trend of women being leaders in agroecology is not unique to Puerto Rico. Women have been identified as key leaders in climate justice and agroecology around the world. Gupta and Moss (2021) explain that 'women are often the backbone of agroecology, leading food systems transformation on the ground. In 2011, the United Nations Food and Agriculture Organization reported that women food producers, given the same resources and access to land, credit, markets, and training as men, could eliminate hunger for 150 million people.' In an article promoting his book, *Animal, Vegetable, Junk*, Bittman (2021) emphasizes that 'agroecology is about not only sane agricultural methods but the empowerment of women and groups of long-exploited people, such as BIPOC, land reform, fair distribution of resources and treatment of labour, affordable food, nutrition and diet, and animal welfare'.

The final theme discussed in this chapter pertains to climate change and ecological healing. Those mobilizing for a more sustainable future in Puerto Rico are well aware of the environmental harm that humans have caused. They are also aware that anthropogenic climate change is predicted to lead to more intense storms in the coming decades, which will undoubtedly be coupled with the socioecological strain of future natural disasters. One person made this clear by saying, 'Ya vemos las simptomas del cambio climático, tenemos que movernos ahora … Huracan Maria no va ser la ultima tormenta' ('We already see the symptoms of climate change, we have to move now … Hurricane Maria will not be the last storm'). Another individual explained that they believed we need to 'buscar una forma de vivir dentro del contexto planetario' ('find a way to live within the context/boundaries of the planet'), and two others described the need to 'heal the island'. Agroecology is a means to do all of these things, as one person said, 'Agroecología es una herramienta para cambiar nuestra relación con el planeta' ('agroecology is a tool to change our relationship with the planet').

9 Discussion

Hurricane Maria laid bare deficiencies in Puerto Rico's political-economic system and food system that stem from a long legacy of colonialism. Yet, the disaster response from FEMA under the Trump administration was poorly handled. In the aftermath of the storm, the people of Puerto Rico needed access to food, water, shelter, and humanitarian aid. Instead, they were treated as second-class US citizens who began to die without medical aid and food. The impact of, and federal response to, Hurricane Maria in Puerto Rico is in line with the critiques of environmental and climate justice scholars. Specifically, the tragedy highlights how negative effects of climate change and environmental disasters are experienced disproportionately, such that those lacking adequate resources and political power face greater repercussions.

In Puerto Rico, collective agency and community-driven solutions are the foundation for a sustainable and just transformation. The Just Recovery initiative, and more specifically agroecological expansion, offers frontline communities in Puerto Rico a path towards resisting extractive, capitalist economic systems while creating climate resilience in ways that are transformational, democratic, and socially just. When faced with unprecedented climatic change, unequal power relationships, and the subsequent injustices faced by frontline communities, agroecology is a tool sovereignty, a tool for ending colonialism, and, most importantly, a tool for healing our relationship with the planet.

While it is clear that agroecology has undeniable, transformative potential for Puerto Rico, there are also tensions between creating an economically profitable business in a capitalist society and working to create sustainable and regenerative economies. More than half of the agroecological enterprises included in this study are non-profits with social justice missions, yet they are also struggling financially. Thus, the cooperative non-profit model provides great social benefits, but also poses challenges for those seeking to be financially sustainable and stable. While I did not address these tensions in my interviews, it would be an interesting future avenue of research. How do the Promotores y Productores agroecológicos define success? Pinpointing what the most successful agroecological enterprises have in common can provide a framework on how to best scale up agroecology in Puerto Rico.

References

Altieri, Miguel A. 'The ecological role of biodiversity in agroecosystems'. In *Invertebrate Biodiversity as Bioindicators of Sustainable Landscapes*, pp. 19–31. Elsevier, 1999.

Altieri, Miguel A., Clara I. Nicholls, Alejandro Henao, and Marcos A. Lana. 'Agroecology and the design of climate change-resilient farming systems'. *Agronomy for Sustainable Development* 35, no. 3 (2015): 869–890.

Belles, Jonathan. 'Atlantic hurricane season recap: 17 moments we'll never forget'. *Weather* (2017). https://weather.com/storms/hurricane/news/2017-11-11-moments-hurricane-season-atlantic-irma-maria-harvey

Bittman, Mark. 'Mark Bittman: We need an agroecological revolution towards a sustainable and equitable system of global food production'. https://lithub.com/mark-bittman-we-need-an-agroecological-revolution/

Cannella, Gaile S., Michelle Salazar Pérez, and Penny A. Pasque, eds. *Critical Qualitative Inquiry: Foundations and Futures.* Routledge, 2016.

Chang, Stephanie E. and M. Shinozuka. 'Measuring improvements in the disaster resilience of communities'. *Earthquake Spectra* 20, no. 3 (2004): 739–755. doi:10.1193/1.1775796

Chase-Dunn, Christopher K. *Global Formation: Structures of the World-Economy.* Rowman & Littlefield, 1998.

Cortés, Jason. 'Puerto Rico: Hurricane Maria and the promise of disposability'. *Capitalism Nature Socialism* 29 (2018): 1–8.

Creswell, John W. *Research Design: Qualitative & Quantitative Approaches.* Sage Publications, 1994.

Dick, Lourdes. 'US tax imperialism in Puerto Rico'. *American University Law Review* 65 (2016): 1.

Dietz, James L. *Economic History of Puerto Rico: Institutional Change and Capitalist Development.* Princeton University Press, 1986.

Dutta, J., S. Tyler, and J. Vázquez. 'Our power Puerto Rico: Moving toward a just recovery'. 2019. https://climatejusticealliance.org/our-power-puerto-rico-report.

Federal Emergency Management Agency. '2017 hurricane season FEMA after-action report'. 2018. www.fema.gov/press-release/20230425/fema-releases-2017-hurricane-season-fema-after-action-report.

Gliessman, Steve, Harriet Friedmann, and Philip H Howard. 'Agroecology and food sovereignty'. *Political Economy of Food* 50, no. 2 (2019): 91–110.

Gupta, Amrita and Daniel Moss. 'Global alliance for the future of food "the future is female"'. https://medium.com/global-alliance-for-the-future-of-food/the-future-of-food-is-female-9bcf3027c54e

Holt Giménez, Eric, and Annie Shattuck. 'Food crises, food regimes and food movements: Rumblings of reform or tides of transformation?' *Journal of Peasant Studies* 38, no. 1 (2011): 109–144.

Jake Price. (2019). https://civileats.com/2020/03/11/farmers-in-puerto-rico-are-growing-a-culture-of-social-justice-and-climate-resilience/

Johnson, David. 'Is this the worst hurricane season ever? Here's how it compares. *TIME Magazine.* 2017. https://time.com/4952628/hurricane-season-harvey-irma-jose-maria

Kelman, I., Gaillard, J. C., Lewis, J. et al. 'Learning from the history of disaster vulnerability and resilience research and practice for climate change'. *Natural Hazards* 82 (Suppl 1) (2016): 129–143. https://doi.org/10.1007/s11069-016-2294-0

Klein, Naomi. *The Shock Doctrine: The Rise of Disaster Capitalism.* Macmillan, 2007.

Klein, Naomi. *The Battle for Paradise: Puerto Rico Takes on the Disaster Capitalists.* Haymarket Books, 2018.

La Via Campesina. 'La Via Campesina (International Peasant Movement): Annual report 2018'. 2018. https://viacampesina.org/en/wp-content/uploads/sites/2/2019/06/2018-LVC-Annual-Report-EN-compressed.pdf.

Mercado, Eliván Martínez. 'Centro de Periodismo Investigativo. The boom of Monsanto and other seed corporations blows in the South of Puerto Rico'. 2017. https://periodismoinvestigativo.com/2017/03/the-boom-of-monsanto-and-other-seed-corporations-blows-in-the-south-of-puerto-rico/

Nelson, Denis. *War against All Puerto Ricans: Revolution and Terror in America's Colony.* Nation Books, 2015.

Organización Boricuá de Agricultura Ecológica. 2020. Accessed January 2, 2020. http://organizacionboricua.blogspot.com/p/quienes-somos.html

Perrett, Connor. 'Puerto Ricans discovered a warehouse full of unused food, water, and supplies from Hurricane Maria, resulting in the firing of the island's emergency manager'. 2020. www.insider.com/puerto-rico-residents-find-warehouse-full-of-supplies-from-maria-2020-1

Quijano, Aníbal. 'Coloniality of power, eurocentrism, and social classification'. In *Coloniality at Large: Latin America and the Postcolonial Debate*, p. 192. Duke University Press, 2008.

Robles, Frances, and Luis Ferré-Sadurní. 'Puerto Rico's agriculture and farmers decimated by Maria'. *New York Times*, 2017. www.nytimes.com/2017/09/24/us/puerto-rico-hurricane-maria-agriculture-.html.

Rodríguez, Clelia O. *Decolonizing Academia: Poverty, Oppression and Pain.* Fernwood Publishing, 2018.

Rosset, Peter M., and Maria Elena Martínez-Torres. 'Rural social movements and agroecology: Context, theory, and process'. *Ecology and Society* 17, no. 3 (2012). www.jstor.org/stable/26269097

Rowsome, Alice. 'Activists and locals are using giant hurricane relief aid ships to rebuild Puerto Rico'. 2017. Vice, November 21. www.vice.com/en_us/article/9kq4ye/activists-and-locals-are-using-giant-hurricane-relief-aid-ships-to-rebuild-puerto-rico.

Saldaña, Johnny. *The Coding Manual for Qualitative Researchers*, Sage Publishing, 2021.

Sbicca, Joshua. 'Solidarity and sweat equity: For reciprocal food justice research'. *Journal of Agriculture, Food Systems, and Community Development* 5, no. 4 (2015): 63–67.

Severino, Kathya, Damayra I. Figueroa, Jennifer Hinojosa, Nashia Roman, and Melendez Edwin. 'Puerto Rico one year after Hurricane Maria'. *Centro for Puerto Rican Studies* (2018). https://centropr-archive.hunter.cuny.edu/sites/default/files/data_briefs/Hurricane_maria_1YR.pdf

Sutter, John D., and McKenna Ewen. 'About 1 million Americans without running water. 3 million without power. This is life one month after Hurricane María'. *CNN*, 20 October 2017.

Tinsley, M. 'Towards a postcolonial critical realism'. *Critical Sociology* 48, no. 2 (2022): 235–250. https://doi.org/10.1177/08969205211003962

Vaccaro, Chris. 'Extremely active 2017 Atlantic hurricane season finally ends'. *National Oceanic and Atmospheric Administration* (2017). www.noaa.gov/media-release/extremely-active-2017-atlantic-hurricane-season-finally-ends

Vélez, Thelma. 'A just recovery: Agroecology and climate justice in Puerto Rico post-Hurricane Maria'. Unpublished doctoral dissertation. The Ohio State University, 2021.

Wallerstein, Immanuel. 'Dependence in an interdependent world: The limited possibilities of transformation within the capitalist world economy'. *African Studies Review* 17, no. 1 (1974): 1–26.

White, Monica M. *Freedom Farmers: Agricultural Resistance and the Black Freedom Movement.* UNC Press Books, 2018.

Willison, Charley E., Phillip M. Singer, Melissa S. Creary, and Scott L. Greer. 'Quantifying inequities in US federal response to hurricane disaster in Texas and Florida compared with Puerto Rico'. *BMJ Global Health* 4, no. 1 (2019).

Xu, Jiuping, Wang, Ziqi, Shen, Feng, Ouyang, Chi, and Yan Tu. 'Natural disasters and social conflict: A systematic literature review'. *International Journal of Disaster Risk Reduction*, 17 (2016): 38–48.

Yeampierre, Elizabeth. '4th Annual Frances Tarlton "Sissy" Farenthold Endowed Lecture Series in Peace, Social Justice and Human Rights'. The Rothko Chapel and Bernard and Audre Rapoport Center for Human Rights and Justice. 2018. http://rothkochapel. org/experience/events/event/elizabeth-yeampierre-sissy-lecture-series.

4

'I WAS POOR BEFORE, BUT CYCLONE AMPHAN LEFT ME DESTITUTE'

Disaster Displacement and Support in Bangladesh

Neil J. W. Crawford, Siddiqur Rahman, Tanzina Nazia, Sennan David Mattar, and Ukegbu Uwa Kalu

> Amphan had a devastating effect on us. We all lost our homes as our homes are very close to the river, the tidal surge took everything away from us. We used to cultivate agricultural lands, we grew crops and life was ok. After the Amphan, we lost everything and we cannot grow any crops due to saline intrusion. We are passing [through] a difficult time now.
>
> (Interview #21, 2021)

1 Introduction

This chapter explores the experiences of people in Khulna District in southwestern Bangladesh following Cyclone Amphan in mid-2020 – the first super cyclone in the Bay of Bengal in the twenty-first century and the largest displacement event of 2020 (IDMC 2020b). Cyclone Amphan formed on 16 May 2020 over the Indian Ocean, moved north-east, and made landfall on 20 May over the coast of India and southern Bangladesh, with wind speeds of 150 km/h (93 mph), causing widespread devastation (IFRC 2021). Bangladesh is among the countries worst hit by disasters and sits at the forefront of the global climate emergency as one of the most climate-vulnerable countries on earth (IDMC 2020a; IEP 2020; UNDRR 2020). Climate-related disasters are on the rise and displace millions of people every year. Bangladesh was ranked in the top ten countries affected by extreme weather events between 1999 and 2018 (Eckstein 2020: 9), and over 4 million people were displaced in the country by flooding in 2022 (Billet 2022). The Sundarbans area, which straddles the border between Bangladesh and India, is a known 'climate hotspot' (Ghosh et al. 2018) for the increasingly frequent and deadly cyclones, as are the parts of southwestern Bangladesh studied here.

DOI: 10.4324/9781003214021-5

In the wake of disasters, large, top-down humanitarian operations to address immediate needs, led by national governments and aid agencies, are common. In Bangladesh, the assessment and coordination of disaster relief is performed by the Humanitarian Coordination Task Team (HCTT), a partnership between the United Nations Resident Coordinator and Bangladesh's Ministry of Disaster Management and Relief (MoDMR). The country's disaster management structure is a two-track hierarchy; a national-level disaster management council oversees the entire structure and reaches down to a ward-level committee, whereas another branch is an interministerial coordination committee which bridges governmental departments down to village-level stakeholders (United Nations Bangladesh 2021). Communities in Bangladesh are already adopting effective adaptation practices, with national and international NGOs assisting local stakeholders to achieve adaptation relevant to their situations, for example, utilizing local knowledge and the planting of coconut, areca nut, and fishtail palm trees for coastal protection from wind and sea surges (Sutradhar et al. 2015). Yet, despite the scale of coordination and examples of good practice by NGOs, poor governance and the lack of evidence-led policymaking have been identified by adaptation scholars as a critical risk to adaptation planning and building resilience to climate change more broadly in Bangladesh (Bhuiyan 2015; Tuihedur Rahman et al. 2018; Islam et al. 2020). Additionally, a limiting factor for good adaptation outcomes is disaster relief funding. Only US$6.5 million, or 26% funding coverage, was allocated to the HCTT humanitarian response plan to Cyclone Amphan by September 2020, whereas the International Federation of Red Cross and Red Crescent Societies (IFRC) was able to meet only 29.3% of their appeal coverage from donors (IFRC 2021). IFRC (2021) has noted that various targets to reach a certain number of people and households were downgraded due to lack of funding. Despite inadequate funding, most donors concentrate their relief aid on providing for immediate needs. For example, 81.8% of UK Aid's £512 million disaster relief budget was allocated to emergency response – and only approximately 13% for prevention and 5% for reconstruction (UK Aid 2022). Given the inadequacy of international and national funding for immediate disaster response, let alone rebuilding, informal and community-driven action is a significant yet underappreciated and underexplored means of responding to emergency situations.

The chapter[1] is based on data gathered in Bangladesh between March and May 2021. Fieldwork took place in three sites in Khulna District, southwestern Bangladesh: Modinabad in Koyra Union and Uttar Bedkashi and Horinkhola in Uttar Bedkashi Union. Khulna District, which sits on the Bay of Bengal, is at a high risk of climate change impacts, with existing mitigation and adaptation strategies inadequate in reducing risks (Islam et al. 2019). Semi-structured interviews were conducted with 25 disaster-affected people across these locations in Khulna District and with eight key informants from civil society and an international organization in Khulna District and the country's capital, Dhaka, between March and May 2021. The 25 participants included 15 women and 10 men, ranging from 25 to 65 years, with an average age of 44. Interviews were conducted in Bengali and translated into English by two of the authors.

The fieldwork took place during the global COVID-19 pandemic. Personal protective equipment (PPE) was used by the research team and distributed to participants, with in-person interviews carried out outdoors where possible. The fieldwork was impacted by the ongoing development of the pandemic: in late March 2021, Bangladesh entered the 'second wave' of COVID-19 and the government imposed the first national lockdown, followed by a stricter lockdown in mid–April 2021 (Bari and Sultana 2021). The fieldwork was truncated to allow the researchers to return home and observe the lockdown, with subsequent key informant interviews conducted online or by phone. The chapter begins with a review of people's vulnerability, migration, and experiences in shelters, before studying the social impacts of Cyclone Amphan. The chapter continues by considering the difficult choices people make, in particular their need to weigh up immediate physical risk with the economic consequences of leaving their homes and possessions. The following sections consider how vulnerability to

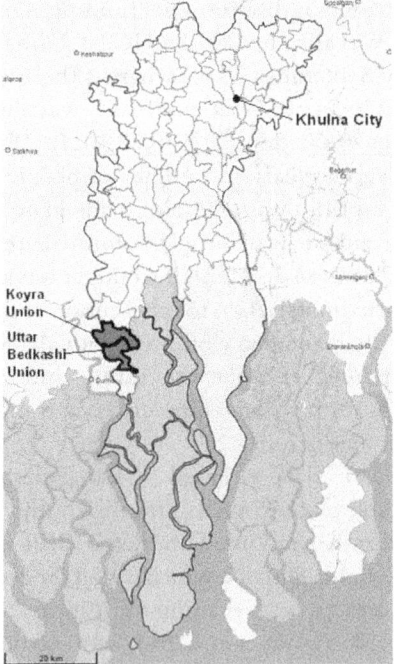

FIGURE 4.1   Khulna District, Bangladesh.

Adapted from ©Thomas Brinkhoff, 2021.

disasters intersects with other issues, namely gender and age, with heightened risks for women and older people. The chapter then considers the avenues for support and assistance that existed for people in the months that followed Cyclone Amphan. This section shows the limitations to the support provided by international and national mechanisms and how many individuals were forced to rely on informal support mechanisms, including family and community loans, to be able to survive. The chapter contributes to scholarship on intersecting forms of vulnerability, including economic and the direct result of disasters, and illustrates the importance of informal support structures in aiding displaced people in post-disaster settings.

2 Cyclones: Displacement, Shelters, and Compounding Vulnerability

Cyclones force people to leave their homes to seek refuge. Sometimes this displacement is short term, and people are able to return once the storm has passed; on other occasions the places they left may no longer be viable to return to, with homes destroyed or land now under water. In other cases, people may lack the resources to return, may be displaced again, or feel compelled to move on to another location in search of new livelihoods or protection from future cyclones. In disaster-prone areas such as southwestern Bangladesh, a short-term response to danger is to seek refuge in a shelter. However, as previous studies have shown, shelters come with their own set of risks. Despite being designed to provide immediate cover, people often have to stay in them for months.

Beyond the immediate threat to people's physical safety, the impacts of disasters in Bangladesh are widely understood as catalysts for compounding existing vulnerabilities. A survey of 478 adults affected by Cyclone Amphan found 55.7% suffered moderate to severe psychological symptoms, whereas this proportion rose to 68.2% among the 13.8% of respondents who had been displaced due to damage to their homes (Hossain et al. 2021). Coastal areas in eastern Bangladesh were largely spared; however, parts of Bangladesh's large refugee population are being moved to areas at a high risk of cyclones (Alam et al. 2020). Public assistance in Bangladesh and India can be extremely limited for individuals who are forced to move or who choose to migrate in the face of cyclones (Dasgupta et al. 2022). The limited assistance afforded to displaced persons reflects a wider international shortcoming as aid interventions often focus assistance on a specific geospatial location for a limited period, that is, where a displacement event occurs, this often fails to reach displaced persons who may have subsequently moved on (Mattar and Mbakwem 2018).

Displacement also carries specific gender- and age-related risks. The loss of housing and existing livelihoods following weather-related disasters is often accompanied by an increase in the risk of gender-based violence (Hasan et al. 2019), human trafficking (Ranjan 2016), child labour (Luetz 2017), and child marriage (McLeod 2019). For example, women forced to migrate to informal settlements in Bangladesh's capital, Dhaka, due to climate change are often forced to rely on criminal organizations to provide access to water because they do not have the class status and political power to demand legal access, exposing them to exploitation (Sultana 2020). The displacement caused by weather-related disasters constitutes a specific injustice for women in Bangladesh as some have reportedly been reluctant to use shelters because of fear of domestic and sexual violence, or social reprisals for leaving their home or taking shelter with strangers (Begum 2017). All of these concerns also followed Cyclone Amphan in Bangladesh. A rapid gender analysis by UN Women (2020a) raised concerns over the safety measures for displaced women and girls in emergency shelter accommodation in Bangladesh, with only 21% of respondents reporting access to gender-based violence support services. Another rapid response assessment by the UN Central Emergency Response Fund (2020) estimated 500,00 women and girls had lost access to sexual and reproductive health services and included reports of increasing domestic violence as a result of Cyclone Amphan.

3   The Social Impacts of Cyclone

Cyclone Amphan caused extensive destruction and displacement primarily in Bangladesh and India. The storm was responsible for the single largest source of new displacements in 2020, with 2.4 and 2.5 million people displaced in India and Bangladesh, respectively (IDMC 2020b). The districts of Satkhira and Khulna in Bangladesh experienced among the worst impacts of the storm; 200,000 houses were fully or partially damaged, and 176,000 hectares of productive farmland were washed away (NAWG 2020). In Satkhira and Khulna District, 60–90 km/h (37–56 mph) wind and high tidal inundation were recorded. Over 851,000 people were affected; over 121,000 houses were damaged and 44,000 destroyed; and over 42,000 hectares of land used for agriculture and aquaculture was damaged (NAWG 2020). One local resident reported that Cyclone Amphan severely impacted farmland as 'the saline flood water destroyed our land quality and we cannot grow any crops now' (interview #5, 2021). Other residents talked about their homes and what the storm did to them. As another resident who works as a day labourer put it, 'Amphan took away everything from me. I only had my clothes that I was wearing that day … The next day, I saw the rooftop of my house sitting 25 metres away from my house. I had to rebuild my house from zero' (interview #18, 2021).

Recovery within the months that followed the cyclone have been hard, particularly for women and older people. Women in Bangladesh hold a disproportionate share of the work associated with rebuilding because 'social customs dictate that women in the household are required to construct and raise the household mud platform (biti), make a new oven (chula), establish a new garden, and, if desired, build new basic infrastructure such as tubewells and latrines' (Sovacool 2018, 190; Lein 2009). However, some women described being unable to rebuild their homes because, as one commented, 'rebuilding houses [is] very difficult in terms of money and labour. Most of the people who lost their houses during Amphan, couldn't rebuild them' (interview #7, 2021). Those with support from either their family or an NGO were most likely to be able to rebuild their homes, as described by a local resident: 'We came back to our old house for the last 3 months or so. We rebuilt the house. My son with his income and savings did that' (interview #14, 2021).

Extreme weather events displace the most vulnerable in Bangladesh, and injustices follow in the months that follow. Despite assumed post-disaster vulnerabilities when rebuilding, academic literature on Cyclone Amphan has primarily focused on the immediate physical impacts of the cyclone, such as its meteorological characteristics (e.g. Ahmed et al. 2021), saline intrusion (e.g. Waheduzzaman and Mizanuzzaman 2021), flooding (e.g. Halder et al. 2021), damage to infrastructure (e.g. Rafa et al. 2021), damage to crops (e.g. Kabir et al. 2020), food insecurity and livelihood loss (e.g. Hossain et al. 2021; Priodarshini et al. 2021), its impact on the COVID-19 pandemic (e.g. Kumar et al. 2021; Pramanik et al. 2021), immediate vulnerability to the cyclone (e.g. Salman 2022), and evacuation behaviour (e.g. Alam et al. 2022; Alam and Chakraborty 2021; Hadi et al. 2021). Some notable studies exist on pre-existing vulnerabilities that offer insight on the future outcomes of cyclones in the region, including Islam et al.'s (2021) examination of coastal community preparedness to cyclones, Parida et al.'s (2021) analysis of the role of media in communicating risks of cyclones, and Priodarshini et al.'s (2021) study of the livelihoods of rural communities before and after Cyclone Amphan and COVID-19 lockdowns. However, there are limited studies that examine the social and political consequences in the months that followed Cyclone Amphan for impacted people. This gap in study is partly a reflection of the slower pace of social and political research. However, it is important to understand people's experiences within the year that followed Cyclone Amphan and what this can tell us to boost longer-term recovery efforts and bolster preparedness for future disasters.

Running parallel to Cyclone Amphan and its aftermath was the global COVID-19 pandemic. By October 2020, Bangladesh had recorded nearly 275,000 confirmed cases of COVID-19 and more than 3600 related deaths (Ober 2020). Exposure to COVID-19 was likely heightened in some shelters due to close proximity and low levels of access to personal protection equipment (PPE). In a survey of evacuees, 92.6% said there was no access to COVID-19 safety kits (including face masks, hand soap, and hand sanitizer) and only 7.14% reported access to sinks with soap (Alam et al. 2022). Long before COVID-19, shelters and refugee camps in Bangladesh and beyond have acted as risk multipliers for the transmission of diseases (Braam et al. 2021; Chan et al. 2018; Kouadio et al. 2014; Hossain 2020; Khan et al. 2020; Loebach and Korinek 2019; Nur et al. 2021; Patwary and Rodriguez-Morales 2022). Cyclone Amphan was observed to compound existing social, physical, and economic vulnerabilities of some of the poorest in Bangladesh, which in turn placed these individuals and communities at a heightened risk of another crisis – the COVID-19 pandemic – further exacerbating an already dire situation.

4 Weighing Physical and Economic Costs

As well as physically, disasters expose people economically. People were forced to make difficult decisions based on weighing up these factors and their relationship to one another. Focusing on people's economic vulnerability helps us to understand why, despite 96.6% of people in coastal areas being issued a cyclone evacuation order, only 42% of people evacuated (Alam et al. 2022). Disaster-impacted people suffer an economic toll as a result of their displacement and destruction of their possessions and must make difficult trade-offs with regard to their physical and economic safety and well-being.

A common theme among those who fled to shelters during Cyclone Amphan was loss of belongings. Reports of income and livelihood loss following the cyclone as well as previous disasters were common (Hossain et al. 2021; Priodarshini et al. 2021). However, some individuals described the economic hardship that resulted from their decision to take refuge in a shelter, as they were unable to bring their belongings with them. As a shopkeeper explained:

> The wind started to blow from [the] early morning. But it became worse … after dusk. At first the dam eroded in my area and the surge started entering. Then I came to the shelter centre with my family. The authorities didn't allow [me] to bring any belongings, so I left them in my house and lost everything.
>
> *(Interview #24, 2021)*

Some residents simply did not have sufficient time to save their belongings. Evacuation orders were issued ten hours before Cyclone Amphan made landfall; however, one study found many residents only received the orders with six hours of notice, and those who complied with the order typically left within up to two hours of landfall (Alam et al. 2022). A local resident describes their experience:

> Previously during cyclones, people used to come to the shelter centre and receive different types of relief support. But this Amphan brought a sudden tidal surge, which has washed away everything, people couldn't save their belongings… I had some chicken and ducks, I lost them too.
>
> *(Interview #23, 2021)*

In response to the destruction of economic livelihoods as well as homes, and to avoid future danger, disasters can spur people to move to new locations. This is one reason for the belief among some scholars that migration is an 'adaptation' strategy in mitigating harm and risk (Black et al. 2011; Vinke et al. 2020), a view that has been communicated as scientific advice to national governments such as the United Kingdom (Foresight 2011). A new trend in this movement is from rural to urban areas, as the latter also offers greater economic opportunities (or the perception of them) much needed following a disaster. Since the 1990s in particular, people who were displaced have been increasingly found in urban areas across the globe (Crawford 2021), prompted in part by attempts to access better services, support, and livelihood opportunities. One respondent, who worked as a mason prior to Cyclone Amphan, described his past decision-making following repeated cyclones. He chose to move to Khulna city, the largest in southwestern Bangladesh and the third largest in the country.

> The biggest impact is the loss of houses, trees and food. Besides, loss of belongings is another problem; I have lost my belongings several times during [Cyclone's] Aila, Sidr, Amphan … Basically, as I was displaced and our area was waterlogged; I had to migrate to Khulna city to overcome the loss.
>
> *(Interview #16, 2021)*

Attempts to escape cyclones through migration were echoed by another respondent, who remarked 'one after another disaster has destroyed us. In the last 10–12 years, we have seen at least six big cyclones … I feel like moving out of this place and migrate (sic) to a new place where we can start a new life' (interview #4, 2021). Despite this, she was uncertain about her ability to support herself elsewhere, as she wondered, 'where can we go and what will we do there?' (interview #4, 2021).

Economic vulnerability renders some individuals immobile as they fear the consequences travelling to a shelter would have for their belongings and what this would mean for them in the longer term. As one resident explained: 'Previously, most of the people were reluctant to go to the shelter centre. You can call us stupid, as we used to stay home to save our belongings. But the fact is if we lose belongings, it becomes very difficult to survive' (interview #17, 2021). As this account suggests, people's ability to migrate to a shelter or onward is often constrained by other concerns, including loss of possessions, livelihoods, and existing economic vulnerability which has been made worse by Cyclone Amphan. The inability to bring cattle to the shelter, for example, is a known concern as half of the interviewees in one study said they would not leave their cattle (Alam et al. 2022).

5 Gender and Age-Related Vulnerabilities of Cyclone Amphan

The research study found that gender and age were key factors in understanding people's experience of the months following Cyclone Amphan. A majority of those interviewed identified greater vulnerability among older people, those living with disabilities, poorer individuals, and women. According to a senior employee of an international NGO in Bangladesh who works with communities impacted by disasters:

> In a patriarchal society, women's movements are restricted, they have less decision-making power, less income and information which make them more vulnerable over men. In addition, we also saw how third gender people, sex workers and people from different caste face difficulty in accessing shelter centres.
>
> *(Interview #32, 2021)*

In advance of Cyclone Amphan, Bangladeshi authorities prepared 12,078 cyclone shelters for over 2.4 million people as part of a four-month $24.6-million response (UN Women2020c). Reaching shelters during heavy rain and flooding was challenging, with women noting that their sarees became heavy and twisted as they became wet, making it hard to run and swim (interview #8, 2021; interview #9, 2021). These shelters were to be equipped with masks and sanitation and intended to provide enough space to allow social distancing. Despite provisions, shelters are very challenging for those who used them, in particular women and girls. As one woman who went to a shelter explains:

> Women with pregnancy and menstruation suffer … a lot during disasters. Menstruation management becomes very difficult for us. Women have to work shoulder to shoulder with men regarding house repairing, taking care of children. We can't think of saving sanitary napkins and face several physical and social problems associated with it. I experienced it during Amphan and face[d] embarrassment and management problems. The whole displaced time, when I was living in the shelter centre with other families; I suffered a lot regarding menstrual management.
>
> *(Interview #25, 2021)*

Concern over belongings and the ability to rebuild were most often communicated by older people. This observation tallies with other studies on disasters and the experiences of older people. Following Hurricane Katrina in the United States in 2005, some elderly people were reported to not have evacuated for fear of losing their belongings, partially explaining why older people accounted for 70% of recorded deaths (Campbell 2007). Following the Fukushima nuclear disaster in Japan in 2011, the mortality risk among older people in care facilities doubled during relocation. Older women accounted for 71% of those who died (Yasumura et al. 2013), demonstrating the risks associated with evacuation and the intersection of age and gender-based vulnerability. Similarly, heightened risk during floods and water-related disease outbreaks in Bangladesh has been explained with reference to a variety of social factors, including gender, age, class, and religion (Sultana 2010). Older people in disaster situations have a fear of losing their belongings, of being unable to rebuild, of the evacuation process, or a combination of these. An older woman with a chronically ill husband detailed her decision-making process:

> I didn't go to the shelter centre. We thought that this storm wouldn't be very powerful. But when the surge entered our area by damaging the dam we took shelter in one of my nephew's half-building[s]. We have two shelter centres around two kilometres away from this area. Most of the nearby people went to the shelter centre. But our area, comprising 25 families is situated far away from the shelter centre. So, none of us went there.
>
> *(Interview #11, 2021)*

Another concern for potential evacuees was the destination itself, with numerous cyclone shelters reportedly overcrowded. Many of the shelters were already being used as COVID-19 quarantine centres (Oxfam International 2020), raising justifiable public health concerns, as crowding made social distancing 'incredibly difficult' (ActionAid 2020, 1). Despite some people's decision not to go to one of the shelters, many did. One individual recounted their experience:

> That day when I arrived at the centre, it was so crowded. There was not an inch [of] empty space at the centre. I saw some people [who] tried to bring their cattle to the shelter centre, but they could not, as it was so crowded with people. Both men and women were on the[ir] feet. We did not sleep that night at all. We were worried and watch[ed] the devastation of the cyclone. I think there were thousands of people at [the shelter] that night. I also heard that the near-by shelter centre was also full. I think, when they built this shelter centre, they did not anticipate this many people.
>
> *(Interview #21, 2021)*

An older resident described overcrowding and the lack of adequate sanitation and living facilities he experienced:

> That day when I arrived at the hospital building, it was very crowded. People like me were everywhere. It was full. We did not have any place to sleep. We all were on our feet all night. There [were] problems with toilets and no place to cook. It was very difficult for us.
>
> *(Interview #13, 2021)*

A similar account was given by another older resident who detailed that the three-floor cyclone shelter was crowded to the point that he saw 'some men sleeping like a chicken on the floor' and, despite there being separate toilets for men and women, 'toilets were a problem. It was always crowded. For women, bathroom facilities were not enough' (interview #20, 2021). Overcrowding in cyclone shelters was not a unique challenge to Cyclone Amphan, nor was the issue an unknown risk to potential evacuees. A survey of people who did not evacuate found that perceptions of overcrowding were the main reported problem with shelters, with 26.94% citing overcrowding as a reason for not evacuating (Alam et al. 2022). Early news reports confirm testimonies in this study that overcrowding was a significant issue (TimesLive 2020; France-Presse 2020; Reuters 2020).

6   Support and Assistance in the Months Following Cyclone Amphan

During the immediate aftermath of Cyclone Amphan, assistance came from national and international aid initiatives in the form of cash support, shelter, WASH support, food, masks, sanitizers, fodder, and other agricultural materials (CERF 2020). Staffed with more than 70,000 volunteers, the Cyclone Preparedness Programme (CPP) and the Bangladesh Red Crescent Society (BDRCS) disseminated cyclone warnings to communities, assisted with finding shelter, and provided medical assistance, aid delivery, and post-disaster recuperation and rehabilitation (American Red Cross 2020; BDRCS 2022). Parallel programmes for disaster recovery by the Government of Bangladesh include the Social Safety Net Programmes (SSNP), which aim to reduce the impact of poverty, hunger, and disasters. The programmes were developed to provide support through personal, domestic, and communal assistance, including social protection and social empowerment (Amhed et al. 2014). SSNPs have reportingly contributed to Bangladesh's poverty reduction campaigns, having provided income security, temporary employment, and medical support for individuals, including those living with disability (World Bank 2019). SSNPs are described as a 'pro-poor' programme targeted at supporting vulnerable people, including those directly impacted by disasters (World Bank 2019). The quality and efficiency of the support provided by the government through the SSNP, however, has been continuingly scrutinized, often with calls for greater efficiency and transparency as a result (Alam and Hossain 2016; World Bank 2020; Maintains 2021).

Despite the seemingly extensive assistance programmes, numerous participants criticized the delivery of aid, citing issues of corruption and nepotism in particular. One described the bias they believed to have occurred:

> Relief distribution was very biased. Basically, people who are involved with the ruling political party and have a linkage with the UP [Union] chairman and [union] members got the relief from the government. In fact, NGOs also did their distribution in coordination with the UP [Union] chairman and [union] members who always preferred some over others. In fact, some well-off

families who didn't face any loss received the support for being relatives of the UP [Union] chairman and [union] members.

(Interview #15, 2021)

Another resident articulated a similar belief and added that 'the UP [Union] chairman and [union] member[s] always favour their relatives and followers over us. Common people have to bribe them to get this support' (interview #17, 2021). A common way corruption happens in post-disaster situations was most clearly explained by a resident, who stated that 'they [Union Chairman] give a wrong list of beneficiaries to get such support from the government by taking bribes' (interview #4, 2021). They did not however believe this bias was 'on the basis of religion' (interview #4, 2021). Nonetheless, in the months following the cyclone, assistance appears to be more elusive for those unaffiliated to the 'right' political party, unrelated to local politicians, or lacking bribery money. As one woman, whose children died and husband had left following Amphan, explains:

Support after any disaster has increased in the last 10 years. But I don't get those, as I can't make any payment to the middleman … sometimes we have to bribe the local leaders to get those support, which I can't manage, so I don't get benefit from the increased support over time.

(Interview #23, 2021)

How then did people access support? One individual felt 'relief distribution through the NGO was okay. But when it is given through the UP [Union] chairman and [union] member[s] … sometimes they vanish half of the relief by themselves' (interview #7, 2021). National and international NGOs with their own distribution channels and volunteers (e.g. CPP and BDRCS) are arguably more likely to engage with households based on needs and risk assessments, strong policies to combat corruption and discrimination, and resourcing limitations. This was the case of the disaster and post-disaster intervention process employed by the IFRC (2021) in response to Cyclone Amphan. However, in the described scenario, NGOs can provide assistance for those who lack sufficient connections to local politicians and decision-makers. NGO interventions after disasters are typically time-limited due to donor-set timelines and being project-based – rather than part of wider development plans, thus lacking in adequate methods to account for community-defined needs – and operationally focused on specific interventions, for example, WASH (von Meding et al. 2009; Islam 2018; He 2019).

The lack of political redress is an underlying barrier to accessing post-disaster support, especially for those whose choices are diminished by a disaster. The issue is strongly illustrated by the case of an older woman who described being displaced by Cyclone Amphan, only to later find her small agricultural plot had been forcefully taken by her sons, forcing her to live on her daughter's balcony.

She explained her experience with housing support after Cyclone Amphan and engaging with political leaders:

> We are badly in need of housing support here. There was a political gathering in our area; political leaders always assure us they will provide housing support, but in reality, we don't get any. I have told the leader that if you want to give us a house, please give something permanently, so that we don't have to hold our roof to stop it being blown away and stop going to the shelter centre. We need buildings with a high basement. If possible, please give this type of housing support, otherwise don't support us at all. As I have spent most of my life facing shocks and stressors, will I be able to survive the rest of my life like this?
>
> *(Interview #17, 2021)*

When it came to the role of their local Member of Parliament (MP) in post-disaster recovery, one resident was clear that 'it is the responsibility of our local MP to rehabilitate us … he should give us food support and repair the damaged embankment' (interview #10, 2021). Yet, support from local MPs did not come. According to another resident, 'we didn't get any remarkable support from the Member of Parliament' (interview #16, 2021), and the support he received only came from a local club and NGOs. While another commented that their local MP had 'never helped us' (interview #7, 2021).

Having the 'correct' political affiliation and its impact on abilities to access support are a known issue for organizations working in the wake of disasters. As a senior employee in a national NGO, which focuses on long-term resilience and livelihoods, explained:

> If you are [a] supporter of the ruling party and have good connection with the local elected body then you have [a] better chance to get government relief or other support … We get the list of potential beneficiaries from the local elected bodies such as UP [Union] Chairman and other elected Members [of Parliament] but that is not final. We double check with ours [list of beneficiaries] and make the final list of beneficiaries. If there is any pressure to include someone's name in the list, we go to the [United Nations Office] and get his/her advice … But it is true that some international NGOs do not bother about it.
>
> *(Interview #26, 2021)*

As a bulwark against entrenching inequalities, one senior employee at a humanitarian response NGO said: 'We always try our best to be as inclusive as possible … We always prioritize PWD [people with disabilty], women, children, [and] elderly while distributing the relief goods' (interview #29, 2021). Women-led and empowerment-focused organizations offered vital disaster response and recovery, as well as being more inclusive (GNWP 2020; UN Women 2020b, 2020c). Most interviewed employees at NGOs referenced their organizations' inclusivity policies, but also acknowledged political bias may affect levels of support to some communities and households. As a senior employee at a community-based organization explained, 'inequality occurs only on the basis of political ground. If you are close to the local representatives and [a] supporter of the chairman, you get more' (interview #27, 2021). A senior employee at an international NGO, while also acknowledging this challenge, explained the background to political bias as follows: 'As we work with the local partners, we have to heavily rely on their list of beneficiaries and sometimes they have to keep elected local government representatives, and other influential local people, happy' (interview #32, 2021).

In 2013, the Government of Bangladesh initiated the first phase of a US$400 million World Bank-supported project, 'Coastal Embankment Improvement Project (CEIP-1)', intended to reduce vulnerabilities of coastal communities to tidal inundation and saline water intrusion and recover agricultural land (BWDB 2022; World Bank 2022). However, according to a senior employee at a national disaster response NGO, 'the progress of that project is not satisfactory … [The] Bangladesh government is spending a lot of money, but due to corruption and lack of monitoring, these are not being effective to reduce the vulnerability of climate change and natural disasters' (interview #33, 2021). They attributed this lack of progress and corruption to the process of funding allocation, with 'the climate change Trust fund … regularly allotted to DC [District Commissioner] for addressing specific climatic need[s] of his working area … In fact, the Department of Environment doesn't have an office at the field level' and, in effect, argued that funding for vulnerable populations was not reaching its intended target (interview #33, 2021). This account is in line with other critiques of impact of corruption on humanitarian assistance in Bangladesh. According to Transparency International Bangladesh (2020), following Cyclone Amphan, there were broad deficiencies in compliance with international pledges, dissemination of misleading information on warnings, lack of accountability and transparency on irregularities, corruption in constructing and maintaining shelters and other disaster-related infrastructure, and a lack of capacity and proper planning to assess actual needs (short and long term) for relief and rehabilitation.

In contrast, residents who received support tended to access it through NGOs. Although unable to recall the names of the NGOs who provided support, one resident remembered that she 'received ready food, dry foods, rice, pulses, vegetables from different NGOs and civil-society organisations' (interview #24, 2021). NGOs mentioned by residents include World Vision, Samadhan, Red Cross, Islamic Relief, Shushilon, and Nobo Jatra, among others. NGO relief was available only in the immediate aftermath of the cyclone and focused on life-saving provisions. As a resident recalls:

> We did not receive much support from the government … We also received food and other support from many NGOs … I must say 90% of the people in

our community received support in one form or another ... But NGO support lasted for a few months say till July/August of 2020, after that the support was almost zero.

(Interview #2, 2021)

There was wide variance in the amount of support received by some residents from NGOs, and this may reflect issues with coordination and tensions between community-based, national, and international organizations. A senior employee of a community-based organization, which partnered with an international NGO, explained how tension can arise:

Historically, there was a significant role of community-based organisation in supporting disaster affected people. But unfortunately, their role is not significant any more. See, what happens, each donor or NGO come[s] to implement their own development project with some specific project goals and with a specific time period. We are too busy and focused with our own project activities and no one cares about the community-based organisation. Some NGOs try to work with them and nurture them for a specific time. Once the project ends, there is no support for the community-based organisation.

(Interview #30, 2021)

There were examples of good practice cited by key informants. Notably, Oxfam's 'REE-CALL' programme is said to strengthen the capacity of CBOs in disaster-prone districts. The programme started in 2017 and a second phase continued until 2021 with the aim to improve disaster and climate resilience through economic empowerment and inclusive leadership (Monash University n.d.; Unnayan Sangha 2022). As a senior employee at a humanitarian response organization explained:

In that program [REE-CALL], we worked with around 800 CBOs [community-based organizations] and we gave them training, provided them [with] equipment in disaster response ... The idea is to provide community people [with] necessary skill sets so that they do not have to wait for CPP [cyclone preparedness programme] or Red Cross volunteer[s] for immediate support.

(Interview #32, 2021)

Community- and family-based networks provided a key source of support for many individuals in this study. This echoes work on the role of community-based groups and organizations in aiding refugees, especially in the absence of national and international assistance (Betts et al. 2020; Pincock et al. 2020a, 2020b). Informal support, including loans, provided by families and community members was essential for people to be able to survive and begin to rebuild after Cyclone Amphan. As one resident explains:

> We are surviving on our own. Sometimes, we take loans from our relatives or neighbours and repay their loans when we make money from our day labour jobs. We also buy goods from the shop on credit. We pay the shopkeeper back when we make money. That is how we are surviving.
>
> *(Interview #22, 2021)*

For some, family networks were a lifeline. One resident described housing her mother following Cyclone Amphan: 'My mother is temporarily living with me, as she has no other place to live. Amphan has displaced her and she could not rebuild her house' (interview #15, 2021). However, respondents also reported limits to the support families can offer each other, such as limits to educational support, made worse by the cyclone. As one resident explained: 'All my daughters have some education. As I am very poor, I could not send them for higher studies. My sons-in-law are also poor. They rarely can help me financially when I need it' (interview #13, 2021). Meanwhile, another resident explained the existence of gender-based wage discrimination in post-disaster projects, how this impacted her family, and the decision-making over who in her family should be involved in working to repair a damaged embankment.

> Payment of men and women labourers varies here. For the same job, a man gets BDT [Bangladeshi Taka] 300 while a woman earns only BDT 200. That's why my son is working on the embankment instead of me. They allowed one member of the family to work in the embankment. So, sending my son is more profitable.
>
> *(Interview #22, 2021)*

The importance of family wealth and support structures was also evident when this support was washed away by Cyclone Amphan. As a resident explains:

> My house was completely washed away in Amphan, so I became displaced. I couldn't save any of my belongings back then. All of my food crops stored in my house were ruined. I had some livestock … All of my goats and chicken got washed away … we didn't have any place to live … we didn't have any source of income … I had land … which we bought for the future of my children. All of my land is outside of embankment coverage now, the breached embankment has taken my land.
>
> *(Interview #10, 2021)*

7 Conclusion

In the words of the UN Secretary-General António Guterres (2022), the world is on a 'fast track to climate disaster' and 'terrifying storms', yet for some in Bangladesh, the disaster has already arrived. As this chapter demonstrates, it is imperative that disaster response and recovery is urgently supported for communities in the Majority World already living with the worst impacts of climate change, but least able to mitigate or adapt to them. It is also important to recognize the myriad ways people show their resilience and find ways to support each other, especially when formal aid mechanisms and structures have failed. Cyclone Amphan has compounded existing vulnerabilities and exposed many to future cyclones and other crises. Beyond a homogenous group, intersecting factors – including age, economic status, gender – influence individuals' and their families' vulnerability to extreme weather events. People have been forced to make painful choices, such as abandoning their homes and running for safety to shelters, but then facing the long-term economic consequences of losing their possessions and animals, or the physical and economic risks that can come from short or long periods spent in an overcrowded shelter. In the months after disasters, life-saving assistance from national and international organizations often dries up, with other forms of assistance limited or inaccessible to those without the 'right' connections. Some people rely on family and community support to help them rebuild, or at least survive. Through providing an essential aid to those with few other recourses, these informal kinds of support can create future challenges, place individuals in debt, and increase already heightened forms of economic marginalization. More is needed across the board – from improved notification systems, improved shelters, assistance beyond life-saving provisions, mechanisms to ensure equitable delivery of support, rebuilding that helps mitigate the destruction of future cyclones, and improved aid delivery. It is also important for research to centre the longer-term experiences and challenges of people who have survived cyclones, including social and political issues, alongside work that focuses on immediate physical threats.

Note

1   The research was made possible by a pump prime grant awarded to Neil J. W. Crawford (principal investigator) from the Global Challenges Research Fund, Scottish Funding Council, via the Global Challenges and Sustainable Development Unit.

References

ActionAid. 'ActionAid provides food support to Cyclone Amphan survivors'. 22 May 2020. https://actionaid.org/news/2020/actionaid-provides-food-support-cyclone-amphan-survivors.

Ahmed, Ishita, N. Jahan, and F. T. Zohora. 'Social safety net programme as a mean to alleviate poverty in Bangladesh'. *Developing Country Studies* 4, no. 17 (2014): 46–54.

Ahmed, Rizwan, M. Mohapatra, Suneet Dwivedi, and Ram Kumar Giri. 'Characteristic features of Super Cyclone "Amphan" observed through satellite images'. *Tropical Cyclone Research and Review* 10, no. 1 (2021): 16–31.

Alam, Akhtar, Peter Sammonds, and Bayes Ahmed. 'Cyclone risk assessment of the Cox's Bazar district and Rohingya refugee camps in southeast Bangladesh'. *Science of the Total Environment* 704 (2020): 135360.

Alam, Md, Torit Chakraborty, Md Hossain, and Khan Rubayet Rahaman. 'Evacuation dilemmas of coastal households during Cyclone Amphan and amidst the COVID-19 pandemic: A study of the southwestern region of Bangladesh'. *Natural Hazards* (2022): 1–31.

Alam, Md Ashraful, and Sheikh Abir Hossain. 'Effectiveness of social safety net programs for poor people in the government level of Bangladesh'. *International Journal of Social Sciences and Management* 3, no. 3 (2016): 153–158.

Alam, Md Shaharier, and Torit Chakraborty. 'Understanding the nexus between public risk perception of COVID-19 and evacuation behavior during cyclone Amphan in Bangladesh'. *Heliyon* 7, no. 7 (2021): e07655.

American Red Cross. 'Cyclone Amphan: In Bangladesh, preparedness paid off Bangladesh'. 30 May 2020. https://reliefweb.int/report/bangladesh/cyclone-

Bangladesh Water Development Board (BWDB). 'Consultancy services for feasibility studies and preparation of detailed design for the following phase of the Coastal Embankment Improvement Project (CEIP)'. Government of the People's Republic of Bangladesh, June 2022. http://ceip-bwdb.gov.bd/Tech_Report/FS/CEIP-2%20June%202022%20Polder%20Screening%20Report.pdf.

Bari, Razmin, and Farhana Sultana. 'Second wave of COVID-19 in Bangladesh: An integrated and coordinated set of actions is crucial to tackle current upsurge of cases and deaths'. *Frontiers in Public Health* 9 (2021). www.frontiersin.org/articles/10.3389/fpubh.2021.699918/full

Begum, Anwara. *Review of Migration and Resettlement in Bangladesh: Effects of Climate Change and Its Impact on Gender Roles.* Bangladesh Institute of Development Studies, 2017.

Betts, Alexander, Evan Easton-Calabria, and Kate Pincock. 'Refugee-led responses in the fight against COVID-19: Building lasting participatory models'. *Forced Migration Review* 64 (2020): 73–76.

Bhuiyan, Shahjahan. 'Adapting to climate change in Bangladesh: Good governance barriers'. *South Asia Research* 35, no. 3 (2015): 349–367.

Billet, Madeline. 'Climate displacement in Bangladesh and India, 4 million displaced'. The Displacement Initiative. www.tdinitiative.com/post/climate-displacement-in-bangladesh-and-india-4-million-displaced.

Black, Richard, Stephen R. G. Bennett, Sandy M. Thomas, and John R. Beddington. 'Migration as adaptation'. *Nature* 478, no. 7370 (2011): 447–449.

Braam, Dorien H., Rafiq Chandio, Freya L. Jephcott, Alex Tasker, and James L. N. Wood. 'Disaster displacement and zoonotic disease dynamics: The impact of structural and chronic drivers in Sindh, Pakistan'. *PLOS Global Public Health* 1, no. 12 (2021): e0000068.

Brinkhoff, Thomas. 'Satkhira (District, Bangladesh) – Population statistics, charts, map and location'. www.citypopulation.de/en/bangladesh/khulna/admin/87__satkhira/.

Campbell, Jenny. 'On belonging and belongings: Older adults, Katrina, and lessons learned'. *Generations* 31, no. 4 (2007): 75–78.

Chan, Emily Y. Y., Cheuk Pong Chiu, and Gloria K. W. Chan. 'Medical and health risks associated with communicable diseases of Rohingya refugees in Bangladesh 2017'. *International Journal of Infectious Diseases* 68 (2018): 39–43.

Crawford, Neil James Wilson. *The Urbanization of Forced Displacement: UNHCR, Urban Refugees, and the Dynamics of Policy Change.* McGill-Queen's University Press, 2021.

Dasgupta, Susmita, David Wheeler, Sunando Bandyopadhyay, Santadas Ghosh, and Utpal Roy. 'Coastal dilemma: Climate change, public assistance and population displacement'. *World Development* 150 (2022): 105707.

Eckstein, David, Vera Künzel, Laura Schäfer, and Maik Winges. *Global Climate Risk Index 2020: Who Suffers Most from Extreme Weather Events? Weather-Related Loss Events in 2018 and 1999 to 2018.* Germanwatch e.V., 2020. https://germanwatch.org/sites/default/files/20-2-01e%20Global%20Climate%20Risk%20Index%202020_15.pdf

Foresight. 'Migration and Global Environmental Change: The Government Office for Science'. (2011). https://assets.publishing.service.gov.uk/government/uploads/system/uploads/attachment_data/file/287717/11-1116-migration-and-global-environmental-change.pdf.

France-Presse, Agence. 'Super-Cyclone Amphan kills up to 20 in India and Bangladesh'. *The Guardian*, 21 May 2020. www.theguardian.com/world/2020/may/21/super-cyclone-amphan-deaths-india--bangladesh

Ghosh, Upasona, Shibaji Bose, and Rittika Bramhachari. 'Living on the edge: Climate change and uncertainty in the Indian Sundarbans'. Working paper 101. STEPS, Institute of Development Studies, 2018.

Global Network of Women Peacebuilders (GNWP). 'Country update: Bangladesh'. 20 June 2020. https://gnwp.org/wp-content/uploads/Bangladesh-COVID-19-Profile.pdf.

Guterres, António. 'Secretary-General's video message on the launch of the Third IPCC Report'. United Nations, 2022. www.un.org/sg/en/content/sg/statement/2022-04-04/secretary-generals-video-message-the-launch-of-the-third-ipcc-report-scroll-down-for-languages

Hadi, Tahmina, Md Sirajul Islam, Denise Richter, and Bapon S. H. M. Fakhruddin. 'Seeking shelter: The factors that influence refuge since Cyclone Gorky in the coastal area of Bangladesh'. *Progress in Disaster Science* 11 (2021): 100179.

Halder, Bijay, Suman Das, Jatisankar Bandyopadhyay, and Papiya Banik. 'The deadliest tropical cyclone "Amphan": Investigate the natural flood inundation over south 24 Parganas using google earth engine'. *Safety in Extreme Environments* 3, no. 1 (2021): 63–73.

Hasan, Md Robiul, Mahbuba Nasreen, and Md Arif Chowdhury. 'Gender-inclusive disaster management policy in Bangladesh: A content analysis of national and international regulatory frameworks'. *International Journal of Disaster Risk Reduction* 41 (2019): 101324.

He, Lulu. 'Identifying local needs for post-disaster recovery in Nepal'. *World Development* 118 (2019): 52–62.

Hossain, Ahmed, Bayes Ahmed, Taifur Rahman, Peter Sammonds, Shamrita Zaman, Shadly Benzadid, and Md Jakariya. 'Household food insecurity, income loss, and symptoms of psychological distress among adults following the Cyclone Amphan in coastal Bangladesh'. *Plos One* 16, no. 11 (2021): e0259098.

IDMC. *Global Report on Internal Displacement.* Internal Displacement Monitoring Centre, 2020a.

IDMC. *Internal Displacement 2020: Mid-Year Update.* Internal Displacement Monitoring Centre, 2020b.

IEP. *Ecological Threat Register 2020: Understanding Ecological Threats, Resilience and Peace.* Institute for Economics & Peace, 2020.

International Federation of Red Cross and Red Crescent Societies (IFRC). *Review of Bangladesh: Cyclone Amphan – Final Report (N° MDRBD024).* Bangladesh Red Crescent

Society, 2021. https://reliefweb.int/report/bangladesh/bangladesh-cyclone-amphan-final-report-n-mdrbd024.

Interview #2, 30 March 2021.

Interview #4, 30 March 2021.

Interview #5, 30 March 2021.

Interview #7, 30 March 2021.

Interview #8, 30 March 2021.

Interview #9, 30 March 2021.

Interview #10, 30 March 2021.

Interview #11, 1 April 2021.

Interview #13, 30 March 2021.

Interview #14, 1 April 2021.

Interview #15, 30 March 2021.

Interview #16, 30 March 2021.

Interview #17, 1 April 2021.

Interview #18, 31 March 2021.

Interview #20, 31 March 2021.

Interview #21, 31 March 2021.

Interview #22, 31 March 2021.

Interview #23, 31 March 2021.

Interview #24, 31 March 2021.

Interview #25, 31 March 2021.

Interview #26, 30 March 2021.

Interview #27, 31 March 2021.

Interview #29, 29 April 2021.

Interview #30, 30 April 2021.

Interview #32, 30 April 2021.

Interview #33, 1 May 2021.

Islam, Md Anwarul, Md Shamsuzzoha, Md Rasheduzzaman, Rajan Chandra Ghosh, and Md Faisal. 'Assessment on climate change adaptation: A study on coastal area of Khulna district in Bangladesh'. *Australian Journal of Engineering and Innovative Technology* 1, no. 6 (2019): 14–20.

Islam, Md Tariqul, Mark Charlesworth, Mohammad Aurangojeb, Sarah Hemstock, Sujit Kumar Sikder, Md Shareful Hassan, Papon Kumar Dev, and Md Zakir Hossain. 'Revisiting disaster preparedness in coastal communities since 1970s in Bangladesh with an emphasis on the case of tropical cyclone Amphan in May 2020'. *International Journal of Disaster Risk Reduction* 58 (2021): 102175.

Islam, Md Zahidul. 'Resourcing for post-disaster housing reconstruction: The case of Cyclones Sidr and Aila in Bangladesh'. PhD diss., London South Bank University, 2018.

Islam, Shafiqul, Cordia Chu, and James C. R. Smart. 'Challenges in integrating disaster risk reduction and climate change adaptation: Exploring the Bangladesh case'. *International Journal of Disaster Risk Reduction* 47 (2020): 101540.

Kabir M., M. Salam, M. Omar, M. Sarkar, A. Rouf, M. C. Rahman, A. Chowdhury, M. Rahaman, L. Deb, S. M. Noman, and M. Siddique. *Impact of Super Cyclone Amphan on Agriculture and Farmers' Adaptation Strategies in the Coastal Region of Bangladesh*. Bangladesh Rice Research Institute, 2020. www.econstor.eu/handle/10419/243315.

Khan, Md Nuruzzaman, M. Mofizul Islam, and Md Mashiur Rahman. 'Risks of COVID19 outbreaks in Rohingya refugee camps in Bangladesh'. *Public Health in Practice* 1 (2020): 100018.

Kouadio, Isidore K., Syed Aljunid, Taro Kamigaki, Karen Hammad, and Hitoshi Oshitani. 'Infectious diseases following natural disasters: Prevention and control measures'. *Expert Review of Anti-infective Therapy* 10, no. 1 (2012): 95–104.

Kumar, Shubham, Preet Lal, and Amit Kumar. 'Influence of super cyclone "Amphan" in the Indian subcontinent amid COVID-19 pandemic'. *Remote Sensing in Earth Systems Sciences* 4, no. 1 (2021): 96–103.

Lein, Haakon. 'The poorest and most vulnerable? On hazards, livelihoods and labelling of riverine communities in Bangladesh'. *Singapore Journal of Tropical Geography* 30, no. 1 (2009): 98–113.

Loebach, Peter, and Kim Korinek. 'Disaster vulnerability, displacement, and infectious disease: Nicaragua and Hurricane Mitch'. *Population and Environment* 40, no. 4 (2019): 434–455.

Luetz, Johannes. 'Climate change and migration in Bangladesh: Empirically derived lessons and opportunities for policy makers and practitioners'. In *Limits to Climate Change Adaptation*, pp. 59–105. Springer, 2018.

Maintains. 'Towards shock-responsive social protection: Lessons from the COVID-19 in six countries'. June 2021. https://www.opml.co.uk/files/Publications/A2241-maintains/maintains-towards-shock-responsive-social-protection-synthesis-report.pdf.

Mattar, Sennan David, and Enyinnaya Mbakwem. 'Climate migration: The emerging need for a human-centred approach'. In *Routledge Handbook of Climate Justice*, pp. 479–493. Routledge, 2018.

McLeod, Christie, Heather Barr, and Katharina Rall. 'Does climate change increase the risk of child marriage: A look at what we know – and what we don't – with lessons from Bangladesh and Mozambique'. *Columbia Journal of Gender and Law* 38 (2019): 96.

Monash University. 'Resilience through Economic Empowerment, Climate Adaptation, Leadership and Learning (REE-CALL)'. n.d. www.monash.edu/__data/assets/pdf_file/0005/994325/REE-CALL-ICT.pdf.

Needs Assessment Working Group (NAWG). *Review of Cyclone Amphan Joint Needs Assessment (JNA)*. Humanitarian Coordination Task Team, 2020. www.acaps.org/sites/acaps/files/key-documents/files/cyclone_amphan_joint_needs_assessment_final_draft_31052020.pdf.

Nur, Most Nusrat Binte, Md Abdur Rahim, and Md Rasheduzzaman. 'Identifying cyclone shelter facilities and limitations for enhancing community resiliency in coastal areas of Bangladesh'. *Asian Journal of Social Sciences and Legal Studies* 3, no. 4 (2021): 107–118.

Ober, Kayly. 'Complex road to recovery & COVID-19, Cyclone Amphan, monsoon flooding collide in Bangladesh and India'. Refugees International. Refugees International, 3 November 2020. www.refugeesinternational.org/reports/2020/10/5/complex-road-to-recoverynbspcovid-19-cyclone-amphan-monsoon-flooding-collide-in-bangladesh-and-india.

Oxfam International. '"Crisis on top of crisis" as India and Bangladesh brace for Super Cyclone Amphan'. 25 May 2022. www.oxfam.org/en/press-releases/crisis-top-crisis-india-and-bangladesh-brace-super-cyclone-amphan-oxfam.

Parida, Debadutta, Sandra Moses, and Khan Rubayet Rahaman. 'Analysing media framing of cyclone Amphan: Implications for risk communication and disaster preparedness'. *International Journal of Disaster Risk Reduction* 59 (2021): 102272.

Patwary, Muhammad Mainuddin, and Alfonso J. Rodriguez-Morales. 'Deadly flood and landslides amid COVID-19 crisis: A public health concern for the world's largest refugee camp in Bangladesh'. *Prehospital and Disaster Medicine* 37, no. 2 (2022): 292–293.

Pincock, K., A. Betts, and E. Easton-Calabria. *The Global Governed? Refugees as Providers of Protection and Assistance.* Cambridge University Press, 2020a.

Pincock, Kate, Alexander Betts, and Evan Easton-Calabria. 'The rhetoric and reality of localisation: Refugee-led organisations in humanitarian governance'. *Journal of Development Studies* 57, no. 5 (2020b): 719–734.

Pramanik, Malay, Sylvia Szabo, Indrajit Pal, Parmeshwar Udmale, Montira Pongsiri, and Susan Chilton. 'Population health risks in multi-hazard environments: Action needed in the Cyclone Amphan and COVID-19-hit Sundarbans region, India'. *Climate and Development* 14, no. 2 (2022): 99–104.

Priodarshini, Rup, Bangkim Biswas, Ana Mariá Sánchez Higuera, and Bishawjit Mallick. 'Livelihood challenges of "double strike" disasters: Evidence from rural communities of southwest coastal Bangladesh during the COVID-19 pandemic and Cyclone Amphan'. *Current Research in Environmental Sustainability* 3 (2021): 100100.

Rafa, Nazifa, Abu Jubayer, and Sayed Mohammad Nazim Uddin. 'Impact of Cyclone Amphan on the water, sanitation, hygiene, and health (WASH2) facilities of coastal Bangladesh'. *Journal of Water, Sanitation and Hygiene for Development* 11, no. 2 (2021): 304–313.

Ranjan, Amit. 'Migration from Bangladesh: Impulses, risks and exploitations'. *Round Table* 105, no. 3 (2016): 311–319.

Salman, Md. 'Assessment of vulnerability and capacity to the cyclone "Amphan" impacts of the southwestern coastal part of Bangladesh: An empirical contextual investigation'. *Natural Hazards* (2022): 1–28.

Sovacool, Benjamin K. 'Bamboo beating bandits: Conflict, inequality, and vulnerability in the political ecology of climate change adaptation in Bangladesh'. *World Development* 102 (2018): 183–194.

Sultana, Farhana. 'Living in hazardous waterscapes: Gendered vulnerabilities and experiences of floods and disasters'. *Environmental Hazards* 9, no. 1 (2010): 43–53.

Sutradhar, L. C., S. K. Bala, A. K. M. S. Islam, M. A. Hasan, S. Paul, M. M. Rhaman, M. A. A. Pavell, and M. Billah. 'A review of good adaptation practices on climate change in Bangladesh'. In *Fifth International Conference on Water & Flood Management*, Dhaka, 6-8 March 2015.

Thomson Reuters. 'At least 14 dead as Cyclone Amphan dumps rain on India, Bangladesh'. *CBC News*, 21 May 2020.

TimesLive. 'Overcrowded shelters: Cyclone Amphan adds to COVID-19 misery in Bangladesh'. YouTube, 2020. www.youtube.com/watch?v=ALuiYitYg3s.

Transparency International Bangladesh. *Governance Challenges in Disaster Response and Way Forward: Cyclone Amphan and Recent Experiences.* Transparency International Bangladesh, 2020. www.ti-bangladesh.org/beta3/images/2020/report/Amphan/Amphan_Study _ES_Eng.pdf

Tuihedur Rahman, H. M., Gordon M. Hickey, James D. Ford, and Malcolm A. Egan. 'Climate change research in Bangladesh: Research gaps and implications for adaptation-related decision-making'. *Regional Environmental Change* 18, no. 5 (2018): 1535–1553.

UK Aid. 'Disaster relief sector breakdown'. Development Tracker, n.d. https://devtracker. fcdo.gov.uk/sector/18.

UN Central Emergency Response Fund (CERF). 'Review of Bangladesh Rapid Response Cyclone Amphan 2020 20-RR-BGD-43537'. Central Emergency Response Fund, 2020. https://cerf.un.org/sites/default/files/resources/20-RR-BGD-43537_Banglad esh_CERF_Report.pdf.

UN Women. *Review of Rapid Gender Analysis Cyclone Amphan*. Gender in Humanitarian Action Working Group, 2020a. www.preventionweb.net/files/73966_rgacycloneamphanbangladeshcompresse.pdf.

UN Women. 'In the aftermath of crisis, rural Bangladeshi women pursue economic security'. ReliefWeb, June 2020b. https://reliefweb.int/report/bangladesh/aftermath-crisis-rural-bangladeshi-women-pursue-economic-security.

UN Women. 'As Bangladesh battles COVID-19 and the aftermath of Super Cyclone Amphan, women's organizations lead their communities through recovery'. UN Women-Headquarters, 15 June 2020c. www.unwomen.org/en/news/stories/2020/6/feature-bangladesh-womens-organizations-in-covid-19-and-cyclone-recovery.

UNDRR. *Human Cost of Disasters: An Overview of the Last 20 Years 2000–2019*. UN Office for Disaster Risk Reduction, 2020.

United Nations Bangladesh. *Review of Strengthening Area-Based Disaster Management Committee (DMC) Coordination Model in Bangladesh*. United Nations Bangladesh, 2021. www.humanitarianresponse.info/en/operations/bangladesh/document/strengthening-area-based-disaster-management-committee-dmc.

Unnayan Sangha. 'Resilience through Economic Empowerment, Climate Adaptation, Leadership and Learning – Ree-Call 2021'. 2022. https://us-bd.org/resilience-through-economic-empowerment-climate-adaptation-leadership-and-learning-ree-call-2021/.

Vinke, Kira, Jonas Bergmann, Julia Blocher, Himani Upadhyay, and Roman Hoffmann. 'Migration as adaptation?' *Migration Studies* 8, no. 4 (2020): 626–634.

Von Meding, J. K., L. Oyedele, and D. J. Cleland. 'Developing NGO competencies in post-disaster reconstruction: A theoretical framework'. *Disaster Advances* 2, no. 3 (2009): 36–45.

Waheduzzaman, M., and M. Mizanuzzaman. 'Impact of salinity on the livelihoods of the coastal people in Bangladesh: An assessment on post Amphan situation'. *Journal of Global Ecology and Environment* 11 (2021): 6–24.

World Bank Group. 'Social safety nets in Bangladesh help reduce poverty and improve human capital'. World Bank, 12 May 2019. www.worldbank.org/en/news/feature/2019/04/29/social-safety-nets-in-bangladesh-help-reduce-poverty-and-improve-human-capital.

World Bank Group. 'Improving the transparency and efficiency of safety nets for the most vulnerable in Bangladesh'. World Bank, 17 November 2020. www.worldbank.org/en/results/2020/11/17/improving-the-transparency-and-efficiency-of-safety-nets-for-the-most-vulnerable-in-bangladesh.

World Bank Group. 'Development projects: Coastal Embankment Improvement Project – Phase I (CEIP-I)'. World Bank, 2022. https://projects.worldbank.org/en/projects-operations/project-detail/P128276.

Yasumura, S., A. Goto, S. Yamazaki, and M. R. Reich. 'Excess mortality among relocated institutionalized elderly after the Fukushima nuclear disaster'. *Public Health* 127, no. 2 (2013): 186–188.

5
THE GREEN CLIMATE FUND AS AN ELABORATE SCHEME OF GENERATING SOCIAL HARMS

Jessica Omukuti and Aidan O'Sullivan

1   Introduction

Unless extremely ambitious plans for mitigation and adaptation are implemented, climate change will push communities to the brink of collapse, for example, through permanent ecosystem losses resulting from a 1.5–2°C rise in temperature (IPCC 2019). International institutions such as the United Nations Framework Convention on Climate Change (UNFCCC) and the Green Climate Fund (GCF) have a critical role to play in climate justice, for example, through provision of high-level governance on climate change (Michonski and Levi 2010). In the Majority World, these institutions influence structuring of responses and allocation of resources for enabling these responses. How these institutions are designed determines the effectiveness of climate action.

Institutional design and operations are political, meaning that how institutions are designed can be based on special interests. Understanding institutions' contribution to climate justice therefore requires interdisciplinarity. However, existing empirical scholarship on climate justice is disciplinary and fails to highlight the cross-scalar nature of climate change and climate change responses (Barrett 2013). Consequently, existing research is not strongly grounded in the everyday realities of people who engage with these institutions. Addressing this gap requires interdisciplinary approaches that enable empirical research on climate justice to inform political decisions (Roser et al. 2015). This chapter uses an interdisciplinary lens to highlight how international institutions determine climate justice outcomes in the Majority World.

Using a case study of the GCF, this chapter critically reflects on how choices of trade-offs within international institutions generate specific design structures with implications for climate justice in the Majority World. The analysis is based on a

DOI: 10.4324/9781003214021-6

review of GCF documents (e.g. GCF operational policies and guidelines) to assess operations, programming, and funding for climate change action. The chapter builds on the concept of social harms from the theoretical lens of zemiology which has emerged from work on critical criminology (Pemberton 2016).

The analysis pays attention to how policy pathways that are supported by international climate change institutions fail to alleviate and, in some cases, generate social harms for communities affected by climate change, particularly those in the Majority World. It presents the concept of 'paradoxical harms', which arises from conscious choices of costs and benefits. These concepts are used to assess how the GCF's design structure justifies actions at different levels of governance that generate social harms (and hence climate injustices) for local communities in the Majority World. The chapter presents the GCF's design structure as a conscious choice by international-level climate change governance stakeholders based on international-level priorities for climate action. However, the prioritization of international interests over local interests depoliticizes climate change and its impacts, resulting in increased drivers and outcomes of climate change risks.

This chapter contributes to literature on climate justice in the Majority World by highlighting the links between institutions for addressing climate change and the role of legalized 'social harms'. It highlights how institutional designs of climate change institutions involve conscious trade-offs that are likely to generate 'acceptable' disadvantage to specific groups in the Majority World. In doing so, it calls for attention to the need for responsibility for climate justice, focusing on how institutions for addressing climate change in the Majority World should be structured to ensure that they deliver climate justice for those disproportionately affected by climate change.

2 The Green Climate Fund

The GCF was established in 2010 and became operational in 2015. It is currently the largest dedicated climate fund globally. Its overall goal is to channel climate finance to 'developing countries' (most of which are in the Majority World) to support low-carbon and climate-resilient development within the framework of sustainable development (GCF 2011). By 2023, the GCF is expected to have channelled over US$18 billion to the Majority World to support essential climate action (Puri et al. 2020), making it a critical facilitator of mechanisms for delivering climate justice.

The GCF design structure is built around three groups of stakeholders: (i) the Board, whose operations are supported by the Secretariat, GCF's donors who are Minority World country governments, and GCF's recipients in the Majority World; (ii) the National Designated Authorities (NDAs) which represent the Majority World and ensure that GCF funding is aligned with country needs; (iii) accredited entities (AEs) which channel funds to Majority World countries by designing and implementing projects that are GCF-funded (see Figure 5.1). AEs can be international institutions such as UN institutions or regional organizations and national institutions from Majority World countries which are direct access entities. The Conference of Parties (COP) monitors the GCF's work against its mandate (GCF 2011).

The GCF's achievement of its mandate strongly hinges on its design structure. Design areas and issues that are particularly important include sources of finance for the GCF which has implications for GCF's capitalization (Cui et al.

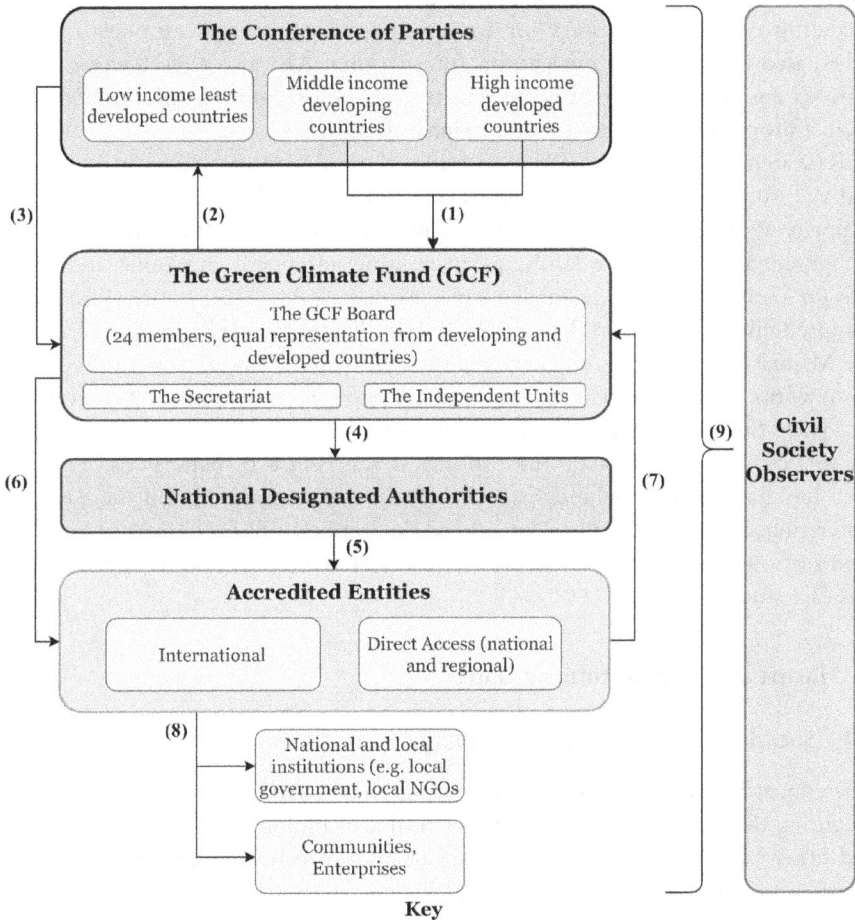

Key

1. Financial pledges and commitments; 2. Annual reports to the COP; 3. COP guidance to the GCF; 4. Support for Readiness activities; 5. Support for project pipeline development; 6. Accreditation; 7. Funding project proposals and self-reporting of outcomes; 8. Implementation of funded activities; 9. Advocacy and capacity building

FIGURE 5.1 Structure of the GCF.

Source: Authors.

2014), decision-making structures (Abbott and Gartner 2011), alignment with international funding principles (Van Kerkhoff et al. 2011), and allocation of finance to the Majority World (Klein and Möhner 2011). However, research on the GCF has identified gaps in the performance against this mandate. For instance, finance allocation within the GCF is found to be skewed towards mitigation, which disadvantages countries in the Majority World whose need for adaptation is higher (Fonta et al. 2018). GCF's slow disbursement of funds and strict compliance requirements also disadvantage direct access entities from the Majority World (IEU 2019). These gaps are usually explained away as necessary for the achievement of the GCF's mandate. For example, the bias towards mitigation financing is seen as necessary for attracting private sector investments through debt-based financing for mitigation. International AEs also have fewer capacity gaps (as compared to direct access entities), which results in easier and quicker project proposal development and implementation.

Research on these gaps fails to consider these design structures as intentional and the outcomes of actions as harms that are imposed on communities in the Majority World. The links between intentions, actions, and outcomes within the global climate governance landscape in framing acceptable outcomes and trade-offs for climate actions are inadequately explored. Assessments therefore fail to comprehensively highlight the role of GCF in shaping climate justice outcomes in the Majority World. This chapter addresses this gap by engaging with theoretical frameworks in critical criminology and concepts in environmental governance that place the social harms that accrue from the normal operations of the market (in this case, climate finance institutions) at the centre of its analysis. The goal is to highlight drivers behind structural designs of institutions and their roles in generating and sustaining social harms and the implications for climate justice. The next section will outline the theoretical bases of zemiology and its application to this case study.

3   Harms and Accountability Gaps

3.1   Social Harms in Critical Criminology

The concept of social harms in relation to climate change can be described using emerging theoretical paradigms from critical criminology, specifically zemiology and green criminology. Zemiology was to address criminology's perceived lack of differentiation of nominally legal and structural 'social harms' from crimes that are legally defined in statutes with an identifiable perpetrator (Hillyard 2015). For zemiologists, the legal status of actions can often have little connection to the level of harm generated. Hence, for many criminal justice systems, many petty acts that are low on the scale of harm receive a disproportionate censure while actions that have harmful outcomes for communities are never punished and are instead legalized or minimized into civil law as opposed to criminal law (White 2013b). Zemiology therefore recognizes that institutional and structural violence can often have further reaching consequences than interpersonal victimization (Pemberton 2016).

Green criminology looks at the influence of human–environment interactions on other humans and non-humans (White and Heckenberg 2014). In the same way as zemiology, green criminology differentiates between green crimes and harms by applying legal-procedural and socio-legal approaches, respectively (Brisman and South 2019). Environmental crimes are actions defined by criminal law as unacceptable, while environmental harms are those considered acceptable within criminal law regulations but cause social harm (White 2018). Harm emerges from state–corporate collusion in pursuit of neoliberal ideals where social harms are considered legally acceptable through state mechanisms (White 2013a; Stoett 2018). Legally defined environmental crimes are addressed by cross-level policing structures (Tomkins 2005). However, these structures cannot sufficiently address environmental harms that are considered legal, thus requiring use of social justice mechanisms to address environmental harms (Hall 2013). Hence, the definition of crime and harm is subject to prevailing political interests resulting in those who are least powerful likely to be disadvantaged.

Social harm theorists grapple with the identification of harms, victims, and perpetrators. Humans and non-humans (i.e. plants and animals) are considered potential victims from legally accepted actions (Halsey and White 1998). Harms should therefore be assessed using a mix of scientific measurement of environmental risk and the social perception of risks by all stakeholders (Gibbs et al. 2009). Victims exist in socially constituted unequal relationships from which harms arise. Harms also affect different groups with different reserves of capital and resources to be able to respond to harm. Hence, zemiology does not merely focus on acts of violence, but also on denial of social and material resources. The identification of perpetrators requires assignment of responsibility arising from omissions or 'abuse of power' (Hall 2013; White 2011).

The perpetrators of social harms range from state to non-state actors. Some research finds that actions that cause social harm 'are frequently not only state sanctioned but are often in fact actively promoted by states pursuant to their … development goals', which complicates identification of harm perpetrators, as links between actions and effects on individuals or groups are not always straightforward (Hall 2014, 130). Intent also becomes a critical factor in assignment of responsibility and identification of perpetrators. For example, when harms are unpredictable, then it is difficult to assign responsibility for harms due to lack of intent (Potter 2013). Social harm theory is therefore more concerned with whether a harm is preventable as opposed to the presence of intent. Hence, accountability (i.e. contributing to the causation of a harm), irrespective of intent, should be used in assigning responsibility for harms (White 2011).

3.2 Paradoxical Harms and Accountability Deficits

Social harms related to climate change emerge in two ways – directly from the impacts and drivers of climate change, and from the responses to climate change, that is, paradoxical harms. First are the direct social harms emerging from the global capitalist systems which encourage extraction of resources at the expense of environmental and human welfare and cause direct and indirect impacts of climate change, for example, food insecurity and cultural heritage loss due to direct climate risks such as sea level rise in Latin America (Ezcurra and Rivera-Collazo 2018). These harms are generally structurally reinforced at different scales and emerge from the slow normalization of marginalization and oppression of some social groups in society (Christie et al. 2008). The harms are linked to 'various forms of "legitimate" coercion exercised by the state and its agents in the effort to maintain existing institutions, contain social conflict, or forcibly pursue and defend a particular definition of the collective interest' (Soron 2007, 10). Second are harms emerging from climate change interventions, for example, development of urban green infrastructure that displaces and dispossesses local communities (Anguelovski et al. 2019). These are paradoxical climate change harms.

Climate change-related paradoxical harms emerge when actions that are framed as solutions for climate change 'in turn, generat[e] new forms of social and environmental harm' (White 2012, 63). Paradoxical harms are dissimilar to unintended consequences because, unlike the latter, harms are well known (White 2012, 63). This means that the harms are considered acceptable trade-offs for the benefits emerging from the implementation of climate change adaptation and mitigation actions. Adaptation literature recognizes that addressing drivers to climate change will require systemic transformations in approaches for addressing societal inequalities to actions and systems that enable inclusion of marginalized populations (Pelling et al. 2015). However, some responses to climate change sustain the state–corporate collusions that create climate change and reinforce inequalities across scales (Kammerbauer and Wamsler 2017). Paradoxical harms can also result from structural violence which evolves over the long term and makes it challenging to link paradoxical harms to the perpetrators (Soron 2007).

Layering the concept of scale over understandings of paradoxical harms provides a multifaceted understanding of social harms. Transnational harms are caused by actors at one scale, but experienced by groups at another scale (White 2011). For example, the development of dams for exclusive hydropower production in the Himalayas (as opposed to multipurpose dams) is likely to dispossess local communities of their water rights (Baruah 2012). Temporally, avoidance of political responsibility for the implementation of the most ambitious adaptation and mitigation actions increases future climate risks. This means that whether (or not) harms are paradoxical depends on the availability of (i) knowledge of sources and nature of adverse risks from climate action and (ii) tools (and systems) to eliminate or limit risks from these climate change actions.

Social harms from climate change and climate change actions result from 'accountability gaps' within global climate and environmental governance (Kramarz and Park 2016; Najam and Halle 2010). Accountability involves a specific set of actors (account holders) demanding that others (agents) report on activities or progress against a given set of expectations, accompanied by rewards or costs for the agents (Biermann and Gupta 2011). Accountability gaps emerge from 'the continued growth of accountability mechanisms without teeth, [leading to] current applications of accountability in … [global environmental governance] run[ning] a real risk of enabling continued environmental degradation rather than protection' (Kramarz and Park 2016, 19). Hence, 'international law remains relatively weak on climate justice, despite hundreds of multilateral environmental agreements and several agencies of global [climate] governance' (Stoett 2018, 2). Accountability gaps also emerge when the distance between decision-makers and the impacts of their actions is too large, which enables them to get away with any negative consequences that may emerge (Newell 2008). Powerful actors leverage these gaps and misuse their power at the expense of marginalized groups (Grant and Keohane 2005).

Drivers of accountability deficits are structural or cultural. Structurally driven deficits emerge when there are changes in structures of governance mechanisms without accompanying allocation of roles and responsibilities, incoherent policies, and guidelines (Lloyd et al. 2008) and are linked to multiple networks of account holders and agents with diverse interests, which generates negative consequences to third-party account holders such as communities (Krahmann 2016). Culturally driven accountability gaps emerge from the absence of an accountability culture and political commitment towards accountability, which results in under-implementation of policies and underperformance of procedures (Lloyd et al. 2008), and are sustained by 'blame avoidance' where actors fail to take responsibility for their inability to reach targeted goals or any negative consequences that emerge and instead blame it on politics (Bache et al. 2015). Addressing these drivers of accountability gaps requires the design and implementation of context-specific accountability tools. Examples of these include ensuring availability of information, participation, and transparency to achieve legitimacy (Koenig-Archibugi 2004). In climate change governance, accountability principles are used for assigning responsibility for climate change action and ensuring participation and representation of those impacted by climate change in decision-making (Newell 2008).

4 Accountability Gaps Created and Used by the GCF

Accountability within the multilateral climate finance landscape is embedded into the GCF's design structure. The GCF's accountability mechanisms are achieved through transparency (e.g. stakeholder participation), certification, monitoring and evaluation, and self-reporting (Scobie 2018). However, although the GCF's structure delivers on international principles of country ownership and fair contribution to climate finance in alignment with the Paris Agreement, it also reinforces structural violence by the Minority World towards the Majority World and perpetuates a culture of blame avoidance within international climate finance. This results in local-level climate injustices. These structures sustain a pre-existing hegemonic relationship between Majority and Minority World countries and between governments and non-government and community representatives, which disadvantages those severely affected by climate change.

4.1 Donor Control through Accountability Mechanisms

The GCF's design structure promotes accountability through its pursuit of egalitarian representation and decision-making in three ways – the GCF's relationship to the COP, the structure of the GCF Board, and priority on country ownership. The GCF is 'accountable to and function[s] under the guidance of the COP' (GCF 2011, para. 4). This aims 'to reduce its perceived dominance by donor state interests' (Vanderheiden 2015, 34). Although the GCF Board has full responsibility for funding decisions, Board members have expressed concern over the unequal relationship between the GCF and the COP, indicating greater control by the COP over the Fund (GCF 2013, para. 182). Researchers question the autonomy of the GCF Board in designing and implementing policies (Recio 2019). This is because the GCF Governing Instrument mandates the GCF to respond to COP guidance (thus making COP guidance legally binding) and grants the COP authority to terminate the GCF (Recio 2019). In the backdrop of the North–South divide in influence at the COP, the power of COP over GCF is worrying. Specifically, Minority World countries have historically been more influential in COP negotiations and their outcomes (Rowe 2015), with some countries having been said to have previously held the COP negotiations hostage (Christoff 2010). This suggests disproportionate Minority World country control over the GCF's operations through the COP.

Second, the GCF Board has equal representation from both 'developed' and 'developing' countries where, until 2019, decision-making was based on consensus. This ensures a fair and balanced representation of countries in GCF policy and resource allocation decisions. The equal representation within the GCF Board has led to a higher preference for the GCF by Majority World countries for channelling of climate finance (Vanderheiden 2015). The seemingly reduced Minority World country decision-making power within the GCF Board contributes to making the GCF unattractive to donor countries, as this reduces the opportunities for donor control as compared to other bilateral funding options (de Sépibus 2014). However, consensus-based decision-making has led to gridlocks in the adoption of key decisions (Bowman and Minas 2019). This means that the equal representation in decision-making design is not delivering results that accelerate climate action in the Majority World.

The COP–Board collaborative approach to accountability, while ostensibly useful for more transparent and equitable decision-making, has enabled subtler forms of donor control of project activities. First, a methodology for allocating responsibility for financial contributions amongst Minority World countries is still missing (Cui et al. 2014). This means that financial commitments and contributions of Minority World countries to the GCF remain voluntary and pledge-based (Schalatek et al. 2012), which fails to fulfil the common but differentiated responsibilities and respective capabilities principle of the UNFCCC (Vanderheiden 2015). This creates leeway for Minority World countries to redirect climate finance to Majority World countries through bilateral Official Development Assistance channels, which have previously exhibited (and perhaps continue to exhibit) greater donor control (Marcoux et al. 2013) and yet still report these flows as climate finance (Michaelowa and Michaelowa 2007). Contribution (or non-contribution) by Minority World countries to the GCF therefore becomes a reputational issue as opposed to a binding commitment, which makes the Fund an 'empty signifier' in advancing transformational climate action (Methmann 2010).

Greater donor control has resulted in the GCF's pursuit to engage the private sector. The private sector is an avenue for the GCF to, first, generate more finance to cover the resourcing gap left by Minority World countries' voluntary and insufficient financial commitments and, second, to finance mitigation actions which are not considered a priority by Majority World countries. Minority World countries, through their disproportionate influence in COP negotiations and in the GCF Board, have made funding commitments and financial allocations conditional on private sector finance mobilization (Bowman and Minas 2019). At the basic level, this is necessary to meet COP guidance to the GCF on engagement of the private sector (UNFCCC 2015, decision 7 para 9). However, even though private sector finance can address the current climate finance deficit in Majority World countries (Bowen 2011), the GCF's reliance on private sector finance correlates with its promotion of concessional tools of financing which are not preferred by Majority World countries. The push for privatization of climate change solutions promotes managerial approaches to environmental governance, which have previously left Majority World countries highly indebted to Majority World countries and institutions, with outcomes benefiting donor countries (Roberts and Parks 2009). Greater engagement of the private sector also conflicts with the need for transparency in funding, operations, and outcomes, as private sector actors are averse to having their operations publicly scrutinized (Kalinowski 2020). Overall, these design structures reflect notions such as those represented in discourses on climate protection, which 'embodies a hegemonic discourse which prevents major changes to the social structures of global capitalism' (Methmann 2010, 350).

4.2 *Blame Avoidance through Hierarchical Accountability*

The GCF's goal is to implement ambitious climate change action by financing programmes that contribute towards countries' Nationally Determined Contributions and National Adaptation Plans (GCF 2019). Pursuance of this goal has led to the GCF's adoption of a design structure that reflects a hierarchical approach to accountability which prioritizes engagement with international and national governments and civil society while overlooking other intrastate civil society actors. These relationships shift the responsibility for implementing and monitoring of projects and programmes to national and regional organizations. For example, national governments, through the NDAs, are the primary decision-makers of the composition of project pipelines. This means that the GCF does not directly engage with communities that are impacted by climate change and which implement the funded projects, for example, non-state actors, subnational governments, and local-level communities (Najam and Halle 2010).

This hierarchical system can also be observed in the GCF's grievance mechanism which individuals and groups can use to report or register concerns for activities that are causing or likely to cause social harms. GCF's AEs are responsible for informing communities about available channels for expressing grievances. The first line of reporting grievances is to the institutions linked to the AEs, for example, see Conservation International (2021). This reflects the intention to promote greater autonomy and country ownership of projects by the AEs. However, the GCF's Independent Integrity Unit and the Independent Redress Mechanism also have mechanisms through which individuals or groups can report misconduct or grievances (GCF 2020a). These are available on the GCF website and recognize that those wishing to file grievances or complaints may be unable to do so through the AEs. However, these are exclusionary as they require that a complainant be literate and have access to a smart device and an internet connection, both of which are not guaranteed for some of the places that GCF's AEs already operate or should be operating. Board observers have noted that the GCF and its AEs' approaches to capturing and recording grievances could be made more proactive to protect the rights of local communities (GCF 2020b, para. 446). The continued use of GCF's design structure for accountability despite these concerns reflects a conscious trade-off between effectiveness and efficiency of operations.

The distancing between the GCF and local institutions gives AEs greater autonomy to prioritize, design, and implement adaptation and mitigation actions. This reflects the GCF's trust in its accreditation processes based on whether AEs meet GCF standards and can implement set environmental and social safeguards (ESS) that protect communities from harms. This ensures that GCF avoids being overly prescriptive and allows countries and actors to track progress that is important to them (van der Ven et al. 2016). This distancing also builds the capacity of these institutions to engage with other funds and donors, which diversifies the climate finance portfolio across the landscape. When viewed from an international level, these design structures generate a bottom-up system of accountability and claims-making within the GCF. However, when viewed from the local level, accountability remains top-down and state-driven, where actors seeking to get their concerns heard must do so through state governments (Newell 2008).

Yet, recent assessments of GCF projects and project documents indicate that even though AEs demonstrate their capacity to implement Environmental and Social Safeguards (ESS) during accreditation, project documents fail to adequately categorize project risk and have limited application of ESS standards and grievance mechanisms (Perrault and Leonard 2017). The reliance on self-reporting for GCF's monitoring and evaluation of project investments and outcomes results in limited timely transparency on project activities and their outcomes, especially for local-level communities (Perrault and Leonard 2017). However, self-reporting has led to misinformation on the (positive or negative) impact of GCF-funded activities (IEU 2018). For example, projects are reported to overestimate the number of beneficiaries (IEU 2021). Although evaluations have recommended that the GCF work out ways to verify these self-reports (IEU 2020), these mechanisms are still missing.

This hierarchical model shifts responsibility for the implementation and results monitoring onto these actors while avoiding engagement with local-level communities. The GCF's focus on international and national institutions assumes a trickle-down of resources to the local level where climate risks are experienced, and a trickle-up of local knowledge from the local communities to inform national and international climate change processes. This keeps climate change action tepid by making the national government of the receiver country responsible for responding to the multilevel ecological disordering of climate change while ignoring other subnational and local actors who can deliver more effective climate change action (Colenbrander et al. 2018). Consequently, 'the overwhelming preference for applying accountability solely to functional, end-of-pipe concerns such as verification, measurement, and compliance ultimately risks doing little to protect the global environment' (Kramarz and Park 2016, 2).

This opens spaces for blame avoidance which is characterized by the emphasis on a superficial bottom-up system of accountability where the GCF is 'motivated primarily by the desire to avoid blame for unpopular actions rather than seeking to claim credit for popular ones' and where 'the political costs of failure tend to outweigh the benefits of success' (Bache et al. 2015, 71). Working through international and national accredited institutions translates country ownership into 'institutional capacity to handle a project', which results in top-down financialization of climate change adaptation and mitigation as opposed to bottom-up climate responses (Bertilsson and Thörn 2020). Blame avoidance is therefore likely to shift the responsibility for emission and vulnerability reductions and failure to achieve these goals onto national-level actors in Majority World countries. The role of international climate change actors, including the GCF, is therefore reduced to that of facilitation.

4.3 The Outcome: Reinforcement of Historical Structural Violence

Together, these design structures contribute to a GCF-led finance system that fails to challenge the structural causes of climate change and provide superficial solutions to addressing vulnerability to climate change. Finance channelled through the GCF has therefore become a veil for sustaining an apolitical approach to addressing climate change and provides an ideological cover for the continued functioning of systems at different scales that cause climate change and disproportionately allocate impacts to Majority World countries. The GCF is therefore complicit in the broader structural 'aggression[s] by the rich [and powerful] against the poor' that drive climate change (see Abbott [2008, 8] citing Ugandan President Yoweri Museveni's speech at the African Union Summit in 2007). It does for the most part take an instrumental stance towards the environment and its degradation in terms of what this means for the humans who inhabit it (White 2013b). This is demonstrative of how reliant these flows are on hegemonic participants in the GCF and are reflective of geopolitical realities.

The structural violence outcomes of climate action by the GCF are created and perpetuated by processes that intentionally generate trade-offs from different decisions and choose outcomes that benefit global climate governance at the expense of Majority World countries. These are the paradoxical harms of international climate finance, where allocation of finance for climate change is based on the 'greater good' and usually at the expense of local communities in Majority World countries that are most affected by climate change. These harms to these communities are socially accepted within international climate finance, and measures instituted to address them are ineffective because they do not address the root causes of the problem – the structural design of the international climate finance institutions. The drivers of these social harms and ultimately injustices are mainly transnational actors who in this case are Minority World countries.

These outcomes from climate actions align with zemiological concerns that the approach of state legislatures towards these structural harms is to treat them as regulatory issues (Hillyard and Tombs 2004). This generates ineffectual solutions that insufficiently highlight or address the injustices of climate change. This is unsurprising due to the unequal power the GCF's contributor countries, which are mainly Minority World countries, have over what form interventions for climate change should take (see Farand 2021). These interventions side-line vulnerable communities and other stakeholders in Majority World countries by prioritizing consultation with central-state actors. For the local level in Majority World countries, actions guided by the GCF's design structure result in actions that fail to deliver on local-level vulnerability reduction. National government-driven allocation of finance in Majority World countries is politically informed as opposed to being based on vulnerability (Barrett 2014). This means that in addition to the failure to recognize perpetrators of climate harm in Minority World countries and corporate backers who provide panaceas for intervention through GCF, there is also a misidentification of victims within Majority World countries, which entrenches climate injustices.

5 Conclusion

Global efforts for addressing climate change will only be successful if the underlying structural drivers of climate collapse are addressed with meaningful input from the most vulnerable populations in the Majority World. This chapter set out to analyse how GCF's design structure caused social harms. It applies the concept of paradoxical harms which was derived from zemiology and critical criminology. The GCF cannot be compared to the fossil fuel industry or Minority World countries that have the greatest responsibility for causing climate change. However, the GCF bears the responsibility for the paradoxical harms that accrue to populations in Majority World countries through its funded activities. This is a common feature of the global climate governance architecture, where actions by legitimate institutions generate conscious trade-off choices that are sanctioned at one level but that cause harms at another.

The GCF should be an instrument for amelioration of the harms caused by climate change, yet it fails to challenge the drivers of these harms. Instead, it builds on and exacerbates the inequalities between countries to advance notions of climate protection that align with global climate change objectives. It obscures those responsible and those harmed and the fundamentally unequal and exploitative relationships between them. In doing so, the GCF fails to adequately define the problem – anthropogenic climate change and multiscalar inequalities – with the gravity and urgency it deserves. This chapter has shown that finance allocated through the GCF demonstrates a symptom of wider emphasis on global liberal governance that is built on the inequalities between Minority and Majority World countries. For example, it frames the market (through private sector financing) as the preferred domain to define and seek for the climate crisis. As countries in the Majority World become more unstable through impacts of climate change, they will continue to rely on international climate finance which will reinforce the unequal power relations between Minority and Majority World countries.

The global climate governance system is complex, with networked actors and diverse account holder–agent relationships (Widerberg and Pattberg 2017). However, how these actors engage with one another has implications for how climate change is addressed and the climate justice outcomes for those in the Majority World. Reduction of trade-offs and the harms that emerge from these governance systems will require an improvement in accountability by climate change adaptation and mitigation institutions. International climate finance institutions, such as the GCF, should be a starting point for accountability improvements. Greater accountability will enable the GCF to justify its actions not just to international-level actors but also to local actors and communities in the Majority World, resulting in improved legitimacy. This can be achieved through the valuation of GCF's impacts that includes both input and output measurements as well as the systemic impacts of interventions (van der Ven et al. 2016). In doing so, responses to climate change can be effective in avoiding the catastrophic effects and losses that communities in the Majority World will face if climate change persists.

International climate finance institutions and communities in the Majority World need to reimagine new pathways through which the climate crisis can be addressed without deepening the harms that we have identified in this chapter. Although research interest on climate finance institutions, such as the GCF, is increasing, future research can focus on understanding how climate finance institutions' design structures can be adjusted to reduce or address the trade-offs that generate social harms for the Majority World. Evidence of the effectiveness of these alternative pathways to climate solutions in the Majority World that address the climate crisis without compromising the Majority World's development ambitions and adaptation capacity is also urgently needed. This can hopefully then be integrated into future climate finance institutional design processes to ensure climate justice in the Majority World.

References

Abbott, Chris. 2008. *An Uncertain Future: Law Enforcement, National Security and Climate Change*. Oxford Research Group.

Abbott, Kenneth, and David Gartner. 2011. 'The Green Climate Fund and the future of environmental governance'. Earth System Governance Working Paper 16.

Anguelovski, Isabelle, Clara Irazábal-Zurita, and James J. T. Connolly. 2019. 'Grabbed urban landscapes: Socio-spatial tensions in green infrastructure planning in Medellín'. *International Journal of Urban and Regional Research* 43 (1):133–156. https://doi.org/10.1111/1468-2427.12725.

Bache, Ian, Ian Bartle, Matthew Flinders, and Greg Marsden. 2015. 'Blame games and climate change: Accountability, multi-level governance and carbon management'. *British Journal of Politics and International Relations* 17 (1):64–88. https://doi.org/10.1111/1467-856X.12040.

Barrett, Sam. 2013. 'The necessity of a multiscalar analysis of climate justice'. *Progress in Human Geography* 37 (2):215–233. https://doi.org/10.1177/0309132512448270.

Barrett, Sam. 2014. 'Subnational climate justice? Adaptation finance distribution and climate vulnerability'. *World Development* 58:130–142. https://doi.org/10.1016/j.worlddev.2014.01.014.

Baruah, Sanjib. 2012. 'Whose river is it anyway? Political economy of hydropower in the eastern Himalayas'. *Economic and Political Weekly*, 41–52.

Bertilsson, Jonas, and Håkan Thörn. 2020. 'Discourses on transformational change and paradigm shift in the Green Climate Fund: The divide over financialization and country ownership'. *Environmental Politics*, 1–19. https://doi.org/10.1080/09644016.2020.1775446.

Biermann, Frank, and Aarti Gupta. 2011. 'Accountability and legitimacy in earth system governance: A research framework'. *Ecological Economics* 70 (11):1856–1864. https://doi. org/10.1016/j.ecolecon.2011.04.008.

Bowen, Alex. 2011. 'Raising climate finance to support developing country action: Some economic considerations'. *Climate Policy* 11 (3):1020–1036. https://doi. org/10.1080/14693062.2011.582388.

Bowman, Megan, and Stephen Minas. 2019. 'Resilience through interlinkage: The Green Climate Fund and climate finance governance'. *Climate Policy* 19 (3):342–353. https:// doi.org/10.1080/14693062.2018.1513358.

Brisman, Avi, and Nigel South. 2019. 'Green criminology and environmental crimes and harms'. *Sociology Compass* 13 (1):e12650. https://doi.org/10.1111/soc4.12650.

Christie, Daniel J., Barbara S. Tint, Richard V. Wagner, and Deborah DuNann Winter. 2008. 'Peace psychology for a peaceful world'. *American Psychologist* 63 (6):540. https:// doi.org/10.1037/0003-066X.63.6.540.

Christoff, Peter. 2010. 'Cold climate in Copenhagen: China and the United States at COP15'. *Environmental Politics* 19 (4):637–656. https://doi.org/10.1080/09644 016.2010.489718.

Colenbrander, Sarah, David Dodman, and Diana Mitlin. 2018. 'Using climate finance to advance climate justice: The politics and practice of channelling resources to the local level'. *Climate Policy* 18 (7):902–915. https://doi.org/10.1080/14693062.2017.1388212.

Conservation International. 2021. 'CI-GCF agency grievance mechanisms'. www.conse rvation.org/gcf/grievances.

Cui, Lian-Biao, Lei Zhu, Marco Springmann, and Ying Fan. 2014. 'Design and analysis of the Green Climate Fund'. *Journal of Systems Science and Systems Engineering* 23 (3):266–299. https://doi.org/10.1007/s11518-014-5250-0.

de Sépibus, Joëlle. 2014. *The Green Climate Fund: How Attractive Is It to Donor Countries?* NCCR Trade Regulation, World Trade Institute, University of Bern.

Ezcurra, Paula, and Isabel C. Rivera-Collazo. 2018. 'An assessment of the impacts of climate change on Puerto Rico's cultural heritage with a case study on sea-level rise'. *Journal of Cultural Heritage* 32:198–209. https://doi.org/10.1016/j.cul her.2018.01.016.

Farand, C. 2021. 'Dispute grips the Green Climate Fund over net zero condition of accessing finance'. www.climatechangenews.com/2021/10/08/dispute-grips-green-climate-fund-net-zero-condition-accessing-finance.

Fonta, William M., Elias T. Ayuk, and Tiff van Huysen. 2018. 'Africa and the Green Climate Fund: Current challenges and future opportunities'. *Climate Policy* 18 (9):1210–1225. https://doi.org/10.1080/14693062.2018.1459447.

GCF. 2011. *Governing Instrument of the Green Climate Fund*. Green Climate Fund.

GCF. 2013. *GCF/B.05/24/Rev.01: Report of the Fifth Meeting of the Board, 8–10 October 2013*. Green Climate Fund.

GCF. 2019. *Turning Ambition into Action: How the Green Climate Fund Is Delivering Results*. Green Climate Fund.

GCF. 2020a. 'Accountability'. www.greenclimate.fund/about/accountability.

GCF. 2020b. *Report of the Twenty-Seventh Meeting of the Board, 9–13 November 2020*. Green Climate Fund.

Gibbs, Carole, Meredith L. Gore, Edmund F. McGarrell, and Louie Rivers, III. 2009. 'Introducing conservation criminology: Towards interdisciplinary scholarship on environmental crimes and risks'. *British Journal of Criminology* 50 (1):124–144. https:// doi.org/10.1093/bjc/azp045.

Grant, Ruth W., and Robert O. Keohane. 2005. 'Accountability and abuses of power in world politics'. *American Political Science Review* 99 (1):29–43. https://doi.org/10.1017/S0003055405051476.

Hall, Matthew. 2013. *Victims of Environmental Harm: Rights, Recognition and Redress under National and International Law*, 1st ed. Routledge.

Hall, Matthew. 2014. 'Environmental harm and environmental victims: Scoping out a "green victimology"'. *International Review of Victimology* 20 (1):129–143. https://doi.org/10.1177/0269758013508682.

Halsey, Mark, and Rob White. 1998. 'Crime, ecophilosophy and environmental harm'. *Theoretical Criminology* 2 (3):345–371.

Hillyard, Paddy. 2015. 'Criminal obsessions: Crime isn't the only harm'. *Criminal Justice Matters* 102 (1):39–41. https://doi.org/10.1080/09627251.2015.1143645.

Hillyard, Paddy, and Steve Tombs. 2004. 'Beyond criminology'. In *Beyond Criminology: Taking Harm Seriously*, 10–29. Pluto Press.

IEU. 2018. *Independent Evaluation of the GCF's Results Management Framework (RMF2018)*. Independent Evaluation Unit of the Green Climate Fund.

IEU. 2019. *Forward-looking Performance Review of the Green Climate Fund (FPR)*. Evaluation Report No.3 (2nd ed.). Independent Evaluation Unit of the Green Climate Fund.

IEU. 2020. *Independent Evaluation of the GCF's Environmental and Social Safeguards and the Environmental and Social Management System*. Independent Evaluation Unit of the Green Climate Fund.

IEU. 2021. *Independent Evaluation of the Adaptation Portfolio and Approach of the Green Climate Fund*. Green Climate Fund.

IPCC. 2019. 'Summary for policymakers'. In *Climate Change and Land: An IPCC Special Report on Climate Change, Desertification, Land Degradation, Sustainable Land Management, Food Security, and Greenhouse Gas Fluxes in Terrestrial Ecosystems*, edited by P. R. Shukla, J. Skea, E. Calvo Buendia, V. Masson-Delmotte, H.-O. Pörtner, D. C. Roberts, P. Zhai, R. Slade, S. Connors, R. van Diemen, M. Ferrat, E. Haughey, S. Luz, S. Neogi, M. Pathak, J. Petzold, J. Portugal Pereira, P. Vyas, E. Huntley, K. Kissick, M. Belkacemi and J. Malley, 3–36. Cambridge University Press, Cambridge, UK and New York, NY, USA. https://doi.org/ 10.1017/9781009157988.

Kalinowski, Thomas. 2020. 'Institutional innovations and their challenges in the Green Climate Fund: Country ownership, civil society participation and private sector engagement'. *Sustainability* 12 (21):8827. https://doi.org/10.3390/su12218827.

Kammerbauer, Mark, and Christine Wamsler. 2017. 'Social inequality and marginalization in post-disaster recovery: Challenging the consensus?' *International Journal of Disaster Risk Reduction* 24:411–418. https://doi.org/10.1016/j.ijdrr.2017.06.019.

Klein, Richard J. T., and Annett Möhner. 2011. 'The political dimension of vulnerability: Implications for the Green Climate Fund'. *IDS Bulletin* 42 (3):15–22. https://doi.org/10.1111/j.1759-5436.2011.00218.x.

Koenig-Archibugi, Mathias. 2004. 'Transnational corporations and public accountability'. *Government and Opposition* 39 (2):234–259. https://doi.org/10.1111/j.1477-7053.2004.00122.x.

Krahmann, Elke. 2016. 'NATO contracting in Afghanistan: The problem of principal–agent networks'. *International Affairs* 92 (6):1401–1426. https://doi.org/10.1111/1468-2346.12753.

Kramarz, Teresa, and Susan Park. 2016. 'Accountability in global environmental governance: A meaningful tool for action?' *Global Environmental Politics* 16 (2):1–21. https://doi.org/10.1162/GLEP_a_00349.

Lloyd, Robert, Shana Warren, and Michael Hammer. 2008. *2008 Global Accountability Report*. One World Trust.

Marcoux, Christopher, Bradley C. Parks, Christian M. Peratsakis, J. Timmons Roberts, and Michael J. Tierney. 2013. 'Environmental and climate finance in a new world: How past environmental aid allocation impacts future climate aid'. WIDER Working Paper.

Methmann, Chris Paul. 2010. '"Climate protection" as empty signifier: A discourse theoretical perspective on climate mainstreaming in world politics'. *Millennium* 39 (2):345–372. https://doi.org/10.1177/0305829810383606.

Michaelowa, Axel, and Katharina Michaelowa. 2007. 'Climate or development: Is ODA diverted from its original purpose?' *Climatic Change* 84 (1):5–21. https://doi.org/10.1007/s10584-007-9270-3.

Michonski, Katherine, and Michael A. Levi. 2010. *Harnessing International Institutions to Address Climate Change*. Council on Foreign Relations.

Najam, Adil, and Mark Halle. 2010. *Global environmental governance: The challenge of accountability. Sustainable Development Insights Series*. Boston University. https://open.bu.edu/ds2/stream/?#/documents/170847/page/1

Newell, Peter. 2008. 'Civil society, corporate accountability and the politics of climate change'. *Global Environmental Politics* 8 (3):122–153. https://doi.org/10.1162/glep.2008.8.3.122.

Pelling, Mark, Karen O'Brien, and David Matyas. 2015. 'Adaptation and transformation'. *Climatic Change* 133 (1):113–127. https://doi.org/10.1007/s10584-014-1303-0.

Pemberton, Simon A. 2016. *Harmful Societies: Understanding Social Harm*. Policy Press.

Perrault, Anne, and Stephen Leonard. 2017. *The Green Climate Fund: Accomplishing a Paradigm Shift? Analysis of the GCF Approach to Safeguards, Indigenous Rights, and Participatory Processes*. Rights and Resources Initiative.

Potter, Gary R. 2013. 'Justifying "green" criminology: Values and "taking sides" in an ecologically informed social science'. In *The Values of Criminology and Criminal Justice*, edited by M. Cowburn, M. Duggan, A. Robinson, and P. Senior, 125–141. University of Bristol.

Puri, Jyotsna, Archi Rastogi, Martin Prowse, and Solomon Asfaw. 2020. 'Good will hunting: Challenges of theory-based impact evaluations for climate investments in a multilateral setting'. *World Development* 127:104784. https://doi.org/10.1016/j.worlddev.2019.104784.

Recio, María Eugenia. 2019. 'Dancing like a toddler? The Green Climate Fund and REDD+ international rule-making'. *Review of European, Comparative & International Environmental Law* 28 (2):122–135. https://doi.org/10.1111/reel.12286.

Roberts, J. Timmons, and Bradley C. Parks. 2009. 'Ecologically unequal exchange, ecological debt, and climate justice: The history and implications of three related ideas for a new social movement'. *International Journal of Comparative Sociology* 50 (3–4):385–409. https://doi.org/10.1177/0020715209105147.

Roser, Dominic, Christian Huggel, Markus Ohndorf, and Ivo Wallimann-Helmer. 2015. 'Advancing the interdisciplinary dialogue on climate justice'. *Climatic Change* 133 (3):349–359. https://doi.org/10.1007/s10584-015-1556-2.

Rowe, Elana Wilson. 2015. 'Locating international REDD+ power relations: Debating forests and trees in international climate negotiations'. *Geoforum* 66:64–74. https://doi.org/10.1016/j.geoforum.2015.09.008.

Schalatek, Liane, Smita Nakhooda, Sam Barnard, and Alice Caravani. 2012. 'Climate finance thematic briefing: Adaptation finance'. Climate Funds Update. https://climatefundsupdate.org/wp-content/uploads/2022/03/CFF3-Adaptation-Finance_ENG-2021.pdf.

Scobie, Michelle. 2018. 'Accountability in climate change governance and Caribbean SIDS'. *Environment, Development and Sustainability* 20 (2):769–787. https://doi.org/10.1007/s10668-017-9909-9.

Soron, Dennis. 2007. 'Accidental environments cruel weather: Natural disasters and structural violence'. *Transformations: Journal of Media & Culture* (14):1–10.

Stoett, Peter J. 2018. 'Unearthing under-governed territory'. In *Just Security in an Undergoverned World*, edited by W. Durch, J. Larik and R. Ponzio, 238–262. Oxford University Press.

Tomkins, Kevin. 2005. 'Police, law enforcement and the environment'. *Current Issues in Criminal Justice* 16 (3):294–306. https://doi.org/10.1080/10345329.2005.12036326.

UNFCCC. 2015. *Report of the Conference of the Parties on Its Twentieth Session, Held in Lima from 1 to 14 December 2014 Addendum Part Two: Action Taken by the Conference of the Parties at its Twentieth Session.* UNFCCC.

van der Ven, Hamish, Steven Bernstein, and Matthew Hoffmann. 2016. 'Valuing the contributions of nonstate and subnational actors to climate governance'. *Global Environmental Politics* 17 (1):1–20. https://doi.org/10.1162/GLEP_a_00387.

Van Kerkhoff, Lorrae, Imran Habib Ahmad, Jamie Pittock, and Will Steffen. 2011. 'Designing the Green Climate Fund: How to spend $100 billion sensibly'. *Environment* 53 (3):18–31. https://doi.org/10.1080/00139157.2011.570644.

Vanderheiden, Steven. 2015. 'Justice and climate finance: Differentiating responsibility in the Green Climate Fund'. *International Spectator* 50 (1):31–45. https://doi.org/10.1080/03932729.2015.985523.

White, Rob. 2011. *Transnational Environmental Crime: Toward an Eco-global Criminology.* Willan Publishing.

White, Rob. 2012. 'Climate change and paradoxical harm'. In *Criminological and Legal Consequences of Climate Change*, edited by Stephen Farrall, 63–77. Hart Publishing.

White, Rob. 2013a. *Crimes against Nature: Environmental Criminology and Ecological Justice.* Taylor and Francis.

White, Rob. 2013b. *Environmental Harm: An Eco-justice Perspective.* Policy Press.

White, Rob. 2018. *Climate Change Criminology*, 1st ed. Bristol University Press.

White, Rob, and Diane Heckenberg. 2014. *Green Criminology: An Introduction to the Study of Environmental Harm.* Routledge.

Widerberg, Oscar, and Philipp Pattberg. 2017. 'Accountability challenges in the transnational regime complex for climate change'. *Review of Policy Research* 34 (1):68–87. https://doi.org/10.1111/ropr.12217.

6

CLIMATE JUSTICE IN LATIN AMERICA

Mapping the Key Emerging Debates

Lira Luz Benites Lazaro, Zenaida Luisa Lauda-Rodriguez, Susanne Börner, Andrea Lampis, and Leandro Luiz Giatti

1 Introduction

In this chapter, we discuss the emerging climate justice debates in Latin America. This region holds the highest biodiversity on the planet and a variety of eco-climatic gradients. This diversity is not only environmental but also human and social. The interaction between them results in several preserved and threatened natural areas, large development projects, growing urban centres, and multiple vulnerable groups exposed to socioeconomic and environmental vulnerabilities, as well as the risk of disasters and adverse effects of climate change (Ramos and de Salles Cavedon-Capdeville 2017). Furthermore, Latin America is home to inequality that has been characterized as the highest in the world (Serna 2017). Despite the progress achieved in the past decade, poverty reduction efforts in the region were impacted as a result of the COVID-19 health crisis. In 2021, extreme poverty in Latin America affected over 86 million people (CEPAL 2022). This represents a setback of 27 years. The health crisis revealed the structural problems of inequality, poverty, informality, and vulnerability that ended up exacerbating the impact of health emergency on the middle-income population.

The region experiences multiple stressors on natural and human systems resulting in part from land use changes (increasing pressure from competing uses such as cattle ranching, food production, and bioenergy) and exacerbated by climate variability. The IPCC *Sixth Assessment Report (AR6)* emphasized that in Central and South America, global warming is altering the intensity and frequency of extreme weather, water, and climate events (IPCC 2021). The 2022 report of the World Meteorological Organization on the state of climate in Latin America, for example, stresses that glaciers in the tropical Andes have lost at least 30% of their area since the 1980s, placing an increased risk of water

DOI: 10.4324/9781003214021-7

scarcity for the Andean population and ecosystems. Extreme rainfall, floods, and landslides induced substantial losses in 2021, leading to hundreds of lives lost, tens of thousands of homes destroyed or damaged, and hundreds of thousands of people displaced. In, 2021, 7.7 million people in Central American countries experienced high levels of food insecurity with contributing factors including continuing impacts from hurricanes and COVID-19 pandemic economic impacts (WMO 2021). At the same time, these natural resources face climate variability and are strongly impacting vulnerable populations. The worst effects of climate change – for example, flooding, droughts, water scarcity, glaciers melting, and crop failures – are disproportionately felt by countries in Latin America. Thus, discussing climate justice deserves attention in this region.

In Latin America, territorial, social, cultural, and climate justice are part of a single complex resistance narrative. This is particularly the case with Indigenous revindication and claims over both land and their cultural heritage in dealing with climate variability from a historical perspective (Feola 2015; Göbel et al. 2014; Rivas 2012). By a similar token, it is the case of those claims regarding extractive and the 'sacrifice' of large extensions of territory to vested economic and corporate interests with the parallel exclusion of human rights (including social, cultural, and environmental rights) of entire subregions and communities (Alimonda 2011; Svampa 2013). It is not a coincidence that in chapter 12 of the recently published contribution of Working Group II of the IPCC to *AR6*, leading scholars have pointed out (a) the inseparability of social issues such as poverty, land grabbing, migration, deforestation, nutrition, as well as several forms of racial and ethnic violence and (b) the great limitation of institutional, policy, and academic approaches to adaptation to climate change for their almost exclusive focus on the biophysical dimension of climate change (IPCC 2021). This focus on the physical effects of climate change ignores the broader societal and cultural implications, including the role of historical colonialism as a root cause of the current crisis. Colonialism, long recognized as a system of violence and exploitation, has also contributed significantly to climate change through its legacy of expropriation and environmental degradation.

In the last decade, Latin American governments were expanding the extractive frontier, promoting megaprojects in the name of national development (Picq 2020) and as part of climate solutions initiatives such as hydroelectric dam construction. Many of these megaprojects generated conflicts in the region as is corroborated by the Environmental Justice Atlas (EJAtlas 2022), which documented cases and information illustrating, for example, the impressive number of socioenvironmental conflicts related to energy generation in Latin America. It reveals contradictions of how projects aimed at mitigating climate change – if planned barely focused on a single sector – can impact other sectors sometimes in unpredictable ways (Benites-Lazaro et al. 2020, Lazaro et.al, 2022a) Renewable energy projects are increasingly contested. The social groups facing energy injustices related to the implementation of these projects face inequalities related to decarbonization policies; on the one hand, they are negatively affected by climate change and, at the same time, face injustices related to the mitigation of climate change (Lehmann and Tittor 2021). In Section 2 we discuss the concept of climate justice from the analysis of published articles and the main topics on climate justice in Latin America.

2 Mapping Climate Justice Literature

The study started from a keyword search of the term 'climate justice' in the journal database Scopus. We identified 4869 articles from 2000 to 2021 worldwide. We narrowed down the search by using the country/territory Scopus tool, which allowed us to single out 155 articles from the Latin American region. As Figure 6.1 shows, there has been a significant increase in numbers over the last ten years, and in 2021, there were 820 articles published across the world, of which 37 on Latin America. This shows that the region is a fertile ground for studies on climate justice. In the region, several climate justice-related issues can be identified that deserve an in-depth study. Climate justice in Latin America has a direct impact on people's lives and livelihoods. Moreover, environmental resistance weaves into existing struggles for social justice since people face environmental threats in all areas of daily lives. Climate injustice is real for the millions who breathe poisoned air in cities and pesticides in agricultural areas. Key issues include soil

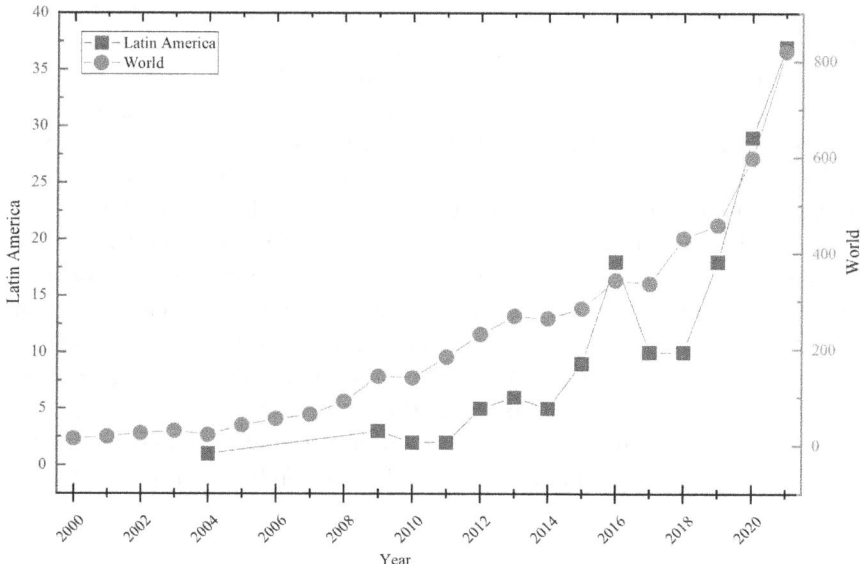

FIGURE 6.1   Number of papers published on 'climate justice' identified via the Scopus database.

Source: Scopus database.

desertification, the disappearance of ancestral forests, snow melting, biodiversity loss, the spreading of infectious diseases, scarcity, as well as frequent and severe floods, droughts, heat waves, and cold waves.

The world literature reveals that the climate justice movement has emerged from global justice protests and that it has been at the forefront of a discourse that is moving away from climate change towards climate justice. Moreover, the climate justice movement constitutes a broader potential for collectivity among communities and social groups engaged in climate change negotiations (Guerrero 2011; Scandrett 2016). Since then, highly specialized environmental NGOs have emerged and participated in climate negotiations as policy experts or, in some cases, as quasi-negotiators, or as members of national delegations of their respective countries. Climate justice movements, together with social movements and the NGOs, have been raising the social justice dimensions of climate change (Guerrero 2011). The literature shows that peasant movements and activists have also played an important role in shaping the climate justice movement by combining issues of biodiversity and climate change with food sovereignty, water depletion, land grabbing, and by embedding the critique of biofuels within the global movement for climate justice (Borras and Franco 2012; Saladini et al. 2018; Svampa 2013). However, the literature also shows that smallholders, plantation workers, and local voices have been conspicuously absent at the transnational level (Kelman 2010), and that the absence of local voices in discussions and decision-making has created a cognitive gap.

Existing literature also focuses on human rights and their connection with climate change, particularly in response to growing scarcity of natural resources and deteriorating climate conditions worldwide. Climate justice literature demonstrates that addressing climate change has proven to be an exceptional space for fighting and building countermovement (Arning and Ziefle 2020; Bebbington 2012; Farrell 2016). On the one hand, there is an intersection of issues, including new land and water and carbon capture programmes, which has blurred the margins of climate change mitigation and resource grabbing. On the other hand, radical movements fighting for agrarian and social justice have encountered resistance to their sector-specific concerns, as evidenced by their intersection with wider issues such as unequal access to resources and conflicts arising from resource scarcity. This resistance can be attributed to the current model of development (Barett 2013; Tramel 2016). In Latin America, the exports of raw commodities have accentuated a violent history of natural resource dispossession and conflicts (Picq 2020). This places Latin America as the deadliest region for environmentalists and land defenders.

3 Climate Justice in Latin America: Main Topics

Figure 6.2 shows the 14 main topics from the analysis of 155 abstracts of articles published with a focus on Latin American countries. They were obtained using the

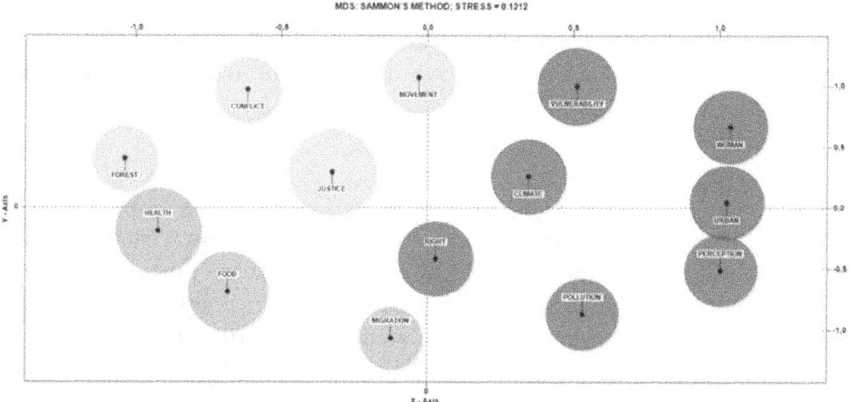

FIGURE 6.2   Results of the topic modelling analysis of articles focused on Latin America.

topic modelling method, which comprises a collection of algorithms that facilitate the organization, understanding, summarization, and visualization of large amounts of textual information (Blei et al 2003). In this study, we used T-Lab's topic model tool that provides a way of identifying, examining, and modelling topics from the textual data. The result is presented by using a multidimensional scaling (MDS) map that allows displaying the relative positions of a number of objects, given only a table of the distances between the topics. In this study, the topics were clustered into four quadrants and colours, according to the similarity among the 14 topics. Previous studies (Lazaro et al. 2022b; Urbinatti et al. 2020) have applied topic modelling to analyse scientific literature and showed the usefulness of the method to identify key topics and their evolution over the years.

In our analysis, those articles draw attention to the fact that increasing extreme climate events, combined with inadequate development and a general lack of preparedness in many countries, can cause serious adverse impacts on communities. Furthermore, these issues can contribute to persistent poverty, social and political exclusion, chronic social vulnerability, poor access to social services, and a lack of opportunities (Kim et al. 2018). Climate justice seeks to link human rights with an understanding of development that safeguards the rights of the most vulnerable, and suggests that the burdens and benefits of climate change as well as its effects and impacts be shared among all countries in an equitable and fair manner (Scandrett 2016).

When comparing climate justice discourses in the international arena with their Latin American equivalent, it becomes evident that there is a tendency of a number of authors to make absolute claims in the name of minorities to whom they neither belong nor legitimately represent. As pointed out by Scandrett (2016), this is reflected by the fact that virtually all interpretations of climate justice seek interventions that protect the poorest and most vulnerable and promote greater equality, or at least do not exacerbate inequality. As it follows, 13 topics are presented and discussed, and the 14th, which is climate, is transversally argued alongside others.

3.1 Social Movements

The climate justice concept originated from within the social movements, and, as such, it is a diffuse and complex one to pin down with sharp precision (Herbert 2020). As mentioned, previously, environmental justice, political ecology, and decolonial approaches can hardly be separated in the Latin American context. Their reading of what climate justice should look like builds around the notions of political and cultural resistance to land grabbing and economic colonization; two preconditions of paramount importance for Latin American movements are debates on land and food sovereignty.

In Latin America, political parties and trade unions from the left have had historical difficulty accepting the environment into their agendas as the latter was not considered a matter of class inequality and, therefore, disconnected from legitimate political struggles (Lampis 2016). The latter has notable exceptions in the peasant movements of the Sem Terra in Brazil (Pegler 2015; de Paulo and Porto 2018; Pegler 2015), in the hundreds of Indigenous struggles related to the extractive practices of the multinationals (Avellaneda 2003; Lalander et al. 2019), and later by the state itself in several countries (Gudynas 2013). Also worth mentioning is women's struggle for environmental justice (Hicks and Fabricant 2016; Muñoz and del Villarreal 2019), an issue discussed later in this chapter. The second peculiarity is a later move towards reconciling struggles based on class and identity under the unifying concept of 'territorio'. *Territorio* (in Spanish) or *território* (in Portuguese) translates neither landscape nor territory, as in English both terms are employed in physical geography. A territory for Latin American social movements is a living construction, which is the foundation of culture, the place that keeps memory alive, and the source of material and symbolic resources. The latter is a central notion in post- and decolonial thinking in the region (Alimonda 2011; Escobar 2010).

Climate justice movements mobilize multiple strands of environmental justice pressure groups, mainly from Indigenous and peasant rights movements. The emergence of climate justice claims is linked to Latin American social movements' narratives around the defence of life and territory. The Latin American and Caribbean Platform for Climate Justice brings together a group of movements, organizations, and social networks in the region to face the serious climate crisis in an articulated way, coordinate the resistance to the current capitalist system and its false solutions to climate change, and promote alternatives for a just transition at the national and local levels, led by the peoples from the territories (PlataformaJusticiaClimática 2022). The transnational social movement that has had adherence in Latin America is the International Indigenous Peoples Forum on Climate Change (IIPFCC), which participates in global climate discussions, particularly in the context of the United Nations Framework Convention on Climate Change. It has sought to increase Indigenous participation in the UN processes and regain control over ancestral territory (Gligo et al. 2020; Martinez-Alier et al. 2018).

Social movements in Latin America have also engaged with new international networks such as Blokadia, a networked movement with three features stemming from a wider set of claims that connect campaigners. First, almost all claims and vindications try to go beyond the protection of the environment, delving into issues of democracy and resource control (Martinez-Alier et al. 2018). Second, campaigns often combine local concerns with a message put forward by those advocating for a global awareness on climate change. Third, frontline activists are local people who attend local village assemblies, council meetings, marches in provincial cities or capitals and are sometimes arrested or subjected to violence (Martinez-Alier et al. 2018).

3.2 Socioenvironmental Conflicts

Latin America is a site of numerous socioenvironmental conflicts, and as such it is also a place of constant resistance and struggle. For the most part, people and communities face violence in response to defending their natural territories against the expansion of extractive processes. However, environmental injustice does not always spark grassroots mobilization, as in the latent conflicts excluded from the main database. That is because environmental or health problems are not easy to identify, or because those affected may be (or feel) too weak or invisible to society (Orihuela et al. 2022).

Heinemann (2018) identifies two major recent urban conflicts in which the direct biophysical impacts of climate change play a major role in unleashing social protests and the strengthening of local social movements in Colombia and Guatemala. Since 2010, the documented increase in precipitation has increasingly impacted the Policarpa Barrio (neighbourhood in Spanish) in Cartagena de las Indias, Colombia. On several occasions, the residents of the neighbourhood confronted the municipality to get their land tenure rights recognized, requesting a solution to growing environmental problems caused by the coupled effect of flooding and industrial pollution. In El Cambray II, an area of Guatemala City, a nearby hill collapsed on the night of 1 October 2015, causing the death of 280 people, while 70 disappeared and hundreds of homes were destroyed. The affected community organized resistance and entered into a conflict with the municipality asking for resettlement and a reform of land use planning, demanding housing security as a matter of human rights in the face of intensifying extreme rains.

The Andean indigenous and social movements along with climate justice activists have come together to address the issue of climate change. The phrase "*sumac kawsay*" is a Quechua term that translates to "living well," and it represents the idea that humans should live in harmony with nature. These groups have mobilized around this concept, calling for action against climate change. In 2014, they organized a joint march to raise awareness about the issue (Hicks and Fabricant 2016). The Bolivian Platform against Climate Change (PACC) has flagged how increasing climate variability hinders the right of Indigenous people to the conservation of their culture and practices.

In Peru, a case study (Caitlin 2016) provides a positive example of the potential for local organizations and people to take charge of their own resiliency efforts where international projects and protocols may otherwise prove ineffective. The paper analyses the efforts of a conservation NGO working with the community to improve resilience to climate change impacts. A local NGO, Asociación Ecosistemas Andinos (ECOAN), works with 21 communities in the Andean highlands outside of Cusco to conserve the endangered Polylepis forests directly affected by climate change.

In south-western Uruguay, governmental decentralization programmes have tried to increase public responsiveness to the adaptation needs of people, proving however to be only partially successful. Thompson (2016) has conducted case studies on four communities, in the same region, who have mobilized against existing governance structures, claiming that the latter do not protect local communities from environmental stresses at the community level.

3.3 Justice

Mainly, environmental and climate justice claims in Latin America are related to the unfair production and distribution of damages and environmental and health risks of the economic development model, which is carried out via the exploitation and privatization of common goods such as water, land, and biodiversity, without any regulatory control over their negative externalities or violating the weak existing protective regulations. In addition, evaluation mechanisms for large energy and infrastructure projects and the impact of the accuracy of environmental impact assessments are often contested. This, in turn, can compromise the free, prior, and informed consent of affected communities, further highlighting the need for more transparent and inclusive decision-making processes (Berger and Mauricio 2012). For example, the construction of gigantic infrastructures to generate electricity through water generated resistance and enormous economic, environmental, and social costs. The justice-related issue regarding its development is the alteration of terrestrial ecosystems and biodiversity, alteration of aquatic ecosystems and biodiversity, influences in fishing, changes in the river regime, alteration of natural cycles of floods, alteration of the landscape and socioeconomic status, especially of Indigenous communities (Hernández-Gutiérrez et al. 2022; Hess 2018).

On the other hand, large-scale food production also creates profound forms of social and environmental disruption. The contradiction of the food industry is possible in part for the fact that producing food seems a naturally benign practice destined to feed people. However, this perception ignores the negative impacts that the industry can have, such as deforestation caused by practices like soybean agriculture and cattle ranching. These issues are further compounded by various injustices, including land expropriation and labour exploitation. As a result, it is crucial to recognize the complex and multifaceted nature of the food industry and work towards more sustainable and equitable practices (Gordillo 2017). For example, Brazil, despite being known as the principal player in food production, 33 million Brazilians go hungry, according to the Brazilian Research Network on Food and Nutrition Sovereignty and Security. When the cut is expanded to people in situations of food vulnerability, more than half (56%) lack food and nutrients because they are not able to pay for food (Mendes and Cristina 2022). The risks faced by Indigenous and peasant communities with small agriculture are particularly acute, given the threats posed by industrial pollution, agribusiness expansion, mega-mining, and massive deforestation, among other serious problems. As a result, various calls for justice have emerged in recent years, including food justice, water justice, and energy justice. These calls demand access to basic resources and participation in their management, highlighting the urgent need for sustainable and equitable practices.

3.4   Forest

Deforestation is a primary contributor to GHG emission. Uncontrolled fires reflect new ecologies of the Anthropocene, driven by interactions of multiple actors and sectors across scales. They threaten the ecological integrity of tropical forests, impact global climate regimes, and importantly cause considerable social and economic burdens. According to the Global Forest Watch, loss in South America rose 2.8% between 2018 and 2019. Colombia, Peru, and Bolivia had particularly big surges in deforestation. In 2021 the losses of primary tropical forests amounted to 3.75 million hectares (375,000 square kilometres). In Brazil, the loss was 1.5 million hectares (Global Forest Watch 2022).

For example, deforestation has a considerable risk of continental impacts in the Amazon rainforest since the biome influences outside territorial domains through atmospheric circulation and precipitation. Increasing deforestation in many places in Latin America has a tragic impact due to its provision of uncountable ecosystem services, such as regulation of water and nutrient cycles, regional climate modulation, carbon storage, and provision of water flows indispensable for agriculture (Börner et al. 2007; Malhi et al. 2008). Also, the constant disturbances and impacts of deforestation in megadiverse ecosystems like the Amazon are strongly associated with the emergence, re-emergence, and spread of numerous infectious diseases (Ellwanger et al. 2020). Together, the ecological disturbances and social instabilities, like insufficient public policies, social conflicts, and the absence of state, can corroborate the emergence of new and threatening diseases (Giatti et al. 2021).

A study conducted by Myers et al. (2018) using data collected from over 700 interviews in five countries, namely Indonesia, Mexico, Peru, Tanzania, and Vietnam, with both Reduction of Emissions from Deforestation and Forest Degradation (REDD+) and non-REDD+ cases, demonstrates that the failure to incorporate political notions of justice into conservation projects such as REDD+ will result in messiness within governance systems, which is a symptom of injustice and illegitimacy. First, conservation, payment for ecosystem services, and the REDD+ project is viewed through a technical rather than political lens, leading to solutions that focus on procedures, such as benefit distribution. Second, the focus on the technical aspects of interventions came at the expense of political solutions such as the representation of local people's concerns and recognition of their rights. Third, the lack of attention to representation and recognition of justices resulted in illegitimacy. This has led to 'messiness' in the governance systems, which was often addressed in technical terms, thereby perpetuating the problem.

3.5   Climate Migrants

In Latin America, the exposure to extreme meteorological and climate conditions exacerbates pre-existing risks and vulnerabilities that can hamper life and labour in certain regions, often resulting in human mobility. In this context, migration can be an adaptation strategy, but it can also be a last-resort survival option in face of adverse environmental conditions (Ramos and de Cavedon-Capdeville 2017). An example of this is the populations that moved in caravans from Central America to the US border after the Hurricanes Eta and Iota in November 2020. Some of the migrants were fleeing the crisis generated by the hurricanes in their countries (OIM 2021). The situation of vulnerability, as well as issues related to the lack of protection, food, water, and refuge, has characterized the mobility of these groups, most of them Indigenous, traditional, or peasant communities and small farmers. The conditions of poverty, violence, food insecurity, and exclusion in their territories are linked to environmental factors, which are aggravated by the effects of climate change. This interaction between vulnerability, exposure to risk, and human mobility reveals the relationship among climate justice and human mobility, especially when those who suffer most from climate change have done the least to cause it. The 99% deaths from weather-related disasters occur in the world's 50 least developed countries, which contribute less than 1% of global carbon emissions (Global Humanitarian Forum 2009).

It is necessary to better understand the factors involved in human mobility linked to environmental events. In territories where it is possible to observe the effects of disasters on human mobility in Latin America, the studies must focus not only on the so-called rapid-onset events, which are more visible because they generate disasters, but also on slow-onset ones, such as sea level rises, temperature rise, or desertification. These studies must privilege local-scale analysis where the impacts of climate (in)justice become more evident.

3.6 Food

Climate change is one of the main threats to food justice. The impacts of climate change in Latin America are increasingly worrying – in particular, those that affect the agricultural and forestry sector, given its high dependence on climatic conditions. This creates a situation of economic, social, environmental, and political vulnerability in the Latin American countries, putting food security, human security, and the basic conditions for poverty reduction. The food sectors are severely affected by the effects of climate change, in particular due to the excessive dependence of agriculture on rain, mainly due to reliance on subsistence agriculture, which requires sustainable adaptation policies to handle this scenario. Related to food are issues such as climate change, pollution of water and soils by pesticides and fertilizers, loss of soil fertility, loss of biodiversity to search potential solutions to achieve food security.

The transnational agrarian movement La Via Campesina (LVC) advocates for food sovereignty – the right of the people to have healthy and culturally appropriate food produced through sustainable practices. This vision has spurred organized social movements in Latin America rooted in the rights of peasants, Indigenous peoples, and women, who want to regain land and food sovereignty as well as representation on the land and agriculture policy (La Via Campesina 2021).

In recent years, the complexity of the effects of climate change on the different dimensions of food security and on agriculture has been recognized – availability (production and trade), access, stability of food sources, and food utilization. In the last decades, Latin American agriculture began to be transformed by market and trade liberalization, the specialization of production, and the domination of the world's food system by agribusiness. The negative ecological impact of agribusiness in Brazil, for example, has been shown by forest deforestation practices, waves of mass dispossession and ecological destruction in the countryside, and the ensuing employment crisis in the rural and urban economies (Alimonda 2011; Rodriguez Acha 2017).

3.7 Health

Climate change is a social justice as well as an environmental issue. The magnitude and pattern of changes in weather and climate variables are creating differential exposures, vulnerabilities, and health risks that increase stress on health systems while exacerbating existing and creating new health inequities. Therefore, health and climate change in Latin America can be strongly related to intertwined ecosystems and socioenvironmental vulnerabilities, exacerbating existing risks and precariousness. A wide range of pathways can relate to heat waves, increased floods and droughts, changes in the spread of vector-borne diseases, ruptures on food systems, scarcity of natural resources, malnutrition, mental health disorders, and non-communicable diseases. Some conditions related to vulnerability can maximize risks, like those related to poor quality or loss of housing and jobs, poverty, mass migrations, violence, and political instabilities. The extreme temperatures in major Latin American cities could lead to an increase in mortality rates (Kephart, et al. 2022). The impact of climate change on the region has been analysed in various sectors such as agriculture, health, transport, and energy (Bárcena et.al. 2020).

Events that affect health conditions, like climate disasters and climate variability, can spread complex typologies of diseases. Acute events, such as floods, provide clear evidence of the direct impacts on health, such as the spread of water-borne diseases. These events typically unfold over a short timescale, measured in days, and are localized in their effects. In contrast, the indirect and secondary repercussions of long-lasting climate events are more difficult to measure, and their impacts can be magnified by socio-environmental vulnerabilities. These impacts can spread across larger territories, affecting a greater number of people over a longer period, making it challenging to accurately forecast the full extent of their effects (Hales et al. 2004).

3.8 Rights

There is no doubt that the adverse impacts of climate change pose a serious threat to the effectiveness of human rights. From the rights to health, food, water, housing, education, to the rights to development and to life itself, all of them are affected, putting our survival at risk. The adverse impacts of climate change are even more serious for people and groups in vulnerable situations, including women (Rodriguez Acha 2017), Indigenous peoples (Pugley 2018), children (Gasparri et al. 2021; Gonzalez-Ricoy and Rey 2019), migrants (Ahmad 2019), people with disabilities, traditional communities, and low-income groups, who can be disproportionately affected by climate variability (CEPAL and ACNUDH 2019).

In Latin America, the Regional Agreement on Access to Information, Public Participation and Justice in Environmental Matters in Latin America and the Caribbean (Escazú Agreement) represents a valuable regional contribution to ensure that environmental and climate actions respect, protect, and make effective human rights and basic democratic principles. It is the first regional treaty on the environment in the region – and the first in the world – to contain specific provisions on human and environmental rights defenders. In a context of the Global South, where countries suffer the most devastating consequences of global warming because of the irrational accumulation system (Facchinelli et al. 2022), there is a need to demand climate justice for the poorest countries and communities. This justice must be in the distributive dimension, denoting the administration of justice processes in the resolution of disputes and the distribution of resources, as well in the restorative dimension, which promotes the commitment to repair the rights of victims of climate change (Torres et al. 2019).

Despite significant advances for the defence of rights in the context of climate justice, such as the approval of the Escazú Agreement. The treaty is very contested, and some countries are still in the process of ratification – for example, Brazil under Jair Bolsonaro, whose government has not been supportive of environmental or human rights mechanisms; and Colombia, which ranks among the top countries in the region for the number of assassinations of environmental defenders (Miguel 2021). Thus, it is still necessary to deepen studies for a better understanding of the dynamic of violations of rights and their structural and root causes, mainly in territories with a high history of socioeconomic vulnerability and highly exposed to adverse effects of climate change, such as Latin America.

3.9 Pollution

According to the World Health Organization (WHO), the highest incidence of disability-adjusted life years (DALYs – a health indicator that expresses lost years of life among a population due to avoidable deaths and disability) is associated with air pollution in regions also known for scoring lower in terms of human development. Except for Bolivia and Guyana – respectively in the 1000–1499 and >1500 DALY ranges – most Latin American countries fall within the 300–679 range. While the regional performance is better than that of Africa or Southeast Asia, the impact is still noticeable (Babatola, 2018; WHO 2019). The human health hazards associated with air pollution are not distributed equally across countries. By studying three Latin American megacities of Mexico City (Mexico), Santiago de Chile (Chile), and São Paulo (Brazil), Bell and colleagues found that introducing air pollution control policy would have vast health benefits for each of the three cities, averting numerous adverse health outcomes including over 156,000 deaths, 4 million asthma attacks, 300,000 children's medical visits, and almost 48,000 cases of chronic bronchitis in the three cities over a 20-year period; the relative economic value of the avoided health impacts was estimated in the order of roughly US$21–165 billion (Bell et al. 2006).

On the whole, the issue with climate justice and air pollution is redistributional, concerning the differential capacities of countries and regions to provide improved environmental conditions and access to healthier environments to large populations, which is reflected by intercountry comparison. Nonetheless, as highlighted in the conclusions, as in the case of air pollution, a climate justice lens is going to increasingly reveal procedural and recognition issues determined by forms of injustice impacting groups with a weaker political voice and power, such as women, Indigenous and peasant groups, children, the elderly, and the disabled both in urban and rural contexts, not to mention the increasing concerns represented by climate migration movements (Fernandez et al. 2021).

3.10 Public Perception of Climate Justice

The concept of climate justice emerges as an unfolding of the environmental justice paradigm and the perception that climate change impacts social groups in different ways and intensities. Although this perception is increasingly frequent, especially among different social movements that have been using the climate justice discourse, it is still not widely understood among the communities and populations most affected by the adverse effects of climate change, nor associated within speeches on the subject (Azócar et al. 2020; Cortés et al. 2021).

Several factors can influence the general public perception of climate change and the risks associated with it. While for scientists these perceptions are strongly based on the process of analysis and results of analytical research, for the public, perceptions are built through processes of association and affectivity, based on the information that individuals have in their attention to the subject and in the confidence in disclosed data and personal experiences (IPCC, 2022). At the community level, several elements, such as the cultural aspects of communities, power disputes, and economic interests, also influence the different ways of characterizing risks, giving them different degrees of perception and importance (Lauda-Rodriguez, Ribeiro, 2019; Hagen et al. 2022).

Milanez and Fonseca (2011) studied the social perception of extreme weather events, analysing the content of articles on disasters (floods and landslides) that mainly affected poor populations in São Paulo (2009) and Rio de Janeiro (2010) in the main newspapers of these cities. They showed that most articles tend to focus on engineering issues and land use problems, and climate change and its injustices are not widely evidenced, which ends up influencing the perception of the topic. It also highlights the lack of information aimed at the general public, especially the populations most affected by these events. This information is essential to guarantee their right of participation – a fundamental pillar of climate justice.

3.11 Urban Issues

Urbanization is increasing in Latin America, with more than 80% of the population living in cities (World Bank 2019). Cities are facing challenges related not only to climate change but also to their accelerated growth and increasing constraints, including urban transport issues, insufficient infrastructure, and the increasing demand for water, energy, and food. Despite their roles as regional economic, demographic, and innovation centres, however, these cities suffer the greatest economic impacts from climatic disasters, for example, hurricanes, floods, droughts, and water, energy and food shortages (GIZ and ICLEI 2014).

For this reason, cities are considered to play a fundamental role in tackling climate change. Many Latin American cities continue to employ traditional planning and management approaches that are unable to assess the complex nature of the impacts and possible effects of climate change on the environment and society (Benites-Lazaro and Giatti 2021; Lazaro et.al. 2022c). Another characteristic of these cities is that they are sites of immense social inequality, with poor communities living in highly populated slums, which mostly have little or no basic sanitation and infrastructure. It is these populations that suffer the most from the consequences of climate variability (Amaral et al. 2021; Silva et al. 2022).

Regarding a megacity and its complexities, the study by Amaral et al. (2021) discusses a vast urban system, the São Paulo Macrometropolis, with more than 34 million inhabitants, through analysing ecosystem services provision and the possibilities of human development. The study identifies a set of peripheral municipalities with a considerable role in developing green infrastructure to support the urban population. From the 180 municipalities analysed, those who were categorized as providers of ecosystem services demonstrate injustices with lower indicators of well-being and human development. That is a considerable context in understanding the demands for substantial urban populations and an antagonistic set of territorial iniquities. Moreover, it brings an appeal for necessary intersectoral approaches dedicated to the territory to conceive adaptive alternatives for the huge urban problems in Latin America.

In the same vein, Smith and Henríquez's (2019) analysis of urban climate injustice in Chile demonstrated the sharp socioeconomic differences in the quality of the environment to which urban residents are exposed, and how high-income areas, located in the peri-urban zones of the city, enjoy better climate, environment, and air quality. Another study of Torres et al. (2019) shows the importance of strengthening decision-makers' capacity within local governments by identifying data and knowledge gaps that limit the implementation of plans and actions regarding adaptation in urban areas. De Silva et al. (2022), by studying the Brazilian city of Salvador's resilience strategy, discussed the importance of including resilience as a guiding theme in urban planning agendas with social inclusion and taking into account vulnerable peoples and communities, emphasizing that all aspects and sectors of social and urban planning that are fundamental to building urban resilience must be involved.

3.12 Vulnerability

Latin America is the most vulnerable region to climate change; this vulnerability is amplified by inequality, poverty, population growth, high dependence of national and local economies on natural resources, land use change, particularly deforestation with consequent biodiversity loss, and soil degradation (IPCC 2022). Vulnerability connects with climate justice through asset depletion. When physical and, more broadly, material assets are hit by extreme climate events, climatic stress, or disasters, the poor and, in general, those who are most vulnerable experience greater challenges in bouncing back (Baptiste and Rhiney 2016; Lampis 2016).

In addition, as shown in the Latindadd (2021) report, the Latin American region is being increasingly affected by extreme weather events such as hurricanes, floods, prolonged droughts, melting glaciers, spread of diseases, and disappearance of species, which put the physical integrity of people at risk, particularly among the most vulnerable groups. In this region, food, water, and energy insecurity is growing as a result of climate change, in addition to migration, infrastructural damage, and economic losses, placing many people in this region in a situation of vulnerability.

The distribution of gains and losses of the developmental processes entailing the exploitation of natural resources, and that of hazard and risks, is a crucial factor in determining poverty and powerlessness outcomes (Adams 2009). Nonetheless, in Latin America, adaptation policies often address climate impact drivers, but seldom include the social and economic underpinnings of vulnerability (IPCC 2022), thus also hindering better chances to achieve significant results in climate justice. The study of Alcántara-Ayala (2019) reveals that disaster studies in Latin America continue to be unbalanced, with much more concentration on threats and technocratic early-warning systems than on exposed vulnerable communities and the problems due to unsustainable practices and uses of the territory as well as disaster risk governance. Obviously, as shown by the various disasters in Latin America, the most vulnerable are those who suffer most consequences, both through the loss of life and loss or damage of their precarious homes (Satriano 2022).

3.13 Women and Young People

Climate change impacts women and girls disproportionately; for example, given the scarcity of water, women and girls are forced to carry out additional tasks, such as walking long distances to collect water for cooking and cleaning, which generates additional physical strain and increases the risk of gender-based violence (Abdenur 2020). Brown and Flores (2021), by studying the Peruvian woman and the impacts of climate change, identified the environment–gender health nexus as problematic in this country. The differentiated impacts of climate change from a gender perspective, as shown by Revelo (2021), reveal the structural knots of socioeconomic inequality and the persistence of poverty, the gender division of labour and the unjust social organization of care, the predominance of the culture of privilege, the patriarchal, discriminatory, and violent cultural patterns, and the unequal concentration of power. Thus, the effects of climate change can deepen existing gender inequalities in the region. This makes the topic of climate justice and gender worthy of in-depth study.

Women and girls from every region of the world created a mass movement for climate justice and joined in the 'Global Call of Women for Climate Justice' declaration (WEDO 2015). At the COP21 in Paris, the movement demanded women's inclusion in the discussions of climate change issues as well as the inclusion of the gender perspective dimension in climate change adaptation debates and policies looked at through climate justice and intersectional lenses.

For example, research by Erwin and colleagues (2021) shows how the inequality in identities, like gender and age, compounds and interacts with systems of power to shape how people adapt to changes at the individual, household, and local scales in Caylloma (Peru). There, local women groups are at the forefront of political struggles and play a leading role in reclaiming environmental justice on issues such as decreasing water supplies and rapidly retreating glaciers. Similarly, Franco-Orozco and Franco-Orozco (2018) depict the crucial role of Colombian women actively engaged in scientific research in addressing some of the world's most urgent problems such as global poverty, health, and climate change from their geographically and politically located perspectives. In discussing how women's rights are still being continuously affected by violence, social exclusion, exploitation, and discrimination in the country, they again reiterate how gender inequalities create a phenomenal barrier towards achieving greater justice in any domain.

Reflections on intersectionality, intra-, and intergenerational climate justice should also seek to widen the focus to include young people and the elderly alongside other marginalized groups such as women or Indigenous populations. Children and youth, for instance, have been recognized as particularly vulnerable to the impacts of climate change; however, so far, debates have often overlooked the inherent capacity of youth to dialogue with and adapt to resource scarcity on an everyday basis, albeit with restrained choices (Börner et al. 2021; Börner 2020). Based on their research with marginalized youth in the urban periphery in Brazil, Börner et al. (2021) argue in favour of a discourse shift from vulnerability to a focus on resilience and a broader understanding of (youth) climate agency. They revisit current understandings of climate activism, beyond 'visible' young climate agents such as Greta Thurnberg who have transformed the international climate arena by introducing more nuanced and pluralistic understandings of agency based on young people's adaptive practices and their everyday (inter)actions with their environment (Börner et al. 2021).

At the same time, the climate justice lens needs to be better tuned to reveal, beyond the most self-revealing distributional climate injustices, a whole set of procedural and recognition justice issues impacting groups with weaker political voice and power. Generally, different forms of misrecognition, exclusion, and a disregard of capabilities interact to shape justice claims (Schlosberg 2007). Moreover, the above analysis has shown how the predatory exploitation of natural resources favours the intensification of environmental risks which are aggravated by climate variability.

The unfolding of climate events may, for instance, contribute to a depletion of conditions of food production, exacerbated by an underlying lack of adaptive capacities as a result of poverty, low literacy, and absence of governmental programmes. Climate change hence exacerbates underlying systemic inequalities and resource insecurities that contribute to socioenvironmental vulnerability. Tackling these complex existing and future challenges requires interconnected solutions to strengthen long-term resilience beyond short-term solutions. This is where nexus thinking around food, water, energy, and security in relation to the impacts of climate change can be useful in identifying interconnections (Lazaro et al. 2022c). Integrated nexus thinking may enable recognizing complex challenges to food–water–energy systems by identifying resource interdependencies and trade-offs, such as the repercussions of deforestation processes on water availability. However, there is a need for more bottom-up research on how different communities – both rural and urban, including marginalized groups such as women and youth – deal with and adapt to resource interdependencies in the context of scarcity and climate change, rather than prioritizing a technocratic understanding of the nexus as well as top-down policy responses (Kraftl et al. 2019; Börner 2021).

4   Conclusion

The discourses and the repertoires of actions deployed by social movements that engage with the climate justice agenda have proved a useful lens through which the mapping of main debates on climate justice can be charted. The analysis based on scientific literature reveals a great diversity of themes on climate justice related to the Latin American region. The chapter found the pervasiveness of the climate justice debate and its relevance in touching upon all major areas or the developmental arena of debate, from urban transformations to food security, up to pollution, migration, and human rights.

Socioenvironmental conflicts and Latin American social movements are directed against the development model, extractivism, the colonial dimension of modernity, and capitalism that do not take into account the environmental dimension and its negative externalities that produce the health of both the population and environment. Moreover, it is paramount to further our understanding of the relationship between climate change, justice, and sustainable development, including revisiting debates on intragenerational as well as intergenerational justice. Multiple interactions and relationships of scales – temporal and spatial, the distribution of social costs, and financial redistribution – are required to understand for a more comprehensive analysis of climate justice.

Successive environmental agreements since the 1970s and climate negotiations have emphasized the need to achieve sustainable development with justice, precisely to highlight the intrinsic connection between justice and sustainability. Studies about climate justice can help address issues directly associated with poverty and the reduction of inequality. There are several critical issues related to justice and access to resources in the region, including the right to water in the face of recurrent water crises in different territories. Ensuring food security and recognizing food as a basic right are also essential. Additionally, there is a growing need for energy justice, which includes access to clean and affordable energy sources, particularly in marginalized communities that often lack access to reliable energy services. By addressing these issues, we can promote more equitable and sustainable development in the Latin America region.|

5 Acknowledgements

The authors acknowledge financial support received from São Paulo Research Foundation (FAPESP), Grant 2017/17796–3, 2019-24479-0 and 2018/17626-3. Dr Susanne Börner acknowledges the Marie Sklodowska-Curie Grant Agreement No. 833401 (NEXUS-DRR).

References

Abdenur, A. 2020. Gender, Climate and Security in Latin America and the Caribbean: From Diagnostics to Solutions. https://climate-diplomacy.org/magazine/cooperation/gender-climate-and-security-latin-america-and-caribbean-diagnostics-solutions.

Adams, D. W. 2009. Easing Poverty through Thrift. Savings Dev., 33, 73–85.

Ahmad, A. N. 2019. Climate Justice and Migration Mobility, Development, and Displacement in the Global South. www.boell.de/sites/default/files/2020-12/Climate_Justice_and_Migration.pdf.

Alcántara-Ayala, I. 2019. Time in a Bottle: Challenges to Disaster Studies in Latin America and the Caribbean. *Disasters*, 43, S18–S27.

Alimonda, H. (ed.). 2011. *La naturaleza colonizada: Ecología política y minería en América Latina*, 1a ed. Ciccus & CLACSO.

Amaral, M. H., Benites-Lazaro, L. L., Sinisgalli, P., Alves, H., and Giatti, L. L. 2021. Environmental Injustices on Green and Blue Infrastructure: Urban Nexus in a Macrometropolitan Territory. J. Clean Prod., 289, 125829.

Arning, K. and Ziefle, M. 2020. Defenders of Diesel. Anti-decarbonization Efforts and the Pro-diesel Protest Movement in Germany. Energy Res. Soc. Sci., 63, 101410.

Avellaneda, A. 2003. Petróleo, ambiente y conflicto en Colombia. Guerr. Soc. y Medio Ambient, 455–501.

Azócar, G., Billi, M., Calvo, R., Huneeus, N., Lagos, M., Sapiains, R., and Urquiza, A. 2020. Climate Change Perception, Vulnerability, and Readiness: Inter-country Variability and Emerging Patterns in Latin America. J. Environ. Stud. Sci., 111(11), 23–36.

Babatola, S. S. (2018). Global Burden of Diseases Attributable to Air Pollution. J. Public Health Afr., 9(3).

Baptiste, A. K. and Rhiney, K. 2016. Climate Justice and the Caribbean: An Introduction. Geoforum, 73, 17–21.

Bárcena I, A., Samaniego, J., Peres, W., and Alatorre, J. E. 2020. The Climate Emergency in Latin America and the Caribbean: The Path Ahead–Resignation or Action?. ECLAC. https://repositorio.cepal.org/bitstream/handle/11362/45678/10/S1900710_en.pdf

Barrett, S. 2013. The Necessity of a Multiscalar Analysis of Climate Justice. Prog. Hum. Geogr., 37, 215–233.

Bebbington, A. 2012. Underground Political Ecologies: The Second Annual Lecture of the Cultural and Political Ecology Specialty Group of the Association of American Geographers. Geoforum, 43, 1152–1162.

Bell, M. L., Davis, D. L., Gouveia, N., Borja-Aburto, V. H., & Cifuentes, L. A. (2006). The Avoidable Health Effects of Air Pollution in Three Latin American Cities: Santiago, Sao Paulo, and Mexico City. Environ. Res., 100(3), 431–440.

Benites-Lazaro, L. L. and Giatti, L. 2021. O nexo água-energia-alimentos- uma abordagem para cidades sustentáveis e o desenvolvimento sustentável. In P. Jacobi and L. Giatti (eds), *Inovação Para Governança Da Macrometrópole Paulista Face à Emergência Climática*, Curitiba: Editora CRV, pp. 79–97.

Benites-Lazaro, L. L., Giatti, L., Sousa Junior, W. C., and Giarolla, A. 2020. Land-Water-Food Nexus of Biofuels: Discourse and Policy Debates in Brazil. Environ. Dev., 33. https://doi.org/10.1016/j.envdev.2019.100491.

Berger, M. 2012. Justicia ambiental en América Latina. Inteligencia colectiva y creatividad institucional contra la desposesión de derechos. E-Cadernos, https://doi.org/10.4000/eces.1128

Blei, D. M., Ng, A. Y., and Jordan, M. I. 2003. Latent Dirichlet Allocation. J. Mach. Learn Res., 3(Jan), 993–1022.

Börner, J., Mendoza, A., and Vosti, S. A. 2007. Ecosystem Services, Agriculture, and Rural Poverty in the Eastern Brazilian Amazon: Interrelationships and Policy Prescriptions. Ecol. Econ., 64, 356–373.

Börner, S. 2020. Vozes silenciosas sobre o nexo água-alimentos-energia e mudanças climáticas: diálogos cotidianos na periferia urbana de Sao Paulo. Diálogos Socioambientais na Macrometrópole Paulista. Nexos: para a sustentabilidade, 4(10), 12–15.

Börner, S., Kraftl, P., and Giatti, L. L. (2021). Blurring the '-ism' in Youth climate Crisis Activism: Everyday Agency and Practices of Marginalized Youth in the Brazilian Urban Periphery. Child. Geogr, 19(3), 275–283.

Borras, S. and Franco, J. 2012. Global Land Grabbing and Trajectories of Agrarian Change: A Preliminary Analysis. J. Agrar. Chang., 12, 34–59.

Brown, L. J. and Flores, E. C. 2021. Perceptions, Priorities, and Project Delivery Problems in Women's and Environmental Health in Peru: A Qualitative Study. Lancet Planet. Heal., 5, S5.

Caitlin, D. 2016. Building Climate Change Resilience through Local Cooperation: A Peruvian Andes Case Study. Reg. Environ. Chang., 16, 2187–2197.

CEPAL. 2022. *Panorama Social de América Latina*. CEPAL.

CEPAL and ACNUDH. 2019. Cambio climático y derechos humanos: contribuciones desde y para América Latina y el Caribe. https://repositorio.cepal.org/bitstream/handle/11362/44970/4/S1901157_es.pdf

Cortés, S., Burgos, S., Adaros, H., Lucero, B., and Quirós-Alcalá, L. 2021. Environmental Health Risk Perception: Adaptation of a Population-Based Questionnaire from Latin America. Int. J. Environ. Res. Public Health, 18(16), 8600. doi: 10.3390/ijerph18168600

de Paulo, A. F. and Porto, G. S. 2018. Evolution of Collaborative Networks of Solar Energy Applied Technologies. J. Clean. Prod., 204, 310–320.

de Silva, A. M. A., Benites-Lazaro, L. L., Andrade, J. C. S., Prado, A. F. R., Ventura, A. C., Campelo, A., and Tridello, V. 2022. Examining the Urban Resilience Strategy of Salvador, Bahia, Brazil: A Comparative Assessment of Predominant Sectors within the Resilient Cities Network. J. Urban Plan. Dev., 148, 1–11.

EJAtlas. 2022. Mapping Environmental Justice. https://ejatlas.org.

Ellwanger, J. H., Kulmann-Leal, B., Kaminski, V. L., Valverde-Villegas, J. M., Da Veiga, A. B. G., Spilki, F. R., Fearnside, P. M., Caesar, L., Giatti, L. L., Wallau, G. L., Almeida, S. E. M., Borba, M. R., Da Hora, V. P., and Chies, J. A. B. 2020. Beyond

Diversity Loss and Climate Change: Impacts of Amazon Deforestation on Infectious Diseases and Public Health. An. Acad. Bras. Cienc., 92, 20191375.

Erwin, A., Ma, Z., Popovici, R., Salas O'Brien, E. P., Zanotti, L., Zeballos, E., Bauchet, J., Ramirez Calderón, N., and Arce Larrea, G. R. 2021. Intersectionality Shapes Adaptation to Social-Ecological Change. World Dev., 138, 105282. https://doi.org/10.1016/j.worlddev.2020.105282.

Escobar, A. 2010. *Territorios de diferencia: Lugar, movimientos, vida, redes*, 2a ed. Envión. https://doi.org/10.1017/CBO9781107415324.004.

Facchinelli, F., Pappalardo, S. E., Della Fera, G., Crescini, E., Codato, D., Diantini, A., Moncayo Jimenez, D. R., Fajardo Mendoza, P. E., Bignante, E., and De Marchi, M. 2022. Extreme Citizens Science for Climate Justice: Linking Pixel to People for Mapping Gas Flaring in Amazon Rainforest. Environ. Res. Lett., 17, 024003.

Farrell, J. 2016. Network Structure and Influence of the Climate Change Counter-movement. Nat. Clim. Chang., 6, 370–374.

Feola, G. 2015. Societal Transformation in Response to Global Environmental Change: A Review of Emerging Concepts. Ambio, 44, 376–390.

Fernandez, M., Harris, B., and Rose, J. (2021). Greensplaining Environmental Justice: A Narrative of Race, Ethnicity, and Justice in Urban Greenspace Development. J. Race Ethn City. 2(2), 210–231.

Franco-Orozco, C. M. and Franco-Orozco, B. 2018. Women in Academia and Research: An Overview of the Challenges toward Gender Equality in Colombia and How to Move Forward. Front. Astron. Space Sci. 5, 24. https://doi.org/10.3389/fspas.2018.00024.

Gasparri, G., El Omrani, O., Hinton, R., Imbago, D., Lakhani, H., Mohan, A., Yeung, W., and Bustreo, F. 2021. Children, Adolescents, and Youth Pioneering a Human Rights-Based Approach to Climate Change. Health Hum. Rights, 23, 95–108.

Giatti, L. L., Ribeiro, R. A., Nava, A. F. D., and Gutberlet, J. 2021. Emerging Complexities and Rising Omission: Contrasts among Socio-ecological Contexts of Infectious Diseases, Research and Policy in Brazil. Genet. Mol. Biol., 44, 1–9.

GIZ and ICLEI. 2014. Operationalizing the Urban NEXUS: Towards Resource-Efficient and Integrated Cities and Metropolitan Regions Case Studies Content. www.thegpsc.org/sites/gpsc/files/urbannexus_publication_iclei-giz_2014_kl_0.pdf.

Gligo, N., Alonso, G., Barkin, D., Brailovsky, A., Brzović, F., Carrizosa, J., Durán, H., Fernández, P., Gallopín, G., Leal, J., Marino de Botero, M., Morales, C., Ortiz Monasterio, F., Panario, D., Pengue, W., Rodríguez, M., Rofman, A., Saa, R., Sejenovich, H., Sunkel, O., and Villamil, J. 2020. *La tragedia ambiental de América Latina y del Caribe*. CEPAL.

Global Forest Watch. 2022. Forest Monitoring, Land Use & Deforestation Trends. www.globalforestwatch.org.

Global Humanitarian Forum. 2009. Human Impact Report Climate Change. The Anatomy of A Silent Crisis. Geneva. www.preventionweb.net/files/9668_humanimpactreport1.pdf.

Göbel, B., Góngora-Mera, M., and Ulloa, A. 2014. Las interdependencias entre la valorización global de la naturaleza y las desigualdades sociales: abordajes multidisciplinarios. Desigualdades socioambientales en América Latina, 13–46.

Gonzalez-Ricoy, I. and Rey, F. 2019. Enfranchising the Future: Climate Justice and the Representation of Future Generations. Wiley Interdiscip. Rev. Clim. Chang., 10, e598. https://doi.org/10.1002/WCC.598.

Gordillo, G. 2017. On the Destructive Production of Food: Some Lessons from South America. J. Polit. Ecol., 24, 797–800.

Gudynas, E. 2013. Debates on Development and Its Alternatives in Latin America: A Brief Heterodox Guide. In M. Lang and D. Mokrani (eds), *Beyond Development*, Fundación Rosa Luxemburg, pp. 15–39.

Guerrero, D. 2011. The Global Climate Justice Movement. In *Global Civil Society*. Palgrave Macmillan, pp. 120–126.

Hagen, I., Huggel, C., Ramajo, L., Chacón, N., Ometto, J. P., Postigo, J. C., and Castellanos, E. J. (2022). Climate Change-related Risks and Adaptation Potential in Central and South America During the 21st Century. Environ. Res. Lett., 17(3), 033002.

Hales, S., Butler, C., Woodward, A., and Corvalan, C. 2004. Health Aspects of the Millennium Ecosystem Assessment. Ecohealth, 1, 124–128.

Heinemann, A. S. 2018. Cambio climático y conflictividad socioambiental en América Latina y el Caribe. América Lat. Hoy, 79, 9.

Herbert, J. 2020. Environmental Justice. In *The Newcastle Social Geographies Collective*. Rowman & Littlefield, pp. 124–132. https://doi.org/10.1016/B978-0-12-373 932-2.00346-X.

Hernández-Gutiérrez, J. C., Peña-Ramos, J. A., and Espinosa, V. I. 2022. Hydro Power Plants as Disputed Infrastructures in Latin America. Water, 14, 277.

Hess, D. J. 2018. The Anti-dam Movement in Brazil: Expertise and Design Conflicts in an Industrial Transition Movement. Tapuya Lat. Am. Sci. Technol. Soc., 1, 256–279.

Hicks, K. and Fabricant, N. 2016. The Bolivian Climate Justice Movement Mobilizing Indigeneity in Climate Change Negotiations. Lat. Am. Perspect., 43, 87–104. https://doi.org/10.1177/0094582X16630308.

IPCC. 2021. Sixth Assessment Report. www.ipcc.ch/report/ar6/wg1/#FullReport.

IPCC. 2022. Impacts, Adaptation, and Vulnerability. Contribution of Working Group II to the Sixth Assessment Report of the Intergovernmental Panel on Climate Change. www.ipcc.ch/report/sixth-assessment-report-working-group-ii.

Kelman, I. 2010. Hearing Local Voices from Small Island Developing States for Climate Change. Local Environ., 15, 605–619.

Kephart, J. L., Sánchez, B. N., Moore, J., Schinasi, L. H., Bakhtsiyarava, M., Ju, Y., ... & Rodríguez, D. A. (2022). City-level impact of extreme temperatures and mortality in Latin America. *Nature medicine*, 28(8), 1700-1705.

Kim, H., Marcouiller, D. W., and Woosnam, K. M. 2018. Rescaling Social Dynamics in Climate Change: The Implications of Cumulative Exposure, Climate Justice, and Community Resilience. Geoforum, 96, 129–140.

Kraftl, P., Balastieri, J. A. P., Campos, A. E. M., Coles, B., Hadfield-Hill, S., Horton, J., Soares, P. V., Vilanova, M. R. N., Walker, C., and Zara, C. 2019. (Re)thinking (Re) connection: Young People, 'Natures' and the Water–Energy–Food Nexus in São Paulo State, Brazil. Trans. Inst. Br. Geogr., 44, 299–314.

La Via Campesina. 2021. Key Documents (La Via Campesina). https://viacampesina.org/en/who-are-we/what-is-la-via-campesina/key-documents-la-via-campesina

Lalander, R., Lembke, M., and Peralta, P. O. 2019. Political Economy of State-Indigenous Liaisons: Ecuador in Times of Alianza PAIS. Eur. Rev. Lat. Am. Caribb. Stud., 108, 193–220. https://doi.org/10.32992/ERLACS.10541.

Lampis, A. (ed.). 2016. *Cambio Ambiental Global, Estado y Valor Público: La Cuestión Socio-Ecológica en América Latina, entre Justicia Ambiental y 'Legítima Depredación'*, 1a ed. CES/UNAL e CLACSO.

Latindadd. 2021. Vulnerability to Climate Change in Latin America and the Caribbean in the Context of the Pandemic. www.latindadd.org/wp-content/uploads/2022/02/Vulnerabilidad-climática_English.pdf.

Lazaro, L. L.B, Soares, R. S., Bermann, C., Collaço, F. M. A., Giatti, L. L., and Abram, S. 2022a. Energy Transition in Brazil: Is There a Role for Multilevel Governance in a Centralized Energy Regime? Energy Res. Soc. Sci., 85, 102404. https://doi.org/10.1016/J.ERSS.2021.102404.

Lazaro, L. L., Bellezoni, R., Puppim de Oliveira, J. A., Jacobi, P. R., and Giatti, L. L. 2022b. Ten Years of Research on the Water-Energy-Food Nexus: An Analysis of Topics Evolution. Front. Water, 53. https://doi.org/10.3389/FRWA.2022.859891.

Lazaro, L. L.B, Giatti, L., Macedo, L., and Puppim, J. 2022c. *Water-Energy-Food Nexus and Climate Change in Cities*. Springer.

Lehmann, R. and Tittor, A. 2021. Contested Renewable Energy Projects in Latin America: Bridging Frameworks of Justice to Understand 'Triple Inequalities of Decarbonisation Policies'. J. Environ. Policy Plan., 1–12.

Lauda-Rodriguez, Z. L., and Ribeiro, W. C. 2019. Risk, Precautionary Principle and Environmental Justice in Mining Conflicts. Desenvolvimento e Meio Ambiente. 51, 154–179. https://doi.org/10.5380/dma.v51i0.59821

Malhi, Y., Roberts, J. T., Betts, R. A., Killeen, T. J., Li, W., and Nobre, C. A. 2008. Climate Change, Deforestation, and the Fate of the Amazon. Science, 80. https://doi.org/10.1126/science.1146961.

Martinez-Alier, J., Owen, A., Roy, B., Bene, D. D. E. L., and Rivin, D. 2018. Blockadia: movimientos de base contra los combustibles fósiles y a favor de la justicia climática. Anu. Int. CIDOB, 41.

Mendes, J. and Cristina, P. 2022. Tragédia brasileira: de celeiro do mundo a um país com 33 milhões de pessoas sem ter o que comer. www.istoedinheiro.com.br/tragedia-brasileira-de-celeiro-do-mundo-a-um-pais-com-33-milhoes-de-pessoas-sem-tem-o-que-comer.

Miguel, T. 2021. Slain Environmentalists: International Agreement Enters into Force to End Killings of Environmental Leaders in Latin America. https://english.elpais.com/usa/2021-04-27/international-agreement-enters-into-force-to-end-killings-of-environmental-leaders-in-latin-america.html.

Milanez, B., and Fonseca, I. F. 2011. Justiça Climática e eventos climáticos extremos: uma análise da percepção social no Brasil. Revista Terceiro Incluído. 1(2), 82–100. https://doi.org/10.5216/teri.v1i2.17842

Muñoz, E. E. and del Villarreal, M. C. 2019. Women's Struggles against Extractivism in Latin America and the Caribbean. Context. Int., 41, 303–325. https://doi.org/10.1590/s0102-8529.2019410200004.

Myers, R., Larson, A. M., Ravikumar, A., Kowler, L. F., Yang, A., and Trench, T. 2018. Messiness of Forest Governance: How Technical Approaches Suppress Politics in REDD+ and Conservation Projects. Glob. Environ. Chang., 50, 314–324. https://doi.org/10.1016/j.gloenvcha.2018.02.015.

Organización Internacional para las Migraciones (OIM). 2021. La movilidad humana derivada de desastres y el cambio climático en Centroamérica. OIM, Ginebra. https://kmhub.iom.int/sites/default/files/publicaciones/la_movilidad_humana_derivada_de_desastres_y_el_cambio_climatico_en_centroamerica.pdf.

Orihuela, J. C., Pérez Cavero, C., and Contreras, C. 2022. Extractivism of the Poor: Natural Resource Commodification and Its Discontents. Extr. Ind. Soc., 9, 100986.

Pegler, L. 2015. Peasant inclusion in Global Value Chains: Economic Upgrading but Social Downgrading in Labour Processes? J. Peasant Stud., 42, 929–956. https://doi.org/10.1080/03066150.2014.992885.

Picq, M. L. 2020. Resistance to Extractivism and Megaprojects in Latin America. Oxford Res. Encycl. Polit. https://doi.org/10.1093/acrefore/9780190228637.013.1742

PlataformaJusticiaClimática. 2022. https://bit.ly/3MUG0ud.

Pugley, D. D. 2018. *Rights, Justice, and REDD+: Lessons from Climate Advocacy and Early Implementation in the Amazon Basin*. Routledge.

Ramos, E. P. and de Salles Cavedon-Capdeville, F. 2017. Regional Responses to Climate Change and Migration in Latin America. In *Research Handbook on Climate Change, Migration and the Law*. Edward Elgar Publishing, pp. 262–287. https://doi.org/10.4337/9781785366598.00018.

Revelo, L. A. 2021. *La igualdad de género ante el cambio climático: ¿qué pueden hacer los mecanismos para el adelanto de las mujeres de América Latina y el Caribe?* CEPAL.

Rivas, D. S. 2012. Explorando algunas trayectorias recientes de la justicia en la geografía humana contemporánea: de la justicia territorial a las justicias espaciales. Rev. Colomb. Geogr., 21, 75–84.

Rodríguez, D., Olivares, J., Sánchez, J. L., Castilleja, Y., Alemán, Y., and Arece, J. 2014. Latin American Perceptions of Climate Change: Methodologies, Tools and Adaptation Strategies in Local Communities. A Review. Rev. U.D.C.A Actual. Divulg. Científica, 17, 73–85.

Rodriguez Acha, M. A. 2017. We Have to Wake Up, Humankind! Women's Struggles for Survival and Climate and Environmental Justice. Dev., 601(60), 32–39.

Saladini, F., Betti, G., Ferragina, E., Bouraoui, F., Cupertino, S., Canitano, G., Gigliotti, M., Autino, A., Pulselli, F. M., Riccaboni, A., Bidoglio, G., and Bastianoni, S. 2018. Linking the Water-Energy-Food Nexus and Sustainable Development Indicators for the Mediterranean Region. Ecol. Indic., 91, 689–697.

Satriano, N. 2022. Com 178 mortos, tragédia em Petrópolis é a maior já registrada na história do município. https://g1.globo.com/rj/rio-de-janeiro/noticia/2022/02/20/tragedia-em-petropolis-maior-registrada-na-historia-o-municipio.ghtml.

Scandrett, E. 2016. Climate Justice: Contested Discourse and Social Transformation. Int. J. Clim. Chang. Strateg. Manag., 8, 477–487.

Schlosberg, D. 2007. *Defining Environmental Justice: Theories, Movements, and Nature*. Oxford University Press.

Serna, N. 2017. Human Mobility, Natural Disasters and Climate Change in Latin America: From Understanding to Action – Haiti. https://reliefweb.int/report/haiti/human-mobility-natural-disasters-and-climate-change-latin-america-understanding-action.

Smith, P. and Henríquez, C. 2019. Public Spaces as Climate Justice Places? Climate Quality in the City of Chillán, Chile. Environ. Justice, 12, 164–173.

Svampa, M. 2013. Resource Extractivism and Alternatives: Latin American Perspectives on Development. In M. Lang and D. Mokrani (eds), *Beyond Development: Alternative Visions from Latin*, Fundación Rosa Luxemburg (Quito – Ecuador) & Transnational Institute (Amsterdam – The Netherlands) pp. 117–143. www.tni.org/files/download/beyonddevelopment_complete.pdf

Thompson, D. 2016. Community Adaptations to Environmental Challenges under Decentralized Governance in Southwestern Uruguay. J. Rural Stud., 43, 71–82. https://doi.org/10.1016/j.jrurstud.2015.11.008.

Torres, P. H. C., Ramos, R. F., and Gonçalves, L. R. 2019. Environmental Conflicts at São Paulo Macrometropolis: Paranapiacaba and São Sebastião. Ambient. Soc., 22, 1–19.

Tramel, S. 2016. The Road through Paris: Climate Change, Carbon, and the Political Dynamics of Convergence. Globalizations, 13.

Urbinatti, A. M., Benites-Lazaro, L. L., de Carvalho, C. M., and Giatti, L. L. 2020. The Conceptual Basis of Water-Energy-Food Nexus Governance: Systematic Literature Review Using Network and Discourse Analysis. J. Integr. Environ. Sci., 1–23.

WEDO. 2015. The Women's Global Call for Climate Justice. https://wedo.org/get-invol ved/women-climate-justice.

WMO. 2021. State of the Climate in Latin America and the Caribbean. https://public. wmo.int/en/our-mandate/climate/wmo-statement-state-of-global-climate/LAC.

World Bank. 2019. Urban Population – Latin America & Caribbean. https://data.worldb ank.org/indicator/SP.URB.TOTL.IN.ZS?locations=ZJ.

WHO. 2019 Disability-Adjusted Life Years (DALYs). www.who.int/data/gho/indicator-metadata-registry/imr-details/158.

7

SOCIOECOLOGICAL CONFLICTS AND RESISTANCES

The Platformization of Climate Justice Activism in Brazil

Caio Penko Teixeira

1 Introduction

Large-scale extractivism in Latin America establishes key forms of both accumulation and displacement and has a direct impact on the socioeconomic reality of local communities and the environment. Recent work by political ecologists and other critical scholars has explored diverse aspects of criminalization of the populations that defend their territories, from the resistance of Indigenous communities to the activism of environmental organizations. The role of digital advocacy in the struggles for socioecological rights and the needs of those who resist and confront extractivism, however, has received relatively little attention. Based on the work of the anti-dam movement in Brazil – The Movement of People Affected by Dams (MAB) – I focus on their digital advocacy work that has been mobilizing resources and relationships to safeguard the rights of people whose lives and livelihoods have been deeply transformed or even destroyed by extractive projects, including dam-related disasters. This leads to conflicts between mining companies and local communities.

The use of networked technologies by activists illustrates the role of digital advocacy generating alternative knowledge, which highlights the massive environmental degradation caused by natural resource extraction projects, including mining activities and the lack of adequate policy response to peoples displaced and dispossessed by mining projects. Under this approach, I argue MAB's digital advocacy sets an agenda to reconsider the lack of control over industrialized modes of resource extraction, highlighting the sociopolitical power of affected people and their environments as a call to action involving reparations measures, policy change, and resistance strategies. In this sense, digital technologies and knowledge practices reshape the ways in which socioenvironmental conflicts are

DOI: 10.4324/9781003214021-8

understood and framed and offer an interrogation of the political implications of such transformations they produce in social struggles for justice. The adoption of social media platforms for digital advocacy by grassroots activism recasts the expertise-based discourse on managing large-scale environmental disasters as a politically neutral approach restricting it to technical and/or managerial issues.

The chapter introduces the Brumadinho mining tailing dam collapse and relates it to mining capitalism. It frames MAB as a socio-territorial movement and explores the alternative knowledge production and diffusion by MAB's digital advocacy. The article concludes by arguing that MAB is an illustrative example of platformization of climate justice activism. It contributes to the debate on grassroots politics through the lens of digital advocacy. Throughout the chapter, we will see how MAB's digital advocacy helps to reconceptualize the climate justice framework as a multifaceted concept, demonstrating how activists have used digital networked technologies to broadcast their movement's political agenda to a larger audience. The platformization of climate justice outlines the nexus between social media platforms and anti-extractivist activism, and how these dynamics produce and diffuse 'alternative knowledge'. The appropriation of social media platforms by MAB allows the 'knowledge curation work', that is, sharing links and creating and spreading original content. Digital advocacy plays a role in articulating grievances and collective identities. As such, digital advocacy provides a strategic tool for contention based on alternative knowledge produced through and by social media platforms and mobile devices via internet networks.

2   Mining Capitalism: Climate Change, Activism, and Justice

The planetary ecological crisis is first and foremost generated by particular social forces within the system of production, profits, accumulation, and concentrated wealth (Foster et al. 2020). This crisis relies on a carbon-based economic model, and the human-induced climate change related to it can be seen clearly in resource extraction. The extraction of land, minerals, fossil fuels, and forest plants has been driven and underpinned by powerful forces combining corporate state and multinational companies in the pursuit of profit and economic growth. Mining operations are the result of a sociohistorical construction of the 'preciousness' of particular minerals (Ferry et al. 2019). Connecting climate change to mining the earth helps illustrate the implications of predatory land use of natural resources on an ever larger scale. Social movements defending natural resources from mining, pollution, and hydroelectric dams inform the debate on today's political-ecological aspects of rapid climate change. This is best illustrated by climate justice struggles including anti-extractivist organizing which disputes state- and corporate-designed mega development projects.

Over the past decade or so, in the Majority World, there is growing mobilization against large-scale extractive projects. This scenario has resulted in research and action for human rights and environmental justice related to extractivism and its effects on communities. People affected by extractive projects or activities resist the 'geographies of extraction', which rely on state power, technological innovation, and capitalist imperialism (Arboleda 2020). Scholars have discussed the implications of nature–capital interactions as socioecological extractivism (Fraser 2021) and how the commodified resource extraction has also been historically framed as a development strategy across the Majority World (Santos 2019; Engels and Dietz 2017; Gudynas 2015). Large-scale extraction projects drive the rising number of socioenvironmental conflicts in Latin America (Schapper et al. 2019; Gómez-Barris and Taylor 2017; Svampa 2015; Zhouri et al. 2016). This chapter draws on the literature exploring grassroots political communication, mobilization, and digital political activity to understand how a climate justice framework plays a part in community-based anti-extractivist organizing (Dahlberg-Grundberg and Örestig 2017; Helmond 2015). An underexplored issue in this discussion of the contemporary impacts of human-induced climate change has been the question of digital advocacy addressing the socioenvironmental impacts of large-scale resource extraction projects.

There is an emerging link between climate justice organizing and anti-extractive struggles. Azadi et al. (2020) explores climate change–mining relationships and highlights the role of mining industry in contributing direct and indirect greenhouse gas emissions. For resource-rich developing countries, the impacts of the mining industry will increase in face of ever-growing demand to produce minerals (lithium, cobalt, and nickel, for instance) as the basis of the clean energy transition (World Bank 2017). Most of the so-called resource-rich developing countries are localized in the Majority World, where big polluting corporations are profiting from extraction of natural resources and exporting it to wealthier countries in the world, or Minority World. Sonter et al. (2020) challenge the clean energy transition discourse and argue that the mining industry represents a global biodiversity threat, including in land and water contamination. As a climate-wrecking industry, the extractive sector impacts the livelihoods of Indigenous and forest communities. Mining activities and the associated human rights abuse play a part in the deforestation dynamics.

Foster et al. (2020) refer to the 'ecological rift' as a result of socioecological conditions and human-generated crises closely related to the capitalist mode of production. Ecological rift exacerbates the irreversible environmental degradation resulting from the anthropogenic drivers of climate change and relates to the historical detrimental effects of the economic system on the environment. The capitalist relations with the environment rely on the (degradation of) land use for exploitation. The key point here is that land use policies and conflicts are part of global political-economic processes shaped by state-capitalist regulation of nature, and, as Kirsch (2014) points out, it is central to look at the critics of the mining industry through the lens of social movements and how they transform the public perceptions of large-scale resource extraction projects. Mining capitalism reveals the approach of today's logic of exploitation of the mining industry and corporation, which is based on large-scale natural resource extraction. Kirsch (2014) studies the tensions and conflicts emerging from the environmental impacts of the mining industry and explores the role of political activism at the local level.

The implications of mining capitalism can be understood through mega projects, resource extraction, and dam-related risks (Aguilar-Pesantes et al. 2021). In Brazil, a growing debate underlines the socioenvironmental impacts associated with dams as large-scale infrastructure projects. Louzada and Ravena (2019) outlined the lack of social involvement in decision-making regarding safety risk governance of dam-related issues. Dams are used not only for hydroelectric projects but also for tailings disposal of solid and liquid toxic waste from mining operations. Every year, mining companies dump billions of tonnes of hazardous mine waste into water bodies, including rivers and lakes (Aguilar-Pesantes et al. 2021). Any possibility of failure of dam structure can be catastrophic for local communities with lasting environmental problems. Displacement caused by the development of dam projects impacts the life of 80 million people worldwide involving land acquisition and expropriation process (IDMC 2017). At the heart of the literature on dam-induced displacement is the perspective that affected communities can experience tensions, violence, and conflict and are likely to become more vulnerable to precarious life situations (Kirchherr 2018; Huber et al. 2017; Kirsch 2014).

2.1 Socioenvironmental Disasters and Advocacy

Contemporary research on environment-related changes caused by human activity points to the environmental and social impacts of the extractives sector as major emitters of greenhouse gases (Azadi et al. 2020). It makes us reflect on the harmful effects of natural resource extraction and mining companies as a major agenda for the critical debate on human-induced environmental destruction. The impact of extractive activities occurs within and outside the mining lease (Lèbre et al. 2020). A growing literature investigates the socioenvironmental disasters caused by catastrophic mining dam failures, underlining the lack of federal open-access database on geotechnical risk or environmental damage potential classification (Azadi 2020; Carmo et al. 2017; Huber et al. 2017) in Brazil and the need to develop effective management of mine waste and environmental governance, combining public policies for solid waste, dam safety, water resources, and protected areas (Carmo et al. 2020).

Advocacy coalitions enable us to understand the politics behind policy change (Fischer 2014). Contemporary progressive movements rely on advocacy coalitions to foster a repertoire of contention in order to make claims and act collectively to produce social, political, and cultural change and, ultimately, influence policy decisions. The advocacy agenda capitalizes on the tools, resources, and knowledge that organizations have at their disposal as a means towards achieving social and lasting policy change. This is done by building widespread public awareness and support for movement demands around which collective action can take shape. Within the social movements field, advocacy relates to the practice of knowledge production in the development of the collective self (della Porta and Pavan 2017). The collective self involves different knowledges and modes of being, so that 'movement knowledge practices are also oriented to develop a critique of the status quo and substantiate alternative proposals to overcome it' (della Porta and Pavan 2017, 307). Johansson and Scaramuzzino (2019) note that social platforms adoption within collective endeavours offers a new path to advocacy. Digital advocacy refers to acts for political influence and acts for political presence through digital access politics, digital information politics, and digital protest politics.

The adoption of digital media – social media platforms – within grassroots advocacy produces and diffuses alternative knowledge on the issues. For Pavan and Felicetti (2019, 3), the production of alternative knowledge refers to 'sets of practices through which local and highly personal experiences, rationalities, and competences get connected and coordinated within shared cognitive systems which, in turn, provide movements and their supporters with a common orientation for making claims and acting collectively'. More broadly, this perspective underpins work on knowledge and skills reflecting the personal experiences of participants/ activists under the movement's vision, developing collectively their capacity to do analytical work and producing their own information, content, and data outside of institutionalized spaces of knowledge production and epistemic settings. This type of alternative knowledge speaks to activists as well as a broader society to engage diverse audiences and inspire change. Often, alternative knowledge is a community-based practice making room for an alternative mode of understanding. It requires storytelling in communicating across audiences and engaging with and thinking about pressing issues on which they mobilize.

2.2 Climate Justice and Grassroots Activism

The growing interest in anthropogenic climate change, activism, and justice as platforms for climate action has led to new ways of conceiving and thinking about the vulnerability to climate change (Derman 2020) and its disproportionate impact on politically marginalized populations, such as Indigenous, peasant communities (Tokar 2019). As civil society organizations reshape existing understandings of what sustainable development is or ought to be, they offer new visions of what recent scholarship has referred to as 'climate justice'. For Gach (2019), climate justice discourse became more politically institutionalized in the Paris Agreement, including aspects of gender equality and Indigenous rights as part of the responses to climate change. Climate justice plays a part in the climate governance system by fostering international influence on domestic climate institutions around climate change (Navroz 2021). Climate justice research and activism shape public policy dynamics by situating the socioenvironmental implications of development at the centre of analysis rooted in equity, fairness, and political emancipation (Jafry et al. 2019). Verlie (2021) expands the notion of climate justice by approaching it through the lens of more-than-human transcorporeality. This perspective embraces the devastating implications for the natural world highlighting issues of ecological harm regarding more-than-human beings, intersecting political interests, and movements from around the world.

Climate justice research and activism inform us about the ethical and human rights dimensions of those most affected victims of the climate crisis but also point to the injustices and oppressions towards marginalized populations. Climate justice research and activism are thus not only relevant in their own right but also offer a powerful analytical lens to investigate broader sociopolitical transformations of a globally shared conception of climate change and its critique of extractive capitalism and challenges for reaching a more democratic renewable energy future. Climate justice research and activism have brought us a profusion of nuances about the heterogeneous character of climate change – and recognition of affected communities and their role in shaping the climate action agenda.

Literature on climate justice highlights how social movements benefit from the power of social media on grassroots activism (Kaun and Uldam 2018; Hestres and Hopke 2017; Şen and Şen 2016). This advocacy work through a climate justice lens combines issues of ecological harm and (human) rights violations. The advocacy work taking place in the digitally networked spatial engagement of political actors and institutions using digital mediums is seen as a way or tactic of addressing conflicts in land use. The MAB's advocacy exemplifies the platformization of climate justice activism. By platformization of climate justice activism, I mean the rise of social media platforms as an instrumental channel of contemporary grassroots advocacy pushing social justice claims within the context of climate-related issues, taking into consideration vulnerable groups, affected communities, and ecosystems.

2.3   The Platformization of Climate Justice Activism

Using MAB's online presence as an example, we learn how digital social networks enable campaigns, community building, petitions, and advocacy practices. In today's digital landscape, the platformization of climate justice activism leads to the production and diffusion of alternative climate knowledges that reflect the logic of social media platforms. Climate justice movements can be understood as a frame, that is, 'climate justice as a frame brings the movement back to its radical roots – this target can be pursued by concrete action at local levels. Second, the climate justice frame is broader in its appeal, allowing the movement to (re)connect to other movements that its framing work identifies as linked and pertinent to its own goals' (della Porta and Parks 2014, 23).

Platformization refers to the digitally mediated collective action (Milan 2015). It is about the rise of mediation of social and mobile media in shaping the sociopolitical dimension of collective organizing and mobilization. KhosraviNik

(2018) points out the affective characteristics of social media communication targeting promotion of content. The role of emotion in social media platform informs a type of advocacy work which considers networked users' digital behaviours and reactions part of the collective organizing, everyday politics, and political engagement that arises within them. Social media platforms allow political organizing and communication, but it is also a source of technological control over the context and conditions for interaction as a web of regulation for users and the public (Törnberg and Uitermark 2020).

This is an important feature of platformization of climate justice activism, that is, it is a digital mode of alternative knowledge production and dissemination. It recognizes the power of social media platforms within contemporary advocacy politics and in so doing, offers a tool for meaningful communication and critical engagement. The platformization of climate justice emerges as a result of growing institutionalization, professionalization, and datafication of advocacy coalitional efforts pursued by social movements utilizing digital tools for long-term organizing. Contemporary digital platforms and their infrastructures operate as content digital intermediaries (Van Dijck et al. 2018). Therefore, digital platforms serve as instrumental tools for social media activism.

3 Brumadinho Tailing Dam Disaster in Brazil

A mining tailings dam is a long-term disposal method of toxic substances and waste from mining activities. There are more than 32,000 tailings facilities globally, containing more than 220 billion tonnes of toxic waste (World Mine Tailings 2020). The high risk and severe consequences of catastrophic tailings failures related to the mining of mineral resources are experienced worldwide. The Brumadinho mining tailings dam collapse on 25 January 2019 is one of Brazil's deadliest environmental disasters. The dam is owned by Vale Mining Company (VALE), Brazil's iron ore miner, and the Anglo-Australian BHP company. Its catastrophic failure killed 270 people, mostly Vale workers, and countless endangered species and other wildlife. According to Silva Rotta et al. (2020), the model used in Brumadinho was the upstream tailings dam, accumulating the tailings through a successive uphill deposition. Further, the massive toxic mud devastated the local flora and fauna as well as the water flows.

The Brumadinho dam was decommissioned in 2015, and the tailings dam failure was caused by structural instability and liquefaction, so that 'almost the entire dam was fully saturated when it failed … The dam rupture directly impacted the ecosystem and human health mainly due to the toxic components of the mud and the velocity of the mudflow that was released and engulfed habited areas, with no chances of survival during the event' (Silva Rotta et al. 2020, 8). Inadequate maintenance and lack of effective disaster risk governance are two key issues underlined by previous studies on dam failures in Brazil. Carmo et al. (2017, 2020) point to the significance of developing a comprehensive approach to post-disaster management to deal with environmental damage linked to dams, and call for the need for transparency in public policies in the extractive sector.

In today's contestations over mining, hydroelectric dams, and other forms of exploitation of resource territories in Latin America, a growing literature focuses on social movements that organize against neoliberal extractive capitalism and its implications for local communities and its territories (Gómez-Barris and Taylor 2017; Zhouri et al. 2016). MAB has been a significant political actor in Brazil's history of social struggles linked to the forms of access to and control over resources, including dam-related effects on local livelihoods, but also in conflicts over mining activities. The territorialized collective action across multiple states in Brazil embraces contentious eco-popular politics. MAB stands up for dam-affected communities' rights and proposes policy alternatives regarding energy systems and natural resources use. MAB disputes the discourse established by the nexus of corporate extractivism and the coercive top-down power of government. The MAB activism agenda engages with gender relations and human rights violations, pushing the claim to hold companies and extractive industries accountable for their social impacts and as perpetrators of environmental damage, destruction, and harm. The critique of the MAB on the commodification of nature with a focus on gendered power relations is illustrated in the following image.

FIGURE 7.1   Arpilleras craft done by MAB activists. The message says: 'Woman, water and energy are not a commodity'.

MAB contests mining- and land-related transformation processes driven by mega projects of dam and resource extraction and pushes its agenda to address dam-induced displacement. For Hess (2018), the development of large hydroelectric projects and the construction of large dams framed the initial background for MAB mobilization. MAB has been playing an instrumental role in mobilizing the local community around the Brumadinho mining tailing dam. This movement started in the 1970s establishing an agenda for a livelihood restoration programme to replace lost housing and land, focusing on compensations and resettlement sites from the government (Movimento dos Atingidos dos Barragens 2011). The grassroots activism of MAB was initially organized by local activists, peasants, and workers mobilizing around controversial hydropower development and its impacts on local communities in Brazil. It emerged in March 1991 during the First Congress of the Dam-Affected People of Brazil held in Brasilia, and as Hess (2018, 267) clarifies:

> As anti-dam coalitions develop over time, they tend to change both their frames and their coalition partners … The movement began in the 1980s as a coalition that used the frame of peasants' rights to land and livelihood, but by the late 1980s, the movement had become national, and the coalition partners had expanded to include environmentalists … By the early 2000s, the local anti-dam movement had expanded its coalition to the national level and included environmentalists, urban residents, and scientific experts … The frame expansion of the anti-dam movements to include the values of environmental sustainability, rights to water and energy, and the democratic reform of governance processes facilitated coalitions with low-income residents and environmentalists in cities and with scientists and lawyers, whose expertise was essential in the environmental assessment process and litigation.

Within the anti-dam literature, Hess (2018) argues that the anti-dam movement refers to an 'industrial transition movement' (ITM) that mobilizes experts (researchers and lawyers, for instance) in shaping the movement's agenda around environment-related harm and human rights violations. Franklin and Oliver (1999) traced the history of the anti-dam struggle in Brazil through the lens of political opportunity structures. They pointed out the influence of liberation theology within other class/land struggle movement organizations. Over the years, MAB's land-centred activism broadened to include a political ecology critique of extractivism and dam-related impacts on local communities (MAB 2021, 2018). Further, the MAB appropriated transnational advocacy networks as instrumental tactics of their social struggles (Scherer-Warren and Reis 2008).

The work of popular advocacy of the MAB as a tool to rethink human rights violations also explores how oppressions of people affected by the dam combines race and gender issues (Masso and Masso 2020; MAB 2011). MAB's activism highlights the role of gender inequalities, 'the affected women need to face, in addition to the patriarchal society and the social inequalities, the impacts of this energetic model on their territories before, during and after the construction' (MAB 2017). On a broader scale, the effort of incorporating gender perspectives into energy projects organizes and shapes an important part of MAB's activism approach. Towards that end, MAB's activism advocates for a specific gender-responsive policy change into energy policy and programmes. In partnership with the Oswaldo Cruz Foundation (Fiocruz) and Polytechnic Health School Saúde Joaquim Venâncio (EPSJV), MAB published a guide to respect, protect, and fulfil rights related to the women affected by dams, stating that 'the right to health includes the right to water, to energy, to land, to healthy food' (2019, 16).

The MAB's digital campaigns that aimed to raise awareness about the lack of accountability related to the Brumadinho mining tailings dam collapse used a combination of data, videos, images, and real stories to move audiences. This study selected MAB's digital campaigns related to the Brumadinho mining tailings dam collapse from February 2019 to April 2021, using specific keywords and hashtags, including #Brumadinho2Anos #DoLutoàLuta #ValeNadaSemReparação #ParaopebaEmLuta #SomosTodosAtingidos.[1] Twitter, as a social media microblogging platform, plays an important role in their activism allowing users to post and interact via direct messages. Data collection was based on a content analysis of MAB's Twitter account (~13,000 followers, 656 social ties, and 6285 posts) and MAB's dam-related online publications. The data collected was published in Portuguese and English.

The MAB's advocacy work via social media platforms connects the human rights violations to socioenvironmental transformation caused by different forms of extractivism. By combining local knowledge with expert knowledge, MAB's advocacy work highlights the impact of Brumadinho mining tailings dam collapse on communities and the ways in which these communities face the contamination of the river. Among other issues, MAB highlighted that 'there was loss of leisure space in the river used by the riverine community, loss of fishing and fertile lands that guaranteed healthy food, in addition to contamination of water, soil, air and plants that influence the entire food chain, and consequently damage health' (2021, 22). It underlines the neoliberal approach logic of profit-making regarding mining industries, which often overlooks long-standing socioenvironmental controversies linked to large dam projects transforming local people's livelihoods (MAB 2018).

The digitally mediated collective action of MAB relies on the leveraged social media to produce and diffuse alternative knowledge on anti-dam activism, its visions, and practices. It articulates a transnational activist network of affected communities by dam projects in Latin America and explores political alternatives to neoliberal regulations of nature exploitation and exportation (MAB 2020). Figure 7.2 illustrates how MAB activists actively construct and mobilize collective memory regarding a disaster-specific event. The usage of social media by MAB outlines the lack of accountability on the part of governments, state institutions

FIGURE 7.2   The Twitter profile page of MAB posting the following: 'Affected people by Vale's crime in the Paraopeba River basin are united and fighting for justice and reparation!', 25 January 2021.

and public authorities. The MAB approach underlines mining corporations' predatory approach of access to, use of, and control over resources. They argue that mining corporations operate locally with impunity when it comes to their impacts on local communities. As a result, they point to the human rights abuses and lack of public participation in the agreement process, arguing that:

> Vale's only objective is profit over life. In the Brumadinho region, the company seeks to avoid paying for its crimes at any cost and wants to avoid further expenses in restoring rights that have been violated. At the same time, it wants to continue exploring the ore in Minas Gerais, sending it abroad without paying the minimum amount of taxes … As in the Rio Doce basin, again an agreement was struck without the participation of the affected population, under the pretext that reconciliation is the best and quickest way to solve the process. However, we condemn the fact that the speed here only serves the objectives of the government of the state of Minas Gerais and of the Vale company, without guaranteeing the rights of those affected, especially the right to participation.
>
> *(MAB 2021)*

The growing power of platforms serves as a key medium to reach not only customers but also citizens, voters, and/or supporters. Their power to use that

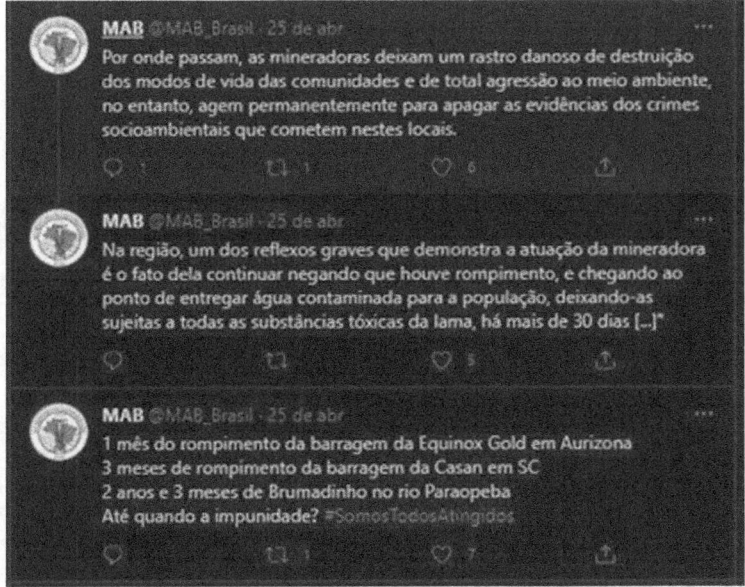

FIGURE 7.3   The Twitter profile page of MAB posting the following:

Wherever they go, mining companies leave a damaging trail of destruction of the communities' ways of life and total aggression against the environment, however, they act permanently to erase the evidence of the socio-environmental crimes they commit in these places. In the region, one of the serious consequences that demonstrates the actions of the mining company is the fact that it continues to deny that there was a breach, and goes so far as to deliver contaminated water to the population, leaving them subject to all the toxic substances in the mud, for more than 30 days. 1 month since Equinox Gold dam broke in Aurizona, 3 months after the Casan dam broke in SC, 2 years, and 3 months of Brumadinho on the Paraopeba river, How long will they go unpunished for?

(#SomosTodosAtingidos, 25 April 2021)

user-generated data and algorithmic tools for persuasion has become increasingly central to public and private life. Social practices related to collective action organized around platforms entail an interactive process that shapes collective action frames, collective interests, and collective identities. The platform-based user interactions among members organizing action, mobilizing resources, or forming collective identities play an important part in the MAB's struggles for rights, needs, and interests of affected populations.

This advocacy work produces accessible alternative knowledge that challenges the official news and suggests the Brumadinho mining tailings dam collapse should be charged as a crime. The figure 7.3, for instance, illustrates the approach adopted by MAB to promote awareness of the aftermath of the environmental disaster in Brumadinho as a result of mining disasters. It reveals how Vale's socioenvironmental crime in Brumadinho devastated the livelihoods of thousands of people. Within its broader action network of people defending natural resources

FIGURE 7.4 The Twitter profile page of MAB posting the following: 'Vale killed 272 people and 11 others are still missing! Our fight for justice continues in Brumadinho, but also in Mariana, no matter how many years pass' (25 January 2021).

from mining and hydroelectric dams, MAB pushes an agenda for reparation measures, accountability, and policy change for greater regulation of extractivism. This suggests that the social media platform of MAB discusses political alternatives regarding extractive governmentality. It shows that alternative knowledge can dispute the narrative over the disaster led by large-scale industrial mines, and that contemporary mining practices are linked to environmental harm and human rights violation, as illustrated by the Figure 7.4.

MAB's advocacy work contests extractivism as a development strategy, challenging its discourses by outlining the lack of regulations. It looks to boost the accountability of institutions that sustain extractivism for their toxic legacy by mining operations. MAB's tactics to hold mining companies accountable for their crimes, including toxic and hazardous waste, involve the production of knowledge as part of their contentious politics. The discourse enacted by and within MAB is mediated, disseminated, documented, and altered through digital social media platforms. Digital advocacy underlies the power relations that shape contestations over extractivism and frames socioterritorial movements as knowledge producers.

4 Conclusion

The social movement's effort to create alternative knowledge via digital media connects to its advocacy strategies for influencing governments and policy change. MAB is one of the clearest examples of platformization of climate justice for which the appropriation of digital tools establishes a relevant strategy for developing and amplifying cooperation through transnational alliances, supporting, and encouraging further mobilization across different actors beyond the zones of their place-based activism. MAB, along with many other environmental organizations and protest groups, continues to push for a more transparent government decision-making and accountable corporate action to tackle the social, environmental, and climate footprint of mining operations affecting already vulnerable communities.

Although the effectiveness analysis of the digital advocacy by MAB goes beyond the aim of this chapter, it is instrumental to point out that they shaped a part of the public debate by making available an alternative understanding of the Brumadinho mining tailings dam disaster, thereby creating free and original content for advocacy campaigns for stronger accountability of the extractive sector and quicker governmental response to affected communities. Ultimately, MAB documented the lack of policy response to people particularly affected by impacts of the extractive economic model. The advocacy campaign served as a reminder of the unaccountable corporate power against land, environment, and people, particularly communities directly affected by mining activities. MAB's advocacy and social media presence enabled a broader media exposure of their struggles and demands, including news published by Globo News, the largest media group in Brazil.

Without the digital advocacy by MAB, the understanding of the Brumadinho mining tailings dam collapse would be shaped mostly by the corporate power and state authorities. MAB generated knowledge alternatives about the Brumadinho dam disaster and produced 'community-led alternatives' to be delivered to political institutions and to underpin the production of policy outputs. For instance, MAB activism requested participation of the movement's representatives in the negotiations for the Preliminary Adjustment Agreement (TAP) with the Vale company, and the 'Federal Public Ministry issued a notice in support of the movement, stating that it hasn't participated in the negotiation meetings without the presence of technical advisors and those affected' (MAB 2020).

The grassroots activism of MAB is not only related to the environmental impacts from large-scale mining projects. They seek ways to advocate for rights in the context of natural water resources and highlight a local struggle to address environmental and social issues within a popular energy project. What we can learn from MAB is that human-caused ecological destruction and environmental injustice can be addressed by new possibilities for organizing grievances and claims on energy. Mining disasters are part of a wider logic of natural resource extraction. The role of social media for community-based activism against corporate extractivism remains an underexplored issue in this literature. It is instrumental to look at the grassroots approach to socioenvironmental conflicts, including dam failure, because of the precarious dam infrastructure in Brazil plagued by insufficient government oversight, limited financial, and technical resources.

The advocacy work of MAB exemplifies how the platformization of climate justice activism may help challenge the dominant discourse of economic growth and development within the extractive global economy. The adoption of digital social media platforms plays an important part in the advocacy work of MAB. In this sense, MAB's digital media presence amplifies their mobilization and resource management strategies to accommodate digital media as something 'necessary' for contemporary protests. MAB's orientation towards contesting intense extractive activities and energy combines collective action and permanent digital communications. This systematic use of digital communications allowed them to reach out to a larger number of activists, supporters, and interested citizens both locally and transnationally. MAB's advocacy has nurtured relationships with the Movement of People Affected by Dams in Latin America (MAR) and Grassroots International, a global grant-making and social action organization.

The significance of the use of social media and digital activism points to the need for greater social control and accountability over mining and extraction processes, including large dam projects, dam planning, and management. Rather than simply a digital manifestation of a pre-existing collective mobilizing, what is at stake is a transformation in the ways in which grassroots movements foster their political agenda. MAB' advocacy refers to a community-building sphere based on advocacy practices that increase the organization's visibility and support base. The advocacy work of MAB points to the lack of public policies and governance frameworks for a stronger regulation of the extractive sector. Grassroot advocacy plays a part in generating policy ideas that address the challenges of environmental degradation such as dam-related displacement and disaster. It sheds light on the implications of mining activities and issues of Brazil's disaster risk governance and accountability.

Clearly, the advocacy work of MAB challenges the neoliberal ideologies of global resource extraction based on techno-economic rationality and market-centred use of and control over resources through exploitative relations. The advocacy work on digitally mediated networks builds alliances for collective action through alternative knowledge. The analysis indicates that digital media do have a role in climate justice activism. It demonstrates the importance of integrating social media platforms with activism on the conflicts over natural resources extraction and contributes to cutting-edge policy thinking on resistance from the grassroots standpoint. The digital advocacy of MAB invites the local community, scholars, policymakers, and ordinary citizens to think critically and pragmatically about the contemporary resource extraction sector and its associated socioecological conflicts in Brazil, the Majority World, and beyond.

Note

1   Free translations: #Brumadinho2years #frommourningtostruggle #NothingWithout Reparations #paraopebainstruggle #weareallaffected

References

Aguilar-Pesantes, A. et al. 2021. A Comparative Study of Mining Control in Latin America. *Mining*, 1, 6–18.

Arboleda, M. 2020. *Planetary Mine: Territories of Extraction under Late Capitalism*. Verso.

Azadi, M. et al. 2020. Transparency on Greenhouse Gas Emissions from Mining to Enable Climate Change Mitigation. *Nature Geoscience* 13, 100–104.

Carmo, F. F. et al. 2017. Fundão Tailings Dam Failures: The Environment Tragedy of the Largest Technological Disaster of Brazilian Mining in Global Context. *Perspectives in Ecology and Conservation*, 15, 145–151.

Carmo, F. F. et al. 2020. Mining Waste Challenges: Environmental Risks of Gigatons of Mud, Dust and Sediment in Megadiverse Regions in Brazil. *Sustainability*, 12(20), 8466.

Dahlberg-Grundberg, M., and J. Örestig. 2017. Extending the Local: Activist Types and Forms of Social Media Use in the Case of an Anti-mining Struggle. *Social Movement Studies*, 16(3), 309–322.

della Porta, D., and E. Pavan. 2017. Repertoires of Knowledge Practices: Social Movements in Times of Crisis. *Qualitative Research in Organizations and Management: An International Journal*, 12(4), 297–314.

della Porta, Donatella, and Louisa Parks. 2014. Framing Processes in the Climate Movement: From Climate Change to Climate Justice. In *Routledge Handbook of the Climate Change Movement*, edited by Matthias Dietz and Heiko Garrelts. Routledge, pp. 19–30.

Derman, B. B. 2020. *Struggles for Climate Justice: Uneven Geographies and the Politics of Connection*. Palgrave.

de Souza Santos, A. A. 2019. Trading Time and Space: Grassroots Negotiations in a Brazilian Mining District. *Ethnography*, 22(2), 184–206.

Dubash, Navroz K. 2021. Varieties of Climate Governance: The Emergence and Functioning of Climate Institutions. *Environmental Politics*, 30(1), 1–25. www.tandfonline.com/doi/full/10.1080/09644016.2021.1979775

Engels, B., and B. Dietz (eds). 2017. *Contested Extractivism, Society and the State – Struggles over Mining and Land*. Palgrave Macmillan.

Ferry, E., A. Vallard, and A. Walsh (eds). 2019. *The Anthropology of Precious Minerals*. University of Toronto Press.

Fischer, M. 2014. Coalition Structures and Policy Change in a Consensus Democracy. *Policy Studies Journal*, 42(3), 344–366.

Foster, J. B. B. Clark, and R. York. 2020. *The Ecological Rift: Capitalism's War on the Earth*. Monthly Review Press.

Franklin, D. R., and P. E. Oliver. 1999. From Local to Global: The Anti-dam Movement in Southern Brazil, 1979–1992. *Mobilization: An International Journal*, 4(1), 41–57.

Fraser, N. 2021. Climates of Capital: For a Trans-Environmental Eco-Socialism. *New Left Review*, 127, 94–127.

Gach, E. 2019. Normative Shifts in the Global Conception of Climate Change: The Growth of Climate Justice. *Social Sciences*, 8(1), 24.

Gómez-Barris, M. 2017. *The Extractive Zone: Social Ecologies and Decolonial Perspectives.* Duke University Press.

Gudynas, E. 2015. *Extractivismos. Ecología, economía y política de un modo de entender el desarrollo y la Naturaleza.* Centro de Documentación e Información Bolivia.

Helmond, A. 2015. The Platformization of the Web: Making Web Data Platform Ready. *Social Media + Society*, 1(2), 1–11.

Hess, D. J. 2018. The Anti-dam Movement in Brazil: Expertise and Design Conflicts in an Industrial Transition Movement, Tapuya. *Latin American Science, Technology and Society*, 1(1), 256–279.

Hestres, L., and J. Hopke. 2017. *Internet-Enabled Activism and Climate Change.* Oxford Research Encyclopedia of Climate Science.

Huber, A. 2017. Beyond 'Socially Constructed' Disasters: Re-politicizing the Debate on Large Dams through a Political Ecology of Risk. *Capitalism Nature Socialism*, 28(3), 48–68.

IDMC. 2017. *Dams and Internal Displacement: An Introduction.* Internal Displacement Monitoring Centre (IDMC), Geneva.

Jafry, T., M. Mikulewicz, and K. Helwig. 2019. Introduction: Justice in the Era of Climate Chante. In *Routledge Handbook of Climate Justice.* Routledge, pp. 1–9.

Johansson, H., & G. Scaramuzzino. 2019. The Logics of Digital Advocacy: Between Acts of Political Influence and Presence. *New Media & Society*, 21(7), 1528–1545.

Kaun, A. and J. Uldam. 2018. Digital Activism: After the Hype. *New Media & Society*, 20, 2099–2106.

KhosraviNik, M. 2018. Social Media Techno-Discursive Design, Affective Communication and Contemporary Politics. *Fudan Journal of the Humanities and Social Sciences*, 11, 427–442.

Kirsch, S. 2014. *Mining Capitalism: The Relationship between Corporations and Their Critics.* University of California Press.

Kirchherr, J. 2018. Strategies of Successful Anti-dam Movements: Evidence from Myanmar and Thailand. *Society & Natural Resources*, 31(2), 166–182.

Lèbre, É. et al. 2020. The Social and Environmental Complexities of Extracting Energy Transition Metals. *Nature Communications*, 11, 4823.

Louzada, A. F., and N. Ravena. 2019. Dam Safety and Risk Governance for Hydroelectric Power Plants in the Amazon. *Journal of Risk Research*, 22(12), 1571–1585.

Masso, T. F., and T. F. Masso. 2020. Onde estão nossos direitos? O campo feminista de gênero bordado pelas mulheres atingidas por barragens. *Revista Brasileira de Políticas Públicas*, 10(2), 490–519.

Milan, S. 2015. When Algorithms Shape Collective Action: Social Media and the Dynamics of Cloud Protesting. *Social Media + Society*, 1(2).

Movimento de Atingidos pelas Barragens (MAB). 2011. *O modelo energético e a violação dos direitos humanos na vida das mulheres atingidas por barragens.* MAB.

Movimento de Atingidos pelas Barragens (MAB). 2017. *Violation of Rights Suffered by Women Affected by Dams and the Struggle for Life.* MAB.

Movimento de Atingidos pelas Barragens (MAB). 2018. *Compromissos com o povo brasileiro para a soberania energética.* MAB.

Movimento de Atingidos pelas Barragens (MAB). 2019. *MAB's Analysis of Vale's Crime in Brumadinho/Minas Gerais*. MAB.

Movimento de Atingidos pelas Barragens (MAB). 2020. *After Occupying Vale's Entrance in Brumadinho, Those Affected Keep Struggling for Reparations*. MAB.

Movimento de Atingidos pelas Barragens (MAB). 2021. *Two Years of Vale's Crime in Brumadinho: The MAB Denounces Agreement without Participation of Those Affected*. MAB.'

Pavan, E., and A. Felicetti. 2019. Digital Media and Knowledge Production Within Social Movements: Insights from the Transition Movement in Italy. *Social Media + Society*, 5(4).

Schapper, A., S. Killoh, and C. Unrau. 2019. Social Mobilization against Large Hydroelectric Dams: A Comparison of Ethiopia, Brazil, and Panama. *Sustainable Development*, 28(2), 413–423.

Scherer-Warren, I., and M. J. Reis. 2008. Do local ao global: a trajetória do movimento dos atingidos por barragens (MAB) e sua articulação em redes. In *Vidas alagadas: conflitos socioambientais, licenciamento e barragens*. Editora UFV, pp. 64–82.

Şen, A. F., and Y. F. Şen. 2016. Online Environmental Activism in Turkey: The Case Study of "the right to water"'. *Global Bioethics*, 27, 1–21.

Silva Rotta, L. H. et al. 2020. The 2019 Brumadinho Tailings Dam Collapse: Possible Cause and Impacts of the Worst Human and Environmental Disaster in Brazil. *International Journal of Applied Earth Observation and Geoinformation*, 90, 1–12.

Sonter, L. J. et al. 2020. Renewable Energy Production Will Exacerbate Mining Threats to Biodiversity. *Nature Communications*, 11, 4174.

Svampa, Maristella. 2015. Commodities Consensus: Neoextractivism and Enclosure of the Commons in Latin America. *South Atlantic Quarterly*, 114(1), 65–82.

Tokar, B. 2019. On the Evolution and Continuing Development of the Climate Justice Movement. In *Routledge Handbook of Climate Justice*, edited by T. Jafry, M., Mikulewicz, and K. Helwig. Routledge, pp. 13–25.

Törnberg, P., and J. Uitermark. 2020. Complex Control and the Governmentality of Digital Platforms. *Frontiers in Sustainable Cities* 2(6).

Verlie, B. 2021. Climate Justice in More-Than-Human Worlds. *Environmental Politics*, 31(2), 297–319.

World Bank. 2017. *Minerals for Climate Action: The Mineral Intensity of the Clean Energy Transition*. World Bank.

World Mine Tailings. 2020. State of Worldmine Tailings 2020. https://worldminetailingsfailures.org.

Zhouri, A., P. Bolados, and E. Castro (eds). 2016. *Mineração na América do Sul: neoextrativismo e lutas territoriais*. Annablume.

8

RESISTING DISPOSSESSION AND DESTRUCTION

Climate (In)justice and Wind Extraction Frontier in the Postcolonial Indian State

David Singh

1 Introducing Energy Transition, Climate Justice, and Resource Extraction

Renewables are omnipresent in the global fight against climate change, particularly as they are acclaimed by a diversity of actors (international organizations, transnational companies, civil society organizations, and states) and well connected to global financial capital and international mechanisms of governance (e.g. Clean Development Mechanism or Green Climate Fund from the UNFCCC). In India, climate policy regime has been mainly oriented towards national missions on solar, wind, or biofuel energy.[1] Since the Paris Climate Agreement in 2015, India has indeed committed itself to produce 40% of its electricity from sources other than fossil fuels by 2030 (United Nations 2015). Electrification projects in postcolonial India historically centred around the large availability of coal resources, entailed from the beginning a profound civilizing narrative of bringing progress and development through science and technology, with the powerful image of 'enlightening' the 'darkness' and 'backwardness' of rural India (Kale 2014). Renewable energy (RE) also advances powerful nationalist narratives, imaginaries, and representations around energy sovereignty, security, and climate leadership (Shidore and Busby 2019); it is constructed as 'green',[2] 'clean', cheap, abundant, infinite, and situated on 'empty' lands in remote areas.

At the same time, calls for climate justice and just transition have been emerging around the world, emphasizing the need to ensure 'a fair and equitable process of moving towards a post-carbon society' (McCauley and Heffron 2018, 2). In the just transition and climate justice rhetoric, the distribution of benefits and burdens related to climate (mitigation and adaptation) actions should be engineered along socially and environmentally just lines. The just transition debate has been recently

DOI: 10.4324/9781003214021-9

mobilized in France by the yellow vest movement who refused to let rural and pauperized sections of the society pay entirely the cost of energy transition (Baber 2019). In more traditional forms of 'Not in My Back Yard' movements, it has been used to oppose the implementation of solar or wind power infrastructures on the basis of aesthetic opposition and landscape changes (Schwenkenbecher 2017). But these appeals coming mainly from industrialized countries do not engage with the specific rights to land, space, and subsistence in the Majority World (Rignall 2016), and they have been largely reappropriated by 'corporate enclosure' to impose an understanding of just transition and just climate actions that is compatible with the 'green growth' rhetoric, sustainable development, and new resource extraction frontiers (Bainton et al. 2021, 627).

Transitions are rarely radical transformations or revolutions. Rather, the energy transition can be analysed as a 'socioecological fix' to interlocking capitalist accumulation and climate crisis (McCarthy 2015), or, as suggested by Newell, a 'trasformismo', a Gramscian term referring to the 'ability to accommodate pressures for more radical and disruptive change and to employ combinations of material, institutional and discursive power to ensure that shifts which do occur in sociotechnical configurations do not disrupt prevailing social relations and distributions of political power' (2018, p. 4). 'Green' energy is not a substitution or disruption to traditional 'black' (coal) or 'brown' (oil) resources, but rather it follows the same complementary and overlapping extractive, exploitative, and destructive logics (Dunlap 2019). Energy transition interventions, such as wind power projects, are embedded in specific land politics and territories where they exercise considerable amount of coercive and discursive power over resources, space, and populations. These projects result in new forms of spatiality and renewed social and land controls as they target so-called 'deserted', 'empty' and 'waste' lands[3] (Baka 2013, 2017), a legal category and classification of land use referring to state-owned land which is neither exploited nor cultivated.

This contentious land politics reveals how much RE infrastructures are entangled with crucial questions of (in)justice in a post-carbon world (Mitchell 2013), as the populations who are currently the most vulnerable and impacted by climate change also carry an important part of the costs and burdens associated to actions aimed at stopping it (Survival International 2009). These violent logics of injustice in the name of climate action are contested by a diverse range of insubordination acts, open resistance, and a renewed repertoire of political (re) actions. Resistance to climate injustice is specifically conducted on the ground of biodiversity and environmental protection, the defence of common lands, and their attached livelihood practices (Temper et al. 2020). It re-energizes traditional agrarian struggles and creates new political outcomes and configurations for alternative development pathways (Scheidel et al. 2018; Temper et al. 2018).

There are already well-established studies on climate (in)justice in the mining sector (Ghosh 2016; Oskarsson, Lahiri-Dutt and Wennström 2019), in conservation and reforestation programmes (Brockington and Igoe 2006; Kabra

2009), and even in wind power or the related rare earth metals extraction in South America (Dunlap 2017; Sanchez-Lopez 2019). Few existing case studies have been conducted on renewables in the Indian subcontinent (Yenneti, Day and Golubchikov 2016; Lakhanpal 2019; Stock and Birkenholtz 2019).[4] We need more research on the dialectics between dispossession-destruction and dissenting political (re)actions or counter-hegemonical movements in the context of climate (in)justice. In this chapter, I investigate the land politics of RE extraction in rural India from the perspective of climate (in)justice, and how it reconfigures resource governance and subordinates' political agency. Drawing from the cases of three wind power projects in western borderland Gujarat, I suggest that the conceptual framework of frontier-making and territorialization (Rasmussen and Lund 2018) helps to illuminate the structural dynamics of domination, colonization, and power underlying the development of wind energy. The concept of 'everyday resistance' (Scott 1989, 1992) is also essential to understand the scale, the scope, and the forms of resistance and contesting voices reacting to wind energy's coercive and discursive expansion. This chapter draws upon critical qualitative research conducted between 2020 and 2021 for a period of height months and relies on analyses of key documents and literature, semi-direct interviews, and informal conversations with around 70 informants as well as (non)participant observation on wind sites, in villages, and other socialization places.

2 Understanding Wind Extraction and Resistance to Renewables

2.1 Frontier-Making and Territorialization

The conceptual foundation of this chapter lies in the assumption that climate change mitigation narratives and energy transition discourses justify a race for 'clean' extraction and a dynamic process of frontier-making and territorialization where long-standing governance instruments enforce patterns of accumulation by dispossession, exclusion, and marginalization.

Energy transition is based on uneven power relations; it is about 'who wins, who loses, how and why' (Newell and Mulvaney 2013, 133) and the outcome of that situation is either just or unjust, depending on the definition of justice we select. The utilitarian definition of justice that is most generally hold in climate change mitigation actions is based on the conception that the further marginalization and exclusion of the few can be accepted and tolerated as long as it helps to achieve the common good for the many. This precise narrow notion of justice is at stake and 'exposes injustices that may be justified in the name of an urgent energy transition' (Bainton et al. 2021, 630). Climate change mitigation narratives help justify the creation of 'sacrifice zones' (Scott and Smith 2017), the extraction of new resources, and a 'thirst for steady, affordable and reliable energy' that in turn 'set[s] in motion land grabs for territories with newfound value' (Newell and Mulvaney 2013, 136).

When these new resources, like wind or solar, are discovered and extracted, Rasmussen and Lund (2018) describe an overlapping cyclical dynamic of both frontier-making and subsequent territorialization. The frontier dynamic refers to the breaking down of existing institutional orders and entails a continuous process of annihilation and destruction of previous property regimes, land uses and rights, claims, and livelihoods (Cons and Eilenberg 2019). The territorialization follows a reverse dynamic as it is about the building of a new institutional order: territorialization or territoriality has been defined by Sack as an 'attempt by an individual or group to affect, influence or control people, phenomena, and relationships by delimiting and asserting control over a geographic area' (2009, 19). It involves a series of governmental techniques where territorial rules, property regimes, and laws are established and performed by both public and private actors. This renewed land control pattern legitimates certain land uses, guarantees property and access rights to certain populations or groups while denying these same rights to others (Sikor and Lund 2009). Frontier-making and territorialization dynamics create new accumulation possibilities based on extensive dispossession, exclusion, and marginalization (Harvey 2004; Levien 2015). When deployed in 'green' energy context, these dynamics generate similar patterns of injustice, violence, and conflict that characterize the carbon economy.

New resource frontiers and territories are then discursively and coercively supported and sustained by a set of concrete governance instruments and development justifications. This process relies first on a practice of rendering energy and climate issues technical, simplified, and depoliticized. This problematization voluntarily eludes questions of dispossession, ownership, property rights, or inclusive development (Li 2007) and makes sure any transformations or transitions are organized 'within the horizons of a liberal–capitalist order' and its sociopolitical status quo (Swyngedouw 2011, 76). Because interventions define what does and does not count as development, and how it can be achieved, they feature an important exercise of power and control (McElwee et al. 2016). Strategies of control are enforced by the power of discourses, knowledge production, science, and data – meaning they are socially constructed and some actors have more capacity to influence this social construction than others (Curran 2012; Bridge 2017). 'Framing' is crucial to this construction, and the mobilization of 'rationalities', 'science', and technology helps to build shared beliefs and representations and impose 'regimes of truth' (Li 2007; Harjanne and Korhonen 2019; McCarthy and Thatcher 2019). Finally, ruling new resource frontiers and territories also implies governing populations, shaping new identities and subjectivities through technologies of government and institutional regulations that infuse new ways of seeing and understanding the world (Agrawal 2005). This leads to the production of 'governmentalizable' subjects, 'i.e. open and amenable to governmental interventions and techniques' (Odysseos 2011, 445).

2.2 Domination and Resistance

The hegemonic discourses of energy transition, climate change mitigation, and their underlying instruments of power, domination, and exclusion are also countered by a large repertoire of resistance and political (re)actions from subordinates. Scholars of resistance and subaltern studies have often seen 'resisting subjects [as] able to protect their consciousness from the colonizing effects of elite policies, dominant cultures, and hegemonic ideologies' as 'always withstand[ing] the powerful, at least in the realm of ideas and beliefs' (Agrawal 2005, 165). Resistance is indeed carried out in reaction to situations of domination and in an oppositional relation to power: Foucault defines resistance as a 'chemical catalyst so as to bring to light power relations, locate their position' (1982, 780). It is therefore particularly useful to look at what kind of resistance practices and discourses emerge from different kinds of power and domination situations (Lilja and Vinthagen 2014). I rely here on Scott's conception of three dimensions of domination and related power (1992, 1989): ideological domination, status domination, and material domination. Given this range of power relations and experiences of domination, we can acknowledge that the reactions to them are equally diverse and multiple. Most of the attention has been oriented towards open, public, and organized resistance movements, but as 'the circumspect struggle waged daily by subordinate groups is, like infrared rays, beyond the visible end of the spectrum' (Scott 1992, 183), it is critical to investigate also the field of everyday insubordination acts that avoid the written and public record and take more ambiguous, dissimulated, and disguised forms of expression.

3 Background: Development Model, RE, and Kutch

The growth model steered by the postcolonial Indian state experienced important shifts and changes since 1947 from a state-led development phase to a more neoliberal and recently 'green' phase, with the common feature of promoting a culture of extraction, primitive accumulation (Marx and Mandel 1990), and accumulation by dispossession (Harvey 2004). Dispossession and enclosure are indeed critical to the internal functioning of capitalism, either as a historical process of enclosing common lands in pre-capitalist societies to create a new class of urban proletariat fuelling the industrialization (Gardner and Gerharz 2016), or as a systemic and cyclical process of global capitalism confronted to crisis of overaccumulation. In the Indian context, Levien (2018) identified several 'regimes of dispossession' endorsed and negotiated by the state and whose outcomes change over time depending on the different class configurations. If the early Nehruvian development model was based on a regime 'that dispossessed land for state-led industrial and infrastructural expansion', 1991 economic liberalization of the country unleashed the genesis of a new regime 'that dispossesses land for private – and increasingly financial – capital' (Levien 2015, 150).

Indeed, India has not been hermetic to the structural changes happening in the world economy, and economic liberalization era has led to the deregulation of previously state-controlled sectors like energy and electricity, the opening to private participation, and a fierce subnational competition where states are compelled to adopt fiscal incentives and 'smooth' land acquisition policies to attract private investors (Joseph 2010; Dubash, Kale and Bharvirkar 2018). RE extraction is fully situated within this neoliberal accumulation and dispossession regime, advancing a new 'green' feature and an even more privatized development pathway as 'the generation of power from RE services is almost 90% in the hands of private parties' (Benecke 2010, 9).

This chapter focuses on Gujarat, as it has been often presented in the literature as a model example of economic liberalization and business-friendly policies (Sud 2012, 2014; Jaffrelot 2018, 2019). Gujarat has been one of the first and most proactive states in liberalizing land policies for private investments since 1987, leading in-state RE development (Phillips, Newell and Purohit 2011). The district of Kutch is of particular interest for this chapter (see Figure 8.1 below), as it shifted from being at the margins of Gujarat towards a 'new resource frontier' and a space for capital accumulation (Mehta and Srivastava 2019). It is the largest district in Gujarat, with a land area of 45,674 square kilometres, constituting 23% of the state. The main occupations have been traditionally pastoralism, agriculture, and fishing in a context of intense droughts; Kutch is increasingly confronted with climate change-related uncertainties, such as cyclones, storms, and land erosion. Pastoral communities are particularly vulnerable to changes in drought patterns, water scarcity, and increased temperatures (Mehta, Bhatt and Joshi 2020). Rehabilitation and resettlement programmes that followed the deadly 2001 earthquake have led to a critical shift in Kutch's development policy, as centre and state governments turned these 'vast drylands' into an attractive space for private investors (Kohli and Menon 2016, 271). The first private Special Economic Zone (SEZ) port in India was created in Mundra and entrusted to Adani Power. The district has been subsequently identified as a wind-rich region by the National Institute of Wind Energy (NIWE) and therefore framed as a major wind corridor, where thousands of windmill projects have been developed by established companies like Adani Power or Suzlon for more than 15 years.

4   Green Climate Injustices: Colonization, Dispossession, and Destruction

This section aims at unravelling the structural, discursive, and physical patterns of injustice, exclusion, and dispossession underlying the development of large-scale land-consuming wind power projects in mainland Kutch. I propose that the construction of 'wasteness' and 'emptiness' is central to the process of frontier-making and that parallel invasive, unlimited, and colonial dialectics are central to wind power territorial expansion and land control.

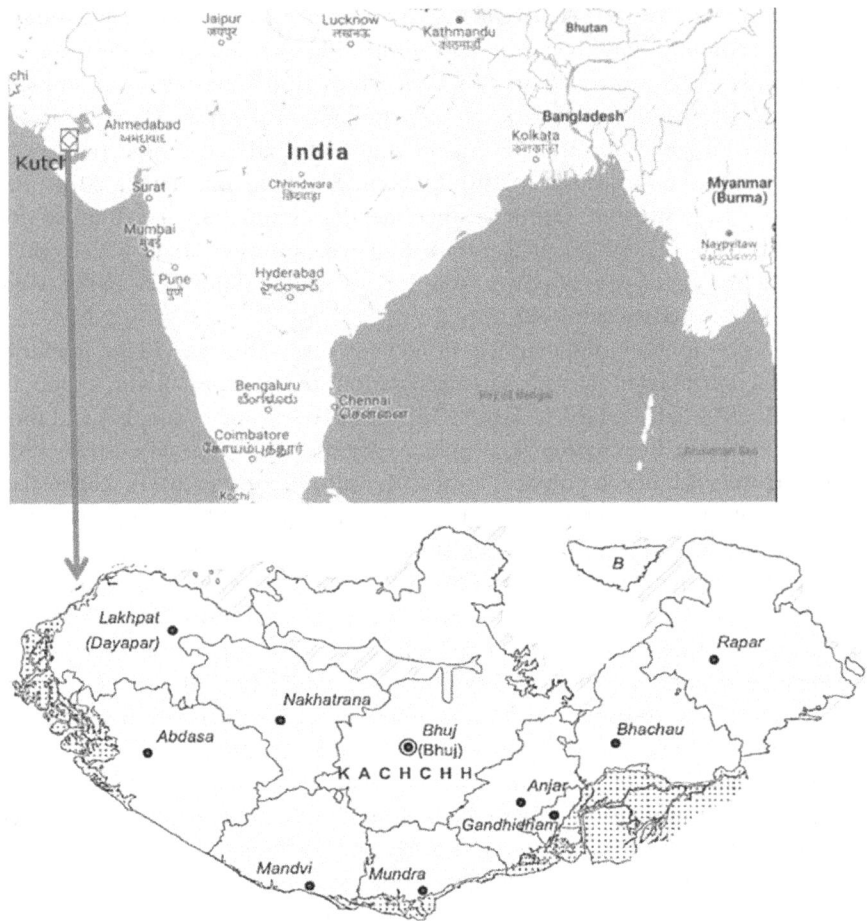

FIGURE 8.1 Modified Google Map of India and administrative map of Kutch/Kachchh District, Gujarat.

Source: www.onefivenine.com/india/villag/Kachchh.

4.1 Constructing 'Wastelands'

Kutch is renowned for its unique ecology and ecosystem biodiversity, hosting within its borders seasonal wetlands, thorn forests, important grasslands, deserts, and even mangrove forests along the coastline (Stanley 2004; Mehta and Srivastava 2017). At the same time, this district is classified as having the highest concentration of 'wastelands' in Gujarat, according to figures from the Wasteland Atlas of India for the years 1992 and 2006 (41.9% and 16.7% of the entire state, respectively) (Gujarat Ecology Commission 2017). But 'wastelands' do not simply appear out of nowhere, nor were they 'discovered' in Kutch with the arrival of wind companies. They follow a political construction that aligns with the long-standing capitalist and colonial expansion over space (Gidwani 2008; Gidwani and Reddy 2011; Harms 2014); 'wasteland' is imagined as an untapped potential for capital accumulation that needs to be integrated into the capitalist discipline of value and private property.

The construction of 'wastelands' in Kutch is rooted within a long history of frontier-making that started in the 1960s where large tracts of common grazing land were opened up by the government for their added agricultural value and are now being converted to industrial programmes and 'green' energy production (Ibrahim 2008; Sud 2020). Policymakers in Ahmedabad (the state capital of Gujarat) have been constantly devaluating and undervaluating lands in Kutch, imagining them as the same single, fixed, disposable, and uniform piece of 'waste' that can be easily taken and appropriated as 'nobody' uses it, or at least not in productive ways, as they define them. These lands, as explained by a wind company representative, 'do not hold any value, do not produce enough. We have a lot of barren lands, a lot of empty lands which don't produce any agricultural value, and even if they do some agriculture it is rainfed crop [of] which production is really low, so we can put them in [sic] a much better use.'[5]

The discursive construction of certain lands as 'waste' leads inevitably to the construction of certain lives, livelihoods, and practices as 'waste' (Gidwani and Reddy 2011; Harms 2014). Traditional pastoral practices in Kutch and the use of common lands for cattle grazing have been historically criminalized and denied legitimacy since colonial times (Mehta and Srivastava 2019), and Gujarat state officials I interviewed still depict pastoral populations in very colonial ways: 'They are uneducated and backward people, the thumb-type who doesn't even know how to write his name, they just move around the jungle all day with their cattle without any true purpose or goal.'[6] Development and industrialization policies promoted by the state have furthered the marginalization of pastoralists as they lose access to pastures and sense of identity and move towards semi-proletarization. Samio,[7] a village I discuss in more detail in the next section, is very illustrative of that phenomenon. In the household survey conducted in January 2021, 80% of the surveyed villagers belonging to the pastoral caste Rabari had partially exited cattle-grazing livelihoods, and half of them were employed outside the village as informal wage labourers.

A powerful instrument in making and creating, both discursively and physically – 'empty spaces' and 'wastelands' – are the mapping and surveying procedures used by companies at early phases of wind extraction. Maps are not neutral representations of an objective reality – they are political choices aimed at legitimizing certain resource uses and claims while marginalizing or erasing others (McCarthy and Thatcher 2019). Wind assessment mapping, satellite images, and remote sensing technologies enforce uniformed and fixed interpretations of land, to a large extent in favour of 'wasteness' and 'emptiness' (Robbins 2016). Engineers and dedicated companies' land teams use the available data on land titles and land records collected from tax officers and land revenue services (district collector, *tehsildar, talati …*), but they also develop their own surveys and microsittings on the ground, visiting villages to identify available lands and potential 'good' locations. During these surveys, land team staff rely on mapping and zoning software, smartphone applications, GPS and maps, and therefore simply do not see nor value the ecosystem surrounding them in the same way local villagers might. As described by a wind energy engineer during a field visit, what he sees are obstacles, costs to reduce, or roads to develop:

> When we survey a location, we have to look for possible roads, where can we bring them and what will be the cost of it. We have to identify areas where we will put the machines and cranes, the assembly platform and the boom-up area, see what additional amount of land will be needed, and what other alternatives do we have.[8]

By doing so, not only do they annihilate and silence the existing uses and appropriations of space, but they also tack and impose a single, rationalized, and ordered perception of space where everything is disposed of for the purpose of wind extraction.

4.2 'We have been completely surrounded by windmills'

That is precisely where the wind frontier-making dynamic is further completed by territorialization, that is, the concomitant imposition of new rules, norms, uses, and claims emanating from wind companies, civil contractors, and local fixers. The territorial expansion and space appropriation of large-scale wind power projects in Kutch follow an extremely invasive, unlimited, and colonial pattern. Unlike other energy extraction projects like solar panels, mining, industries, or SEZ which are usually situated at one single and delimited location, windfarms do not have any visible or physical boundaries, walls, or limits. Windmills are dispersed around large tracts of lands, across hundreds of locations in a 20-km radius, grouped in little clusters, at times covering dozens of villages. The private takeover of these large tracts of mostly revenue lands does not happen 'in one go', at one single time, but follows different temporal and spatial lines. Indeed, wind companies acquire land on a cluster basis of 15–20 locations – once they have overcome the long title-clearing process and started construction, they move on to the next cluster and start again looking for land. This specificity leads to a continuous, cumulative, and ongoing process of land-grabbing, stretched across space and time (Oskarsson, Lahiri-Dutt and Wennström 2019). One company acquiring land in a certain village at a certain time does not forestall any other company, or indeed the same one, to return later and start the construction of new locations. The second village I surveyed, Haroma, which is situated in the outskirts of the subdistrict city of Dayapar, 100 km west of the district capital of Bhuj, has been completely invaded by four different companies in less than three years. A first cluster of five windmills was initially implemented in 2018 by an Indian energy company. Since then, the villagers recount '[they] have been completely surrounded by windmills'[9] as 55 more locations were constructed from three different companies, mostly foreigners this time. When I last visited Haroma in December 2021, 15 new windmills were under construction by the original Indian company. In that same invasive and colonial dynamic, the windmill location site functions as a boundaryless and changing zone whose delimitation and shape gets distorted and stretched according to construction stages. Wind companies and contractors rarely restrict themselves to the one-hectare land allotment per location they are given by the Revenue Department and usually occupy a large amount of land illegally: as explained by a villager from Haroma[10] whose land was partially encroached by companies:

> They clear everything in a four hectares radius, and then they store all their material, all the machines and cranes wherever they want, wherever it best suits them, without asking anyone. There is a whole field of wind energy expansion that is illegal, without permission or consent.[11]

The physical clearing, cleaning, and (un/re)making of space during early construction stages is the most visible and tangible element of territorialization, whereby land is modified, ordered, reorganized, and even domesticated. This entails a profound physical and material process: in Haroma, since windmills have arrived, the grassland-rich areas situated on hilly points have been cut, the paths used by villagers with their cattle have been flattened and gravelled, watercourses used for farming irrigation have been diverted for the sake of road development, groundwater sources have been completely depleted when they were situated on a company's location, while massive tree species, plants, and vegetation used for different religious or medicinal purposes have been simply destroyed. During this territorial (un/re)making, wind companies and civil contractors exercise hegemony and omnipresence: they occupy the land 24 hours per day, with SUVs, trucks transporting raw materials, JCBs, cranes, water tanks, and tractors moving constantly in and around space, day and night, from location to location, from village to village. Companies are able to exercise control and authority over large tracts of space via clustering it, as each staff is being assigned a specific zone to supervise (usually around ten locations). In this process, they develop personal and individual relations with what becomes 'their' zone, with the villages falling under their responsibility, with the influential and powerful men and women, like the village head or other high-caste landowners. Companies also mark their footprint and impose their hegemony on the terrain of sound as windmills' loud noises and regular alarms rhythm the field of everyday life: during my first evening visit on the farmland of a villager in Haroma, I remember being disturbed and puzzled by windmills' rotation as the discussion with my host got interrupted several times because of the heavy noise. When I asked if this was common, my interlocutor smiled at me and responded quite undisturbed while cooking the dinner, 'it's the windmills making their usual sound, at nights you have more wind, so they make more noise'.[12]

The destructive and invasive logics of large-scale wind farms described in this section are precisely what defines climate injustice, or in other words the socioeconomic and territorial marginalization and dispossession of already vulnerable communities in the name of 'green' energy and climate change mitigation. They generate multiple situations of domination: an ideological domination with the 'wasteland' and 'green' energy discourse, a status domination with the use of violent means of persuasion and the strengthening of caste hierarchies, and a material domination with the dispossession of land, the private takeover of common property resources, and the physical (re/un)making of space for windmills. These dialectics of dispossession and domination generate in return unique dialectics of resistance, repertoires, and cultures of contention that range from petitioning, protesting, and rallies to subtler, individual everyday reactions.

5   The Micropolitics of Resistance and Everyday Political (Re)actions

Contrary to coal- or dam-related dispossession as in the Narmada Valley (Nilsen 2010), there is no district-wise mass opposition to wind power in Kutch: the overall lack of resistance to dispossession in Gujarat is linked to a long-standing absence of anti-caste or labour movements and the predominance of Hindu nationalist politics and the Hindu-Gujarati pride discourse (Sud 2012, 2014). Combined with this anti-resistance sociopolitics, the specific material and geographical organization of wind farms, described in the previous section as dispersed around different spatial and temporal lines, certainly does not favour the emergence of pan-district resistance (re)actions, unlike some other parts of India (Lakhanpal 2019) or the world (Dunlap 2018a). What exists are spontaneous, eruptive, and at the same time well-organized public and collective village-level movements trying to oppose and contain the expansion of wind projects and related electricity infrastructures in their specific areas. Central to these movements is the use of traditional tools of contestation (petitions, protests and rallies, hunger strikes), common mobilization factors around the defence of the environment, the defence of livelihoods and identity as legitimate alternatives for achieving climate justice, and a specific social or caste configuration favouring the emergence of unity and collective action. I have come across or heard about no less than five actions in central Kutch where village committees decided to block, protest, and petition against the construction of windmills or the erection of transmission lines and electricity towers. For the purpose of this chapter, I will focus on one of them, the case of Samio, briefly mentioned in the previous section.

5.1   A Case of Collective and Organized Resistance Movement: Samio

Samio is a village situated in the suburb of the subdistrict city of Nakhatrana, 50 km west of Bhuj. The village is populated by 250 households divided in four castes: the land-owning farming Patel community, the pastoral Rabari caste, the Dalits who are numerically dominant and practice informal wage labour, and very marginally the Koli tribe whose members are employed as labourers on Patel's lands. In 2015, a renowned Indian wind company established six preliminary windmills on so-called government 'wastelands', but when 40 more locations were allocated to three other companies in 2018–19, the villagers unanimously opposed the new project, physically blocked the different sites where construction had already started, and decided to challenge all the companies on the basis of illegal and unauthorized cutting and felling of trees. This led to a halt of the different windmill projects and the beginning of a long judicial process that was still pending at the time I visited the village.

The environmental concern due to wind power extraction has been a powerful mobilization factor driving villagers to resist it, and particularly for the pastoral Rabari relying on common and free property resources for their livelihood (Mishra and Mishra 2017). Similarly, non-pastoralist groups in the village united in an intercaste Patel-Rabari-Dalit opportunity alliance as they shared common fears about the future. The leader of the resistance movement in Samio summarized these fears during our first interview:

> Where will we go? And what will we do? If they put all these 40 windmills, there will be nothing left for us in our own village. Wind power is a business, but there is nothing for us, it will never feed us, we will have to go for wage labour. For now, everyone one is surviving around land, either farmers, agricultural labourers or pastoralists. We are all related to the jungle and its land here, all the castes, that is precisely what united us.[13]

While the pastoral Rabaris were at the forefront of the struggle because their livelihood is more dependent on the availability of grasslands than other castes, everyone is related in a way to the space they call the 'jungle', which is actually labelled as 'wastelands' and not forest lands by the government. Indeed, almost every household in the village possesses at least one or two cattle for personal use, providing milk to the family, and for this reason everyone relies at some point on *gauchar* (a Gujarati term referring to common grazing lands owned by the village council) and government-owned 'wastelands'. Villagers have contested the core of the interrelated 'wasteness' and 'greenness' discourse that was mobilized by the state to justify the private takeover of revenue grasslands:

> They were claiming that these lands are waste, it is all dry lands where nothing can grow, where you have no vegetation at all. But this was completely wrong, they were lying, so we got all the maps and records, and with the help of environmental NGOs and social activists[14] we showed to the district collector that these lands have actually a unique ecosystem, with species of trees, vegetation, birds and animals living here. They say that wind is pollution free, that green energy is carbon free, which might be true. But green energy is only

a brand, this same green energy is creating the wide cutting of trees, vegetation and jungle, it is provoking fires and technical incidents.[15]

The way they contested climate injustice and the related global discourse of RE 'greenness' and 'cleanness' is precisely via mobilizing environmental and biodiversity arguments. This is their strategy to resist on the discursive terrain by using the same weapon against the hegemonic interest groups in a 'strategic reversal of the process of domination' (Bhabha cited in Butz 2002, 21). Their fight against climate injustice is profoundly political; it is situated at the intersection of social, cultural, and economic dimensions and re-energizes traditional agrarian struggles around the defence of common lands and attached livelihood practices (Temper et al. 2018). Villagers from Samio express a strong attachment to what they understand to be 'our' lands, 'our' common property resources which have been providing livelihood sources for generations. What is at stake here is the defence of their cultural identity of who they are and what they do (as pastoralist or as farmer) as well as where they belong (their village). In this discursive fight, they hope to legitimize their independent ways of life outside informal wage labour.

5.2 Everyday Insurgency and Insubordination

Complementary to these open and collective resistance movements with powerful empowerment outcomes, I have encountered a tremendous terrain of everyday insubordination acts, of dissimulated, individual, and undeclared contestation taking place when existing power relations did not allow the emergence of cross-caste configurations like in Samio (Scott 1989, 1992). These resistance forms amount to a complex system of insurgency, whereby isolated individuals conduct non-coordinated, spatially and temporally dispersed, multiple and unpredictable attacks on the companies' locations and projects (Dunlap 2018a, 2018b). During my numerous visits to wind sites, I witnessed countless situations of physical blockades, cable or petrol thefts, sabotage, and destruction of windmills done by anonymous individuals. If one isolated situation of blockade or theft could certainly not harm the company, its addition, multiplication, and overlap constitute a real threat to the stability of wind power development, as companies have to deal with multiple tension spots erupting unpredictably at different places and different times. The most common phenomenon I witnessed was the trespassing into public and common lands when projects would start. Pastoral communities occupied the space with large numbers of cattle in the direct vicinity of construction areas, passing between the heavy vehicles, machines, and cranes, sometimes even slowing down the work when a camel, a goat, or a cow suddenly decides to halt in front of a truck.

The notion of 'quiet encroachment' advanced by Bayat seems appropriate here as it refers to 'the silent, protracted but pervasive advancement of the ordinary people on the propertied and powerful in order to survive and improve their lives' (2016, 545). This 'middle-ground patrolling' (Turton 1986) resistance mode purposely exploits the weaknesses of the wind system, the grey zones, and the small spaces left by companies. It acts as a powerful reminder that the villagers are still here, that they still exist, and that this space, even though completely reshaped and modified by wind extraction, is still theirs. It acts as an indirect way to reappropriate dispossessed resources, (re)claim political agency, dignity, and honour after the humiliation of dispossession.

Resisting climate injustice and related dispossession and exclusion takes multiple forms, and there is no need to juxtapose individual, everyday resistance to collective and more exceptional contestation. They actually both respond to different dialectics of domination; they are complementary and inform each other: the silent multiplication of small-scale insurgency acts and low-profile insubordination can feed more organized resistance forms (Scott 1989) and vice versa. Resistance in that sense is always flexible and oscillates within a much more nuanced and complex political terrain (Lilja et al. 2017).

6   Conclusion: Perspectives for Achieving Climate Justice in Energy Transition

From the discussion in the last two sections, the perspectives for achieving climate justice in energy transitions in a context of rural, marginal, and already climate-vulnerable Kutch seem at least very problematic, perhaps even compromised. Indeed, considering the ways windmills are coercively, physically, and discursively deployed around Kutch, 'green' infrastructures will 'easily replicate the patterns of violence and dispossession inherent in traditional extractive industries and operate according to the same logics that prioritize private profits over social and environmental concerns' (Temper et al. 2020, 17). They run the risk of reproducing the same exclusive extraction spaces or zones where private actors are governing 'meta-infrastructures' (Easterling 2016), enclaved, disconnected, 'punctuated [and] discontinuous' (Bridge 2009, 4) territories. These new spaces of 'green' and carbon-free capital accumulation have therefore a lot in common with their fossil fuel counterparts: they are structured on the same patterns of marginalization, enclosure, injustice, and both further the violent absorption of rural communities into capitalist modes of production. Wind energy is certainly not environmentally friendly by nature as it dangerously harms the unique biodiversity of Kutch and the human–ecosystem relations based on mutual services. Nor is it people-friendly as it openly declares certain areas and their populations as 'waste', 'empty' and 'sacrifice zones' (Scott and Smith 2017).

There are still some important perspectives to be drawn. None of them will solve climate injustices alone, but together they might take a further step towards rebalancing the climate burden between North and South, the centre and the margins, the Minority World and the Majority World. First, this involves completely reconsidering the (neo)colonial discourse of 'emptiness', 'wastelands' and 'wasted' populations, which has been justifying and sanctioning legal enclosure, dispossession, and exclusion for centuries. This discourse is now linked to more recent notions of 'greenness' and 'cleanness' in the context of RE development. The assumptions that wind or solar power plants are by nature socially and environmentally sustainable need to be firmly questioned, particularly since issues around rare earth metals extraction, biodiversity, and social impacts have been raised (Avila 2018; Dunlap 2018b; Sanchez-Lopez 2019). It is therefore essential that RE infrastructures clear traditional Social and Environmental Impact Assessment (SEIA) procedures, as they are still now exempted from social and environmental clearances in the Indian legislation. We need to think and develop alternatives to the existing models of industrial and extractive RE expansion as well as privilege bottom-up approaches where impacted communities are included in the decision process via participative and democratic procedures. If free, prior, and informed consent procedures have been certainly criticized for legitimizing land acquisitions and pacifying resistance (Dunlap 2018a), they constitute a first and non-exhaustive step in Kutch, where villagers are still not informed or consulted when wind projects start on villages' 'wastelands'. This raises the final important point of community ownership and small-scale initiatives: ensuring that villages targeted by windmills have access to electricity and get a fair share in the value produced, that new land uses and claims are not enforced at the expense of villagers but are rather made compatible with their needs, constitute basic and elementary steps towards more distributive justice.

This chapter was concerned by the question of climate (in)justice in the context of energy transition and post-carbon world, from the perspective of frontier spaces and frontier communities who face the double cost of climate-related uncertainties and climate change mitigation interventions. As the northern-imposed consensus around climate change and 'green growth' is becoming hegemonic in policymaking and finance, RE projects and interventions will flourish and proliferate in the coming decades, particularly in marginal and peripheral areas of the Majority World where land politics is less contentious. It therefore becomes indispensable to develop critical analyses of RE projects from the lens of political geography and political ecology as initiated in this piece; these perspectives shed light on the territorial processes, the persistence of class–caste relations, and the legacy of extractive and colonial logics in so-called green projects. By drawing on the interrelated concept of frontier-making and territorialization in resource governance as well as everyday resistance in subordinates' political agency, this chapter hopes to make a novel contribution to the growing field of climate justice, particularly by emphasizing the unique dialectics between dispossession and resistance in the context of wind energy extraction in rural India. Beyond the case study examined in this chapter, detailed studies on resistance (re)actions, acts of insurgency, and counterhegemonic movements to climate injustices have the exciting potential to show the way towards future energy transitions that are truly just and fair in nature.

Notes

1 India's climate change policy regime has been mostly dominated by techno-economic considerations around imperatives of growth, development, and security. These imperatives have been justifying the development of large-scale, centralized, and grid-connected renewable infrastructures through national missions on solar, wind, and biomass aimed at increasing the country's installed capacity while also attracting lower energy tariffs (Dubash 2011; Mohan and Topp 2018; Behuria 2020).
2 The term 'green' refers to the political construction of renewables as environmentally sustainable; quotation marks are used in the whole chapter to dissociate with that construction.
3 Quotation marks are used here as a reminder of the political construction underlying that land classification.
4 This chapter was written in 2021. I acknowledge that since then a growing body of literature has emerged on the resistance and justice dimensions of energy transition in India. I am highly indebted to the contributions of Sareen and Shokrgozar (2022) and Stock (2022).
5 Interview, 24.3.21.
6 Interview, 13.2.21.
7 Fictional name.
8 Discussion, 27.1.21.
9 Discussion, 25.4.21.
10 Fictional name.
11 Discussion, 31.3.21.
12 Discussion, 25.10.21.
13 Interview, 26.1.21.
14 He refers here to particular social activists belonging to a Dalit rights organization which was implicated in the movement as the Dalit community is numerically dominant in the village.
15 Villager of Samio, interview, 11.3.21.

References

Agrawal, Arun. 2005. 'Environmentality: Community, Intimate Government, and the Making of Environmental Subjects in Kumaon, India'. *Current Anthropology* 46 (2): 161–90. https://doi.org/10.1086/427122.
Avila, Sofia. 2018. 'Environmental Justice and the Expanding Geography of Wind Power Conflicts'. *Sustainability Science* 13 (3): 599–616. https://doi.org/10.1007/s11625-018-0547-4.
Baber, Zaheer. 2019. 'Climate Change and the Yellow Vest Movement'. *Economic and Political Weekly* 54 (34): 7–8.
Bainton, Nicholas, Deanna Kemp, Eleonore Lèbre, John R. Owen, and Greg Marston. 2021. 'The Energy-Extractives Nexus and the Just Transition'. *Sustainable Development* 29 (4): 624–34. https://doi.org/10.1002/sd.2163.
Baka, Jennifer. 2013. 'The Political Construction of Wasteland: Governmentality, Land Acquisition and Social Inequality in South India'. *Development and Change* 44 (2): 409–28. https://doi.org/10.1111/dech.12018.
Baka, Jennifer. 2017. 'Making Space for Energy: Wasteland Development, Enclosures, and Energy Dispossessions'. *Antipode* 49 (4): 977–96. https://doi.org/10.1111/anti.12219.
Bayat, Asef. 2000. 'From "Dangerous Classes" to "Quiet Rebels": Politics of the Urban Subaltern in the Global South'. *International Sociology* 15 (3): 533-557. https://doi.org/10.1177/026858000015003005.

Behuria, Pritish. 2020. 'The Politics of Late Development in Renewable Energy Sectors: Dependency and Contradictory Tensions in India's National Solar Mission'. *World Development* 126 (February): 104726. https://doi.org/10.1016/j.world dev.2019.104726.

Benecke, Gudrun. 2010. 'Stakeholder Networks in Carbon Governance: The Role of State–Market Relations in the Indian Renewable Energy Sector'. Governance of Clean Development Working Paper 007. School of International Development, University of East Anglia.

Bridge, Gavin. 2009. 'The Holeworld: Scales and Spaces of Energy Extraction'. *New Geographies 2: Landscape of Energy*, 43–51.

Bridge, Gavin. 2017. 'Resource Extraction'. In *International Encyclopedia of Geography*, 1–13. American Cancer Society. https://doi.org/10.1002/9781118786352.wbieg1047.

Brockington, Daniel, and James Igoe. 2006. 'Eviction for Conservation: A Global Overview'. *Conservation and Society* 4 (3): 424.

Butz, David. 2002. 'Resistance, Representation and Third Space in Shimshal Village, Northern Pakistan'. *ACME: An International Journal for Critical Geographies* 1 (1): 15–34.

Cons, Jason, and Michael Eilenberg. 2019. 'Introduction: On the New Politics of Margins in Asia'. In *Frontier Assemblages*, 1–18. John Wiley & Sons. https://doi.org/10.1002/978111 9412090.ch0.

Curran, Giorel. 2012. 'Contested Energy Futures: Shaping Renewable Energy Narratives in Australia'. *Global Environmental Change* 22 (1): 236–44. https://doi.org/10.1016/j. gloenvcha.2011.11.009.

Dubash, Navroz K. 2011. 'From Norm Taker to Norm Maker? Indian Energy Governance in Global Context'. *Global Policy* 2 (s1): 66–79. https://doi.org/10.1111/j.1758-5899.201 1.00123.x.

Dubash, Navroz K., Sunila S. Kale, and Ranjit Bharvirkar. 2018. *Mapping Power: The Political Economy of Electricity in India's States*. Oxford University Press. https://doi. org/10.1093/oso/9780199487820.001.0001.

Dunlap, Alexander. 2017. ' "The Town Is Surrounded": From Climate Concerns to Life under Wind Turbines in La Ventosa, Mexico'. *Human Geography* 10 (2): 16–36.

Dunlap, Alexander. 2018a. 'Counterinsurgency for Wind Energy: The Bíi Hioxo Wind Park in Juchitán, Mexico'. *Journal of Peasant Studies* 45 (3): 630–52. https://doi. org/10.1080/03066150.2016.1259221.

Dunlap, Alexander. 2018b. 'The "Solution" Is Now the "Problem": Wind Energy, Colonisation and the "Genocide-Ecocide Nexus" in the Isthmus of Tehuantepec, Oaxaca'. *International Journal of Human Rights* 22 (4): 550–73. https://doi. org/10.1080/13642987.2017.1397633.

Dunlap, Alexander. 2019. 'Wind, Coal, and Copper: The Politics of Land Grabbing, Counterinsurgency, and the Social Engineering of Extraction'. *Globalizations*, 1–22. https://doi.org/10.1080/14747731.2019.1682789.

Easterling, Keller. 2016. *Extrastatecraft: The Power of Infrastructure Space*. Verso.

Foucault, Michel. 1982. 'The Subject and Power'. *Critical Inquiry* 8 (4): 777–95.

Gardner, Katy, and Eva Gerharz. 2016. 'Introduction. Land, "Development" and "Security" in Bangladesh and India'. *South Asia Multidisciplinary Academic Journal* 13. https://doi.org/10.4000/samaj.4141.

Ghosh, Devleena. 2016. ' "We Don't Want to Eat Coal": Development and Its Discontents in a Chhattisgarh District in India'. *Energy Policy* 99 (December): 252–60. https://doi. org/10.1016/j.enpol.2016.05.046.

Gidwani, Vinay. 2008. *Capital, Interrupted: Agrarian Development and the Politics of Work in India*. University of Minnesota Press. http://ebookcentral.proquest.com/lib/kbdk/det ail.action?docID=334224.

Gidwani, Vinay, and Rajyashree N. Reddy. 2011. 'The Afterlives of "Waste": Notes from India for a Minor History of Capitalist Surplus'. *Antipode* 43 (5): 1625–58. https://doi. org/10.1111/j.1467-8330.2011.00902.x.

Gujarat Ecology Commission. 2017. *State of Environment Report: Land Resources*. Government of Gujarat.

Harjanne, Atte, and Janne M. Korhonen. 2019. 'Abandoning the Concept of Renewable Energy'. *Energy Policy* 127 (April): 330–40. https://doi.org/10.1016/j.enpol.2018.12.029.

Harms, Erik. 2014. 'Knowing into Oblivion: Clearing Wastelands and Imagining Emptiness in Vietnamese New Urban Zones'. *Singapore Journal of Tropical Geography* 35 (3): 312–27. https://doi.org/10.1111/sjtg.12075.

Harvey, David. 2004. 'The "New Imperialism": Accumulation by Dispossession'. *Actuel Marx* 35 (1): 71–90. https://doi.org/10.3917/amx.035.0071.

Ibrahim, Farhana. 2008. *Settlers, Saints and Sovereigns: An Ethnography of State Formation in Western India*. Routledge India.

Jaffrelot, Christophe. 2018. 'Le Capitalisme de Connivence En Inde Sour Narendra Modi'. *Les Études Du CERI* 237: 1–47.

Jaffrelot, Christophe. 2019. 'Business-Friendly Gujarat under Narendra Modi: The Implications of a New Political Economy'. In *Business and Politics in India*. Oxford University Press: 211–233. https://doi.org/10.1093/oso/9780190912468.003.0008.

Joseph, Kelli L. 2010. 'The Politics of Power: Electricity Reform in India'. *Energy Policy* 38 (1): 503–11. https://doi.org/10.1016/j.enpol.2009.09.041.

Kabra, Asmita. 2009. 'Conservation-Induced Displacement: A Comparative Study of Two Indian Protected Areas'. *Conservation and Society* 7 (4): 249. https://doi. org/10.4103/0972-4923.65172.

Kale, Sunila S. 2014. 'Structures of Power: Electrification in Colonial India'. *Comparative Studies of South Asia, Africa and the Middle East* 34 (3): 454–75. https://doi. org/10.1215/1089201X-2826037.

Kohli, Kanchi, and Manju Menon. 2016. 'The Tactics of Persuasion: Environmental Negotiations over a Corporate Coal Project in Coastal India'. *Energy Policy* 99 (C): 270–76.

Lakhanpal, Shikha. 2019. 'Contesting Renewable Energy in the Global South: A Case-Study of Local Opposition to a Wind Power Project in the Western Ghats of India'. *Environmental Development* (January). https://doi.org/10.1016/j.envdev.2019.02.002.

Levien, Michael. 2015. 'From Primitive Accumulation to Regimes of Dispossession'. *Economic and Political Weekly* 50 (22): 146–57.

Levien, Michael. 2018. *Dispossession without Development: Land Grabs in Neoliberal India*. Oxford University Press.

Li, Tania Murray. 2007. *The Will to Improve: Governmentality, Development, and the Practice of Politics*. Duke University Press. http://ebookcentral.proquest.com/lib/uea/detail.act ion?docID=1170540.

Lilja, Mona, Mikael Baaz, Michael Schulz, and Stellan Vinthagen. 2017. 'How Resistance Encourages Resistance: Theorizing the Nexus between Power, "Organised Resistance" and "Everyday Resistance"'. *Journal of Political Power* 10 (1): 40–54. https:// doi.org/10.1080/2158379X.2017.1286084.

Lilja, Mona, and Stellan Vinthagen. 2014. 'Sovereign Power, Disciplinary Power and Biopower: Resisting What Power with What Resistance?' *Journal of Political Power* 7 (1): 107–26. https://doi.org/10.1080/2158379X.2014.889403.

Marx, Karl, and Ernest Mandel. 1990. *Capital: Critique of Political Economy*, vol. 1. Translated by Ben Fowkes. Penguin Classics.

McCarthy, James. 2015. 'A Socioecological Fix to Capitalist Crisis and Climate Change? The Possibilities and Limits of Renewable Energy'. *Environment and Planning A: Economy and Space* 47 (12): 2485–502. https://doi.org/10.1177/0308518X15602491.

McCarthy, James, and Jim Thatcher. 2019. 'Visualizing New Political Ecologies: A Critical Data Studies Analysis of the World Bank's Renewable Energy Resource Mapping Initiative'. *Geoforum* 102 (June): 242–54. https://doi.org/10.1016/j.geoforum.2017.03.025.

McCauley, Darren, and Raphael Heffron. 2018. 'Just Transition: Integrating Climate, Energy and Environmental Justice'. *Energy Policy* 119 (C): 1–7.

McElwee, Pamela D., K. Sivaramakrishnan, and K. Sivaramakrishnan. 2016. *Forests Are Gold: Trees, People, and Environmental Rule in Vietnam*. University of Washington Press. http://ebookcentral.proquest.com/lib/uea/detail.action?docID=4649050.

Mehta, Lyla, Mihir Bhatt, and Pankaj Joshi. 2020. 'How Pastoralists in Kutch Respond to Social and Environmental Uncertainty'. May. STEPS Center.

Mehta, Lyla, and Shilpi Srivastava. 2017. 'The Social Life of Mangroves: Resource Complexes and Contestations on the Industrial Coastline of Kutch, India'. STEPS Working Paper. STEPS Center.

Mehta, Lyla, and Shilpi Srivastava. 2019. 'Pastoralists without Pasture: Water Scarcity, Marketisation and Resource Enclosures in Kutch, India'. *Nomadic Peoples* 23 (2): 195.

Mishra, Saswat Kishore, and Pulak Mishra. 2017. 'Do Adverse Ecological Consequences Cause Resistance against Land Acquisition? The Experience of Mining Regions in Odisha, India'. *Extractive Industries and Society* 4 (1): 140–50. https://doi.org/10.1016/j.exis.2016.11.004.

Mitchell, Timothy. 2013. *Carbon Democracy: Political Power in the Age of Oil*. Verso. http://ebookcentral.proquest.com/lib/uea/detail.action?docID=5176984.

Mohan, Aniruddh, and Kilian Topp. 2018. 'India's Energy Future: Contested Narratives of Change'. *Energy Research & Social Science* 44 (October): 75–82. https://doi.org/10.1016/j.erss.2018.04.040.

Newell, Peter. 2018. 'Trasformismo or Transformation? The Global Political Economy of Energy Transitions'. *Review of International Political Economy*, 1–24. https://doi.org/10.1080/09692290.2018.1511448.

Newell, Peter, and Dustin Mulvaney. 2013. 'The Political Economy of the "Just Transition"'. *Geographical Journal* 179 (2): 132–40. https://doi.org/10.1111/geoj.12008.

Nilsen, Alf Gunvald. 2010. *Dispossession and Resistance in India: The River and the Rage*. Routledge.

Odysseos, Louiza. 2011. 'Governing Dissent in the Central Kalahari Game Reserve: "Development", Governmentality, and Subjectification amongst Botswana's Bushmen'. *Globalizations* 8 (4): 439–55. https://doi.org/10.1080/14747731.2011.585845.

Oskarsson, P., K. Lahiri-Dutt, and P. Wennström. 2019. 'From Incremental Dispossession to a Cumulative Land Grab: Understanding Territorial Transformation in India's North Karanpura Coalfield'. *Development and Change*. https://doi.org/10.1111/dech.12513.

Phillips, Jon, Peter Newell, and Pallav Purohit. 2011. 'Governing Clean Energy in India'. Governance of Clean Development Working Paper 017. School of International Development, University of East Anglia.

Rasmussen, Mattias Borg, and Christian Lund. 2018. 'Reconfiguring Frontier Spaces: The Territorialization of Resource Control'. *World Development* 101 (C): 388–99.

Rignall, Karen Eugenie. 2016. 'Solar Power, State Power, and the Politics of Energy Transition in Pre-Saharan Morocco'. *Environment and Planning A: Economy and Space* 48 (3): 540–57. https://doi.org/10.1177/0308518X15619176.

Robbins, Paul. 2016. 'Fixed Categories in a Portable Landscape: The Causes and Consequences of Land-Cover Categorization': *Environment and Planning A* (December). https://doi.org/10.1068/a3379.

Sack, Robert. 2009. *Human Territoriality: Its Theory and History*. Cambridge University Press.

Sanchez-Lopez, Maria Daniela. 2019. 'From a White Desert to the Largest World Deposit of Lithium: Symbolic Meanings and Materialities of the Uyuni Salt Flat in Bolivia'. *Antipode* 51 (4): 1318–39. https://doi.org/10.1111/anti.12539.

Sareen, Siddharth and Shayan Shokrgozar. 2022. 'Desert geographies: solar energy governance for just transitions'. *Globalizations* 1–17.

Scheidel, Arnim, Leah Temper, Federico Demaria, and Joan Martínez-Alier. 2018. 'Ecological Distribution Conflicts as Forces for Sustainability: An Overview and Conceptual Framework'. *Sustainability Science* 13 (3): 585–98. https://doi.org/10.1007/s11625-017-0519-0.

Schwenkenbecher, Anne. 2017. 'What Is Wrong with Nimbys? Renewable Energy, Landscape Impacts and Incommensurable Values'. *Environmental Values* 26 (6): 711–32. https://doi.org/10.3197/096327117X15046905490353.

Scott, Dayna, and Adrian Smith. 2017. '"Sacrifice Zones" in the Green Energy Economy: Toward an Environmental Justice Framework'. *McGill Law Journal / Revue de Droit de McGill* 62 (3): 861–98. https://doi.org/10.7202/1042776ar.

Scott, James C. 1989. 'Everyday Forms of Resistance'. *Copenhagen Journal of Asian Studies* 4 (May): 33–33. https://doi.org/10.22439/cjas.v4i1.1765.

Scott, James C. 1992. *Domination and the Arts of Resistance: Hidden Transcripts*. Yale University Press.

Shidore, Sarang, and Joshua W. Busby. 2019. 'What Explains India's Embrace of Solar? State-Led Energy Transition in a Developmental Polity'. *Energy Policy* 129 (June): 1179–89.

Sikor, Thomas, and Christian Lund. 2009. 'Access and Property: A Question of Power and Authority'. *Development and Change* 40 (1): 1–22. https://doi.org/10.1111/%28ISSN%291467-7660/issues.

Stanley, Oswin D. 2004. 'Wetland Ecosystem and Coastal Habitat Diversity in Gujarat, India'. *Journal of Coastal Development* 7 (2): 49–64.

Stock, Ryan, and Trevor Birkenholtz. 2019. 'The Sun and the Scythe: Energy Dispossessions and the Agrarian Question of Labor in Solar Parks'. *Journal of Peasant Studies*, 1–24. https://doi.org/10.1080/03066150.2019.1683002.

Stock, Ryan. 2022. 'Triggering Resistance: Contesting the Injustices of Solar Park Development in India'. *Energy Research & Social Science* 86: 102464.

Sud, Nikita. 2012. *Liberalization, Hindu Nationalism and the State: A Biography of Gujarat*. Oxford University Press. www.oxfordscholarship.com/view/10.1093/acprof:oso/9780198076933.001.0001/acprof-9780198076933.

Sud, Nikita. 2014. 'The State in the Era of India's Sub-National Regions: Liberalization and Land in Gujarat'. *Geoforum* 51 (January): 233–42. https://doi.org/10.1016/j.geoforum.2013.06.002.

Sud, Nikita. 2020. 'The Actual Gujarat Model: Authoritarianism, Capitalism, Hindu Nationalism and Populism in the Time of Modi'. *Journal of Contemporary Asia*, 1–25. https://doi.org/10.1080/00472336.2020.1846205.

Survival International. 2009. 'The Most Inconvenient Truth of All: Climate Change and Indigenous People'. https://assets.survivalinternational.org/documents/132/survival_climate_change_report_english.pdf.

Swyngedouw, Erik. 2011. 'Whose Environment? The End of Nature, Climate Change and the Process of Post-Politicization'. *Ambiente & Sociedade* 14 (2): 69–87. https://doi.org/10.1590/S1414-753X2011000200006.

Temper, Leah, Sofia Avila, Daniela Del Bene, Jennifer Gobby, Nicolas Kosoy, Philippe Le Billon, Joan Martínez-Alier, et al. 2020. 'Movements Shaping Climate Futures: A Systematic Mapping of Protests against Fossil Fuel and Low-Carbon Energy Projects'. *Environmental Research Letters* 15 (12): 123004. https://doi.org/10.1088/1748-9326/abc197.

Temper, Leah, Mariana Walter, Iokiñe Rodriguez, Ashish Kothari, and Ethemcan Turhan. 2018. 'A Perspective on Radical Transformations to Sustainability: Resistances, Movements and Alternatives'. *Sustainability Science* 13 (3): 747–64. https://doi.org/10.1007/s11625-018-0543-8.

Turton, Andrew. 1986. 'Patrolling the Middle-ground: Methodological Perspectives on "Everyday Peasant Resistance"'. *Journal of Peasant Studies* 13 (2): 36–48. https://doi.org/10.1080/03066158608438290.

United Nations Framework Convention on Climate Change. 2015. *Adoption of the Paris Agreement*. 21st Conference of the Parties.

Yenneti, Komali, Rosie Day, and Oleg Golubchikov. 2016. 'Spatial Justice and the Land Politics of Renewables: Dispossessing Vulnerable Communities through Solar Energy Mega-Projects'. *Geoforum* 76 (November): 90–99.

9

THE MARGINALITY OF THE PLAINLAND INDIGENOUS COMMUNITIES IN CLIMATE CHANGE PLANS AND FINANCE IN BANGLADESH

Siddiqur Rahman and A. K. M. Mamunur Rashid

1 Introduction

One of the key topics in the scholarly literature on the politics of climate change is burden-sharing principles for costs associated with reducing greenhouse gas (GHG) emissions, adaptation practices, and damages resulting from climate-induced disasters (Maltais 2016; Singer 2002; Gosseries 2004; Caney 2005) – more specifically, the principles according to which burdens are shared between states and, over time, between generations. Parks and Timmons (2010) have noted that climate negotiations have failed for their narrow understanding of the concepts of burden-sharing principles. They argue that the broader structural inequality in the economic relations between the Majority and Minority Worlds hinder the possibilities of bringing justice in climate negotiations and stress the need to address the issues such as trade, investment, debt, and intellectual property. Harlan et al. (2015) emphasize that climate injustice is caused by structural inequality and argue for social science perspectives, local community engagement, cultural ecosystems, and integrated adaptation policies to address the issues of climate injustice. In a similar fashion, we argue that structural inequality that exists within Bangladesh has amplified suffering from climate-induced disasters in the most marginalized communities, including Indigenous communities living in plainland northern Bangladesh. In terms of numbers, Bengalese/Bangladeshis are the ethnic majority population in Bangladesh who mostly live in the country's plainlands. Minority ethnic communities, also known as Indigenous communities, live mostly in the hilly eastern areas and, to a lesser extent, in the plainlands. The latter are generally referred to as 'plainland Indigenous people'.

This chapter aims to explore the extent to which global and national climate finance tackles the negative impacts of climate change through adaptation practices

DOI: 10.4324/9781003214021-10

that are being justly distributed at the national level. In recent years, national and international attention has been focused on the impacts of sea level rise and other climate change issues on the coastal and delta areas in southern Bangladesh. In contrast, there has been little attention on the climate hotspots in the northern parts of the country, and in particular on the specific challenges faced by the plainland Indigenous people. This chapter will first shed light on the specific climate change vulnerabilities faced by this group. Section 2 will highlight the potential, abilities, and existing experiences of Indigenous communities and their traditional adaptation strategies. Finally, the chapter will argue how some Indigenous communities in Bangladesh remain invisible in and left behind by climate change policy, plans, and finance. Meanwhile, their existing adaptation practices should be recognized and be better supported by international and national climate finance.

2 Climate Vulnerability among Indigenous People in Northern Bangladesh

Bangladesh has been repeatedly listed as one of the countries that are most vulnerable to climate change impacts around the globe (Germanwatch 2021; IPCC 2014). The country's susceptibility and vulnerability in this context are shaped by its geographic and climatic characteristics, such as variations in air and ocean temperatures, changes in precipitation patterns, and intensification of extreme weather phenomena such as cyclones, drought, and sea level rise, all exacerbated by the socioeconomic situation of the population living in poverty along with income and social inequalities (GoB 2012).

According to a World Bank (2019) study, the proportion of Bangladesh's population living on less than US$1.90 per day is estimated to have fallen from 99% to 9% between 1972 and 1998. Despite the country's impressive achievements in economic development and poverty alleviation (World Bank 2019), there are still significant spatial differences in the pace and quality of development (Raihan 2018; Rahman 2022). The income disparity holds for different population groups living in Bangladesh, including the Indigenous people living in the hilly areas of eastern Bangladesh (Chakma, Marma, Tripura, Murong, Bom, Khumi, Pankhoya, Tanchanga, Khumi) and in the plainlands (Santal, Orao, Pahan, Mahato, Turi, Malo, Mali, Mahali, Pahari, Munda, Kol, Rakhain, Garo, Hazong, Khashi, Monipuri) (BBS 2015). Indigenous people are among the most marginalized groups in the country due to social and cultural exclusion, insecure land tenure and landlessness (in some cases), and economic discrimination, which exacerbate their vulnerability to climate change (Hasan 2015). The Indigenous people in hilly areas live within a particular territory, while those residing in the plainlands do not have a specific region; they are primarily landless and share a typical village with Bangladeshi people. In this chapter, we are going to focus on the state of climate justice as it relates to the plainland Indigenous people living in the highly drought-prone Barind areas of the country.

The Barind region, in north-western Bangladesh, is characterized by very high temperatures and low annual rainfall. The average yearly temperature and precipitation are between 8°C and 44°C and 1500–2000 mm, respectively, with 80% of the rain falling during the monsoon season (June to September), resulting in extreme water scarcity in the area for most of the year (GoB 2018b; Hasan 2015). Out of 11 districts in the Barind region, five have higher exposure to climate-induced droughts than other districts (BBS 2015), and most plainland Indigenous people also live in these areas. The agricultural sector is also the leading provider of employment for the 15.33 million people residing in the Barind area (65% for Bengali people and 97% for the Indigenous people), of which many are either subsistence farmers or work as day labourers on larger farms (Reza et al. 2011; Paul 1998).

The impacts of intensified drought on Indigenous people in the Barind region include reduced freshwater availability, stress on crop agriculture, and heat- and water-induced health issues. The impacts of droughts on the agricultural sector and associated livelihoods are significant and pose the biggest threat to the agro-based economy and food security in the Barind area. Communities experience direct crop and income loss damage, which leads to increased vulnerability (Shahid & Hazarika 2010). Apart from the loss of agriculture, drought also causes abnormal increases in prices, reduces the availability of jobs for day labourers (on which poor individuals, including many Indigenous communities, depend), and poses a threat to the food security of local communities (GoB 2016; Reza et al. 2011; Paul 1998). Many croplands are converted into low labour-intensive horticulture as an adaptation strategy. However, this adaptation strategy pursued primarily by wealthy landowners negatively affects the livelihoods of plainland Indigenous people who are primarily dependent on day labourer jobs.

Plainland Indigenous populations are disproportionately vulnerable to climate change impacts as they are socially, economically, and politically discriminated to a higher extent than both Indigenous people living in the hills in the south and the dominant Bangladeshi people. They generally lack knowledge about climate change and its impacts and have limited access to adaptation technologies. Academic studies (Garai et.al, 2022; Reuters 2019; Gunter et al. 2008) have identified a range of vulnerabilities experienced by plainland Indigenous people and how they are largely absent from key policy and investment planning.

Due to lower social status, their access to education, health services, social security, infrastructure, the justice system, and the labour market is fundamentally different from mainstream society. Their right to choose where to live, where to work, who and how to worship, and who to marry is often challenged by the dominant ideologies and local elites. They have a distinctive cultural trait (communal mode of production, customary laws, hunting, crop cultivation method, undistilled wine production and consumption, intramarriage system), which is very different from the Bangladeshi majority. They are often reluctant to express their cultural norms both in personal and public life. Most of the plainland Indigenous communities live in rural areas and are landless or own only tiny pieces of land. A feasibility study by the HEKS/EPER Foundation (2018, 93) captured a micronarrative of an Indigenous woman named Sonali Pathan in Naogaon: 'We want to try new crops or other alternative ways, but that requires land which we do not own. The landowners will suggest [to] us what they think is profitable for them, we have to convince them that the new alternatives will be profitable for them. But we ourselves are uncertain about whether it will be profitable or not.'

Plainland Indigenous people depend mainly on their income as day labourers in the agricultural sector. If employed, they are being paid up to 50% less than dominant Bangladeshi Muslim day labourers (IRI 2020), and they have little bargaining power to demand equal wages. They face significant problems with no access to land, ongoing land grabbing, and related disenfranchisement.

Whatever little land plainland Indigenous people might own, it is threatened to be taken from them by fraud, bribery, and force, often further enhanced by historical and religious stereotypes (IRI 2020). Meanwhile, land conversion from agriculture to horticulture by large-scale farmers reduces the need for day labourers in the target regions. As the majority of the plainland Indigenous people were dependent on this seasonal income, they are suffering the most from this climate-induced land usage conversion which left many of them jobless or with significantly reduced income opportunities (i.e. working days per year). Out of sheer desperation, it has become standard practice for some plainland Indigenous people to 'sell' their labour days before the season (also known as 'bonded labour') for as low as 50% of the wages they would get during the season (IRI 2020; HEKS/EPER 2018). As the livelihoods and associated skills of many ethnic minority communities are solely based on agriculture and livestock raising, they often have a lower adaptive capacity than the general society and take up diversified (non-agriculture-based) employment opportunities without additional learning opportunities.

Plainland Indigenous populations face constrained access to the public health system due to cultural stigmatization and financial barriers. Despite the healthcare system being publicly financed, healthcare recipients are expected to contribute to their treatment. Indigenous community members often have difficulties making even marginal contributions, which excludes them from otherwise available health care. Furthermore, if they enter public health facilities, they are often treated disrespectfully or do not receive the same quality service as other patients due to their cultural and social status. This has led to negative attitudes of Indigenous communities towards public health services, as a result of which they often turn to natural healers (HEKS-EPER 2018).

Moreover, the national curriculum is in the Bengali language, and there is no textbook in Indigenous languages. Though the national curriculum has some lessons on climate change, such as at the pre-primary, primary, and secondary levels, the curriculum on Bangla, Bangladesh, and global studies includes some environmental and climate change issues (Shohel et al. 2021). However, these do not address climate vulnerability specific to Indigenous people, nor is this knowledge accessible in their own language. Moreover, children of plainland Indigenous people face language barriers and cultural stigmatization in the public education system. In all public schools, the teaching language is Bangla, which Indigenous populations are not fluent in; therefore, Indigenous children are frequently deprived of formal learning at the entry level, which negatively impacts their motivation to learn general subjects as well as about climate change issues. Educating the children of Indigenous population is not a priority for stakeholders (parents, teachers, supervisors) and education authorities. The lack of cultural sensitivity training/orientation of staff at the decision-making level on the needs and priorities of plainland Indigenous children works to increase their vulnerability.

3 Climate Justice for Plainland Indigenous People of Bangladesh

We will present the situation of climate injustice of plainland Indigenous people from three perspectives: (1) the extent to which national climate change policies, plans, and budgets address the disproportionate vulnerabilities of plainland Indigenous people, (2) the extent to which different projects or regional adaptation investment recognize Indigenous group-specific priorities and participation in their investment/projects, and finally (3) how climate justice for plainland Indigenous people should be based on shared but differentiated responsibilities of Bangladeshi and plainland Indigenous people.

4 Climate Change-Related Policies, Plans, and Budgets

Plainland Indigenous people constitute 1.5% of the total population of Bangladesh (IMLI 2018). As mentioned earlier, they live an extremely marginal social, economic, and cultural life. In the Barind region, around 1.5 million plainland Indigenous people are highly exposed to drought. Another million live in other parts of the country and are exposed to cyclones, saltwater intrusion, floods, and flash floods (BBS 2015). The major climate change-related policies, plans, and budgets are blind to the vulnerability of plainland Indigenous people caused by drought, floods, cyclones, sea level rise, and flash floods. The National Climate Change Budget (2017–18, 2018–19, 2019–20, 2020–21, 2021–22) lacked a fair allocation to plainland Indigenous people, making them invisible within major climate change policies. The other reason is lack of socioeconomic vulnerability assessment and of people-centric approaches in adaptation plans and strategies that are inclusive of marginalized communities. Section 5 illustrates the key climate policy, plans, and budgets of the Government of Bangladesh and assesses the extent of their recognition of the vulnerability of the plainland Indigenous people and of affirmative action necessary to address their climate vulnerabilities.

An assessment of the policy, plans, and strategies on climate change in Bangladesh shows that the Indigenous people and their vulnerabilities to climate change did not receive proper attention by policymakers. This section highlights how, if at all, key policy documents and climate budgets over the years have recognized the climate change vulnerabilities of the plainland Indigenous people. For example, the Nationally Determined Contributions (NDC) document lays out adaptation and mitigation strategies to increase the country's climate resilience. This important government policy document does not recognize Indigenous people as a special vulnerable category; rather, there was an implicit mention of climate-vulnerable communities/groups where Indigenous people were considered. Like the NDC, the Bangladesh Climate Change Strategy and Action Plan (BCCSAP), the country's key climate change plan, provides an overall framework for climate action, recognizing the need for adaptation and highlighting Bangladesh's willingness to follow a low-carbon development pathway. The climate budget is also allocated based on the six priority investment areas. A review of the Plan shows that there is an implicit mention of Indigenous people as a climate-vulnerable group in selected climate hotspots. Moreover, in 2010, the Government of Bangladesh decided to formulate a long-term Bangladesh Delta Plan 2100 (BDP-2100) to integrate the short- to medium-term aspirations of Bangladesh to achieve upper-middle-income (UMIC) status and eliminate extreme poverty by 2031. Like the previous two documents, there is also no clear reference to plainland Indigenous people. The national Climate Change and Gender Action Plan (CCGAP) aims to ensure that both adaptation and mitigation actions benefit men and women equitably without creating gender gaps. Despite being formulated through a participatory process, however, this document also fails to recognize the plainland Indigenous people as a particularly vulnerable group, though it does implicitly mention plainland Indigenous people as vulnerable (MoEF 2013).

The Climate Fiscal Framework 2020 is an important government document which provides principles and tools for fiscal climate policy-making, helping to identify the demand and supply sides of climate funds and ensure that such policies are transparent and sustainable in the longer term. Although the Fiscal Framework mentions Indigenous people among climate-vulnerable groups, the document lacks the justice lens in distributing climate finance within Bangladesh. The Climate Change Country Investment Plan (2016–21) has been prepared by the Ministry of Environment and Forests through a process of review, joint effort, and consultation that included multiple ministries, agencies, experts, NGOs, and civil society organizations as well as various stakeholders at the divisional, district, and community levels. Despite wider participation, this investment plan also did not recognize plainland Indigenous people, and no affirmative actions for them were considered.

The Country Programme of Bangladesh to the Green Climate Fund (GCF) provides a synopsis of the observed and projected climate change impacts in Bangladesh, the climate change governance structure, and an overview of the main actors in the country's climate change response activities. In contrast to the previous policies and plans, this document includes a recognition of Indigenous people as a group vulnerable to climate change impacts and a concrete investment project for plainland Indigenous people. Thus, with the exception of the GCF country programme document, there is a significant lack of recognition of climate injustice experienced by plainland Indigenous people in national policies, plans, and budgets.

5   Participation of Plainland Indigenous People in Climate Finance

The funding flow to address climate vulnerabilities depends on the level of influence Indigenous people have on the budget-setting entities and also on the degree of sensitivity of public policymakers to their adaptation needs. In Section 4, we discussed how Bangladeshi climate policies are largely oblivious to the adaptation needs of Indigenous people, as a result of which the funding flow towards this goal is limited. The participation of Indigenous people both in policymaking and budget-setting is largely absent, which undermines their development as well as adaptive capacity to combat increasing droughts. Key development and climate change adaptation barriers of plainland Indigenous people are recognized by several assessments as detailed below, but Indigenous people did not have any space at the policy table, nor were considered an important actor in the policymaking process.

Key development barriers include constrained access of plainland Indigenous youths to job markets and non-traditional career paths due to poverty and cultural stigmatization, constrained access to public and private services due to social discrimination, weak community leadership and representation of interests of ethnic communities in public governance structures, and access to finance (formal banking finance and microfinance) and insurance (Gunter et al. 2008; HEKS/EPER 2018).

Key climate change barriers include (1) limited reliable data on long-term climate variabilities and associated risks based on localized climate change impacts, (2) lack of access to and awareness about climate change impact scenarios, (3) cultural constraints to interpret risk information, (4) lack of technical and financial capacities to identify, prioritize, and take up climate-resilient livelihoods, (5) lack of awareness about climate change-induced health impacts, (6) insufficient technical skills and knowledge, (7) lack of coordination among key institutions to identify and support the uptake of climate-resilient livelihoods in a culturally sensitive manner, and (8) lack of capacity of ethnic minorities to protect cultural heritage in light of intensifying climate change impacts (Alam et al. 2013; Bhuyan et al. 2018; Dastagir 2015; Etzold et al. 2014; Gunter et al. 2008; Habiba & Shaw 2014).

In order to address development and climate change barriers, Indigenous people require support from national and international stakeholders. For instance, addressing climate change barriers requires reliable national and global finance. However, the current climate change allocation for the Indigenous people is the lowest among all vulnerable social groups in Bangladesh (see Table 9.1).

Major development projects have similarly failed to address the needs of Indigenous people. A very limited number of climate change projects (UNDP-HEKS partnership projects for adaptation, Shiree project of NETZ, Action Aid, Bangladeshi government projects on drought resilience, UNEP projects on drought resilience) address the disproportionate vulnerabilities of plainland Indigenous people in the Barind region. Most of these projects are led by INGOs. While it appears that the plainland Indigenous people were not consulted during the climate-related policy formulations, a few international NGO projects have employed more consultative approaches. The international non-profit organization Action Aid in 2007 piloted an adaptation project (Adaptation and Resilience to Drought in Barind Region) for plainland Indigenous people, where they engaged the population in participatory risk and vulnerability assessment and put their own priority needs to support through the programme. Similarly, for the

TABLE 9.1 Relative share of climate finance for Indigenous people in Bangladesh, National Budget Report, 2017–22

Budget year	Total budget for climate change	% of national budget	% of GDP	% of budget responsive to the needs of Indigenous people	% of budget responsive to the needs of plainland Indigenous people	% of plainland Indigenous people within the broader Indigenous population
2017–18	146336m (US$1.72b)	8.56	0.73	0.02	0.003	15.0
2018–19	189487.6m (US$2.229b)	8.82	0.75	0.024	0.009	37.5
2019–20	237485m (US$2.794b)	7.8	0.80	0.04	0.01	25.0
2020–21	242257m (US$2.85b)	7.5	0.80	0.035	0.015	42.8
2021–22	251249.8m (US$2.955b)	7.26	0.64	0.035	0.02	57.0

Source: Government of Bangladesh.

last 15 years, the HEKS/EPER Foundation has been working deliberately with plainland Indigenous people following a participatory project management cycle (HEKS/EPER 2018). However, these projects provide capacity-building support such as awareness raising, social mobilization, skill development, education and health information, etc. The total budgets of such NGO projects are very small (less than US$3m for five years) in comparison to the government investment or large-scale (more than US$15m for five years) donor-funded projects (HEKS/EPER 2018). The significance of NGO investment is very limited in comparison to investment by the Government of Bangladesh and international financial institutions.

Initiatives by UNDP, UNEP, Action Aid, NETZ, CARE, HEKS/EPER Foundation, and many local NGOs attempt to enhance the resilience of communities and the agricultural sector to droughts, including re-excavation of ponds, improvement of irrigation systems, rainwater harvesting, crop diversification, and afforestation (HEKS/EPER 2018). A stakeholder assessment conducted by UNDP in 2017 showed that over 200 initiatives were carried out by development partners in Chapainawabganj, Rajshahi, Naogaon, and Dinajpur (HEKS/EPER 2018). There are many initiatives providing access to credit programmes for poverty reduction, integrated rural development, and food security. There are 40 active microcredit organizations in the area (HEKS/EPER 2018). However, not all interventions focus on climate change and only a few of them explicitly target plainland Indigenous people. The absence of plainland Indigenous people from avenues of decision-making in climate change-related projects and policies invisibilizes and marginalizes them, aggravating climate injustice.

6 Shared but Differentiated Responsibilities and the Role of the State

The average annual per capita GHG emissions generated by plainland Indigenous people is much lower than the Bangladeshi average. Most plainland Indigenous people live in their traditional houses owned by others. Only 1.8% have agricultural lands and 98.2% are landless (IMLI 2018). The contribution of Bangladesh to global GHG emissions is 0.6% (WRI 2020), and if we count the total contribution of the agriculture sector in the country's GHG profile, the figure is 27.35%. This means 0.16% of the global GHG contribution is from the agriculture sector of Bangladesh (GoB 2021b; WRI 2020). The average per capita emission of Bangladeshi people is 1.02 tCO_2e (GoB 2021b; BBS 2022), whereas the average per capita emission of a plainland Indigenous person in the Barind region is 0.28 tCO_2e (GoB 2021b; WRI 2020; GoB 2018b). Despite Indigenous people living a nature-based life with minimal carbon emission, the toll of loss and damage and chronic vulnerabilities due to climate change among them is much higher. One of the reasons why the average per capita emission of a Bangladeshi citizen

(1.02 tCO₂e) as a whole is lower than many other countries in the Majority World such as Vietnam (2.20), Indonesia (2.03), Argentina (4.61), Azerbaijan (3.45), and Fiji (1.95) (Worldometer 2022) is the fact that most Indigenous people, as many poor rural Bangladeshis, live an almost net-zero-emission life. Yet, global climate finance does not recognize these differentiated responsibilities of different types of people living in a country. The investments are channelled to large infrastructure development and to vulnerable Bengali people, particularly in coastal and delta areas, rather than plainland Indigenous people in the Barind region. Therefore, middle- and high-income Bengali people's higher carbon-intensive life is not discussed in climate talks, nor are the stark differences in terms of climate vulnerabilities that exist within the country. This internal injustice is often ignored in national discussions, while national climate negotiators focus on the global scale where Bangladesh as a whole is disproportionately impacted by the effects of climate change. It is essential to acknowledge and address intracountry climate injustices and how these injustices relate to the national as well as global climate finance.

If we examine the development and well-being situation of Indigenous people in comparison to the economic growth of Bangladesh, it is rather negative. Plainland Indigenous people are gradually becoming more marginal due to loss of land entitlement, cultural marginalization, and relatively lower education. Less than 10% of eligible enrolled Indigenous children in primary school actually successfully completed the secondary school education certificate, in comparison to 60% of the majority Bangladeshi people. There are relatively more jobless youth (90%) among the plainland Indigenous people than the national average (45%) (UNDP et al. 2019). The development deficit (Gunter et al. 2008) and climate deficit (Alam et al. 2013; Bhuyan et al. 2018; Dastagir 2015; Etzold et al. 2014; Habiba & Shaw 2014) of the Indigenous people in the Barind region is a triple injustice. They have the lowest per capita emission levels, highest climate vulnerability, and they receive the lowest amounts of climate finance (see Table 9.1). Therefore, the principle of differentiated responsibilities is grossly ignored in the case of plainland Indigenous people in drought-prone areas. Table 9.1 shows the proportions of the climate budget allocated to Indigenous people as a whole and plainland Indigenous people specifically.

The percentage of the total national budget for climate change allocated to Indigenous people in Bangladesh was 0.02% (GoB, 2017), 0.024% (GoB, 2018), 0.04% (GoB, 2019), 0.035% (GoB, 2020) and 0.035% (GoB, 2021), respectively. In the same budget years, the share allocated to plainland Indigenous people was 0.003%, 0.009%, 0.01%, 0.015% and 0.02%, respectively. This budget allocation shows that Indigenous communities in general get less finance compared to other regions and communities. Comparing the climate finance among the Indigenous people, specifically, plainland Indigenous people get less than the Indigenous people living in the hilly areas in the south-east of Bangladesh. Meanwhile, data show that plainland Indigenous people's per capita GHG emissions are far lower than that of an average Bangladeshi citizen (GoB 2021b; WRI 2020; GoB 2018b).

This injustice is further exacerbated by a poor approach to adaptation in the Barind region. The lack of water is encouraging large landholders to convert cropland into horticulture, which requires less water and does not need intensive labour, which was the major source of income for landless plainland Indigenous people. While horticulture is technically a successful adaptation, it is benefitting only 18% of the people, while the remaining 82% are not getting any benefits from this land use change (HEKS 2018). Thus, this is arguably a case of maladaptation due to lack of human rights and climate justice perspectives in adaptation planning at the regional level which fails to consider the culture and ethnicity of the local population in adaptation planning and financing.

7 Conclusion

Bangladesh is perceived as one of the most vulnerable countries to climate change, while it contributes very little to global GHG emissions. However, the issue of climate injustice also exists *within* Bangladesh, when considering those most affected (i.e. plainland Indigenous people) and those with highest GHG emission rates (mainstream Bangladeshi). It is evident that within Bangladesh, plainland Indigenous people in the Barind region are particularly vulnerable due to their lower capacity to adapt to environmental change (drought), while their relative contribution to national GHG emissions is much lower than that by mainstream Bangladeshi people. The participation of the plainland Indigenous people in climate policymaking is nearly absent, a result of which is their exclusion from climate plans and policy. Plainland Indigenous people's needs are therefore not heard, understood, or reflected on in any climate policy process and documents.

This chapter provided an account of climate vulnerability with a focus on plainland Indigenous communities and argued they are more vulnerable to climate change impacts as a result of other forms of marginalization in Bangladesh – social, economic, and political. Secondly, the chapter shed light on the negative impacts of climate change on their lives and livelihoods. Thirdly, it discussed intracountry climate injustice, and it is not adequately reflected in the allocation of climate finance. Climate finance data in Table 9.1 show that Indigenous people in general are allocated a lower portion of the climate budget, but what is received by the plainland Indigenous people, in particular, is extremely low, and certainly lower than the allocations to other Indigenous communities in the hilly areas in the south-east. This chronically low investment in plainland Indigenous communities by the government and donors further inhibits their ability to adapt to climate change. Finally, the chapter argued that to address climate change vulnerabilities, national climate adaptation plans, policies, strategies, and finance need to be aware of the needs of plainland Indigenous communities, who arguably deserve the most assistance given their low contribution to global increases in temperature and their heightened vulnerability. Through the lens of climate justice, it is equally important that a just distribution of climate finance for adaptation is ensured simultaneously at inter- and intracountry levels.

References

Alam, J. A., Rahman, S. M., & Saadat, A. H. M. 2013. Monitoring meteorological and agricultural drought dynamics in Barind region Bangladesh using standard precipitation index and Markov chain model. *International Journal of Geomatics and Geosciences*, 3(3), 511–517.

Bangladesh Bureau of Statistics (BBS). 2015. *Population & Housing Census 2011 Volume I, II, III, IV*. Ministry of Planning.

Bangladesh Bureau of Statistics (BBS). 2022. *Population & Housing Census 2021*, Ministry of Planning.

Bhuyan, M. D. I., Islam, M. M., & Bhuiyan, M. E. K. 2018. A trend analysis of temperature and rainfall to predict climate change for northwestern region of Bangladesh. *American Journal of Climate Change*, 7(2), 115–134.

Caney, S. 2005. Cosmopolitan justice, responsibility, and global climate change. *Leiden Journal of International Law*, 18, 747–775.

Dastagir, M. R. (2015). Modeling recent climate change induced extreme events in Bangladesh: A review. *Weather and Climate Extremes*, 7, 49–60.

Etzold, B., Ahmed, A. U., Hassan, S. R., & Neelormi, S. 2014. Clouds gather in the sky, but no rain falls. Vulnerability to rainfall variability and food insecurity in northern Bangladesh and its effects on migration. *Climate and Development*, 6(1), 18–27.Re

Garai, Joydeb, Hok Bun Ku, Yang Zhan. 2022. Climate change and cultural responses of indigenous people: A case from Bangladesh. *Current Research in Environmental Sustainability*, 4, Elsevier. https://doi.org/10.1016/j.crsust.2022.100130 [Accessed 17 April 2023].

Germanwatch. 2021. *Global Climate Risk Index*. Germanwatch.

Gosseries, A. 2004. Historical emissions and free-riding. *Ethical Perspectives*, 11, 36–60.

Gunter, B. G., Rahman, A., and Rahman, A. F. M. 2008. How Vulnerable are Bangladesh's Indigenous People to Climate Change? Bangladesh Development Research Working Paper Series (BDRWPS), BDRWPS 1, April, Bangladesh.

Government of Bangladesh (GoB). 2016. Third national communication of Bangladesh to the United Nations Framework Convention on Climate Change. Ministry of Environment and Forests and Climate Change.

Government of Bangladesh (GoB). 2017. Climate financing for sustainable development: Budget report 2017–18. Finance Division, Ministry of Finance.

Government of Bangladesh (GoB). 2018. Climate financing for sustainable development: Budget report 2018–19. Finance Division, Ministry of Finance.

Government of Bangladesh (GoB). 2019. Climate financing for sustainable development: Budget report 2019–20. Finance Division, Ministry of Finance.

Government of Bangladesh (GoB). 2020. Climate financing for sustainable development: Budget report 2020–21. Finance Division, Ministry of Finance.

Government of Bangladesh (GoB). 2021a. Climate financing for sustainable development: Budget report 2021–22. Finance Division, Ministry of Finance.

Government of Bangladesh (GoB). 2021b. Nationally Determined Contributions (NDCs) 2021. Ministry of Environment, Forest and Climate Change.

Gunter, B. G., Rahman, A., & Rahman, A. F. M. 2008. How vulnerable are Bangladesh's Indigenous people to climate change? Bangladesh Development Research Working Paper.

Habiba, U. and Shaw, R. 2014. Drought Scenario in Bangladesh, *Water Insecurity: A Social Dilemma (Community, Environment and Disaster Risk Management, Vol. 13)*, Emerald Group Publishing Limited, Bingley, pp. 213–245. https://doi.org/10.1108/S2040-7262(2013)0000013016

Harlan, Sharon L. et al. Climate justice and inequality. In Riley E. Dunlap, and Robert J. Brulle (eds), *Climate Change and Society: Sociological Perspectives* (New York, 2015; online edn, Oxford Academic, 20 August 2015), https://doi.org/10.1093/acprof:oso/9780199356102.003.0005, accessed 17 Apr. 2023.Bottom of Form.

Hassan, S. 2015. Drought vulnerability assessment in the high Barind tract of Bangladesh using MODIS NDVO and land surface temperature imageries. *International Journal of Science and Research*, 4(2), 55–60.

HEKS-EPER. 2018. *Climate Induced Drought and Its Impact on Plainland Ethnic Minorities in Northern Bangladesh*. HEKS-EPER.

International Mother Language Institute (IMLI). 2018. *Indigenous People Survey – Plainland Indigenous People*. Government of Bangladesh.

International Republican Institute (IRI). 2020. *The Challenges Facing Plainland Ethnic Groups in Bangladesh: Land, Dignity and Inclusion*. IRI.

IPCC. 2014. *Climate Change 2014: Synthesis Report. Contribution of Working Groups I, II and III to the Fifth Assessment Report of the Intergovernmental Panel on Climate Change*. IPCC.

Maltais, A. 2016. A climate of disorder: What to do about the obstacles to effective climate politics. In *Climate Justice in a Non-Ideal World* (ed. C. Heyward and D. Roser, pp. 43–60). Oxford University Press.

MoEF. 2013. *Bangladesh Climate Change and Gender Action Plan*. Ministry of Environment of Forest, Government of the People's Republic of Bangladesh.

Parks, B. C. & Roberts, J. T. 2010. Climate change, social theory and justice. *Theory, Culture & Society*, 27 (2–3). http://doi.org/10.1177/0263276409359018.

Paul, B. K. 1998. Coping mechanisms practiced by drought victims (1994/95) in North Bengal, Bangladesh. *Applied Geography*, 18, 355–373.

Raihan, S. 2018. An anatomy of 'jobless growth' in Bangladesh. *Daily Star*, 12 May. www.thedailystar.net/opinion/economics/anatomy-jobless-growth-bangladesh-1572829.

Rahman, S. 2022. Career strategies of public university students. In *Millennial Generation in Bangladesh: Their Life Strategies, Movement, and Identity Politics* (ed. Kazuyo Minamide), pp- 74-91, University Press.

Reuters. 2019. Indigenous and women's rights can boost climate fight - U.N. Thomson Reuters foundation, Barcelona.

Reza, S., Mazumder, Q. H., and Ahmed, M. 2011. Groundwater balance study in the High Barind, Bangladesh. *Journal of Sciences*, 39, 11–26.

Shahid, S. and Hazarika, M. K. 2010. Groundwater drought in the northwestern districts of Bangladesh. *Water Resource Management*. http://doi.org/10.1007/s11269-009-9534-y

Shohel, M., Roy, G., Alam, S., and Chowdhury, T. 2021. Climate change adaptation and sustainability in the Bangladeshi school curriculum. In *Handbook of Research on Environmental Education Strategies for Addressing Climate Change and Sustainability* (pp. 261–285). IGI Global. http://doi.org/10.4018/978-1-7998-7512-3.ch013.

Singer, P. 2002. *One World*. Yale University Press.

UNDP-HEKS-EDM-GCF. 2019. *Pre-feasibility Study. Strengthening Drought-Resilience of Ethnic Minority Communities in North-western Bangladesh*. UNDP-HEKS-EDM-GCF.

World Bank. 2019. The Bangladesh development update April 2019: Towards regulatory predictability. https://openknowledge. worldbank.org/handle/10986/31504.

World Resource Institute (WRI). 2020. GHG emissions. climatewatchdata.org/ghg-emissions.

Worldometer. 2022. CO_2 emissions per capita. www.wprldometers.info/co2-emissions/co2-emissions-per-capita.

10

ETHICAL DIMENSIONS OF CLIMATE AND ENVIRONMENTAL ISSUES IN PAKISTANI MEDIA

Shafiq Ahmad Kamboh, Muhammad Ittefaq, Sadia Jamil, and Bushra Hameedur Rahman

Climate justice and climate ethics debates are quite interrelated. According to Juliet (2021), the use of the term climate justice has helped frame global warming as an ethical and political issue instead of viewing it as purely environmental or physical in nature. Caney (2021) argues that while facing the disproportionate costs of climate change between Majority and Minority World countries, climate justice debates do seem relevant, but many would argue for the relevance of other kinds of moral consideration, particularly the climate ethics discussions.

Climate change is the biggest threat facing humankind in the twenty-first century. Considering the significance of this emergency, much of the existing scholarship focuses on exploring the multifaceted dimensions of climate change, the consequences of such changes for human beings and infrastructure, and eventually the potential solutions (Graham, 2015). Although such understandings of the crisis are important, many studies highlight the need to incorporate a 'new science' of climate change (O'Brien et al., 2010, p. 19). While explicating this more holistic approach to the problem, Graham (2015) stated that the new science emphasizes equally recognizing climate change as an ethical and a scientific issue, and it further calls for establishing strong collaboration between natural and social scientists to find more concrete solutions.

To meet this need, the United Nations Educational, Scientific and Cultural Organization (UNESCO) has devised and adopted a Declaration of Ethical Principles in Relation to Climate Change (DEPCC) (UNESCO, 2017). This comprehensive document of 18 Articles assigns responsibilities to various stakeholders to address the challenge while setting ethics as the focal point of the climate change discussion. For instance, in a bid to highlight the significance of enhancing public awareness through various communication channels, Article 12 aptly assigns media, among others,

DOI: 10.4324/9781003214021-11

certain responsibilities such as to 'promote awareness regarding climate change and the best practices for responding to it, through strengthening social dialogue, and communication by the media, scientific communities, and civil society organizations, including religious and cultural communities' (UNESCO, 2017, p. 132). This study considers DEPCC's articles in the context of climate and environmental ethics issues and assesses whether contemporary Pakistani journalism is complying or violating these principles. However, despite having this clearly defined normative role, there is a dearth of research examining how and to what extent contemporary journalism has been reminding nation-states and other relevant actors of their ethical responsibilities to formulate and implement policies and actions regarding climate change mitigation, adaptation, and environmental degradation.

Recent research suggests that Pakistan is the world's fifth most affected country by climate change (Eckstein et al., 2018). Therefore, any human activity that becomes a reason for emitting GHGs, increasing pollution or ecological imbalance, and eventually contributing to climate change or environmental degradation was considered a climate or environmental ethics issue by this chapter.

From a media advocacy perspective, we argue that editorial journalism carries a strong potential to publicize cases of poor compliance to climate and environmental ethics to policymakers. To have a meaningful impact on public policy-making processes, such issues – usually underrepresented in mainstream media – need either adequate news coverage or to be highlighted as part of a special journalistic genre (Kamboh & Yousaf, 2020). In this regard, due to its distinctive format, a newspaper editorial has a strong influence on readers, intra- and inter-media news agendas, and on political or policy agendas (Firmstone, 2019, p. 2). However, a survey of print media treatment of climate change issues reveals that newspaper editorials have barely been studied for their role in advocating climate and environmental issues, particularly in the Majority World (Das, 2019, p. 65; Kamboh et al., 2022).

To address these gaps in existing literature, this chapter analysed the editorial coverage of local and regional climate and environmental issues in the mainstream Pakistani Urdu and English language newspapers. We contribute to climate journalism research in two ways: firstly, by measuring the extent of editorial coverage given to climate and environmental ethics issues versus all other types of issues; secondly, by evaluating the editorial coverage priorities of both the local and regional climate and environmental ethics issues. To meet these objectives, two research questions were posed. First, was there a difference between editorial coverage of climate and environmental ethics issues and other issues among Pakistani newspapers? And second, what were the advocacy priorities of selected newspapers in the context of local and regional climate and environmental injustice issues?

1 Media Advocacy via Editorial Journalism

There are a number of journalistic forms that are opinion-based and make no claim of objectivity. Among them are columns, editorials, and editorial cartoons.

While defining editorials as an essential and regular feature of contemporary newspaper journalism and highlighting some of their main objectives, Firmstone (2019) contends that editorial journalism is exclusive to newspapers and includes articles, often called leaders. Unlike columns and articles that include the name of their authors, editorials are anonymously written. Editorials represent the collective opinion or the public voice of a newspaper. Editorials help newspapers support and oppose individuals, speak in the best interest of their readers, speak to their readers, and speak to politicians, policymakers, and other organizations.

For the sake of distinguishing editorials from other forms of opinion-based writings, Pimentel et al. (2021, p. 3) assert that 'editorials have a noble place in the opinion section, which differentiates them from other genres, such as columns'. A few authors consider editorials as the writings by a group of people (namely, the editorial board) as opposed to columns or letter to editors that are written by individual authors. On a given day, the number of published editorials in a newspaper may differ from country to country. While revealing different editorial writing practices across the globe, Firmstone (2019) notes that editorials are usually written by veteran journalists known as editorial writers. In the United Kingdom, newspapers publish three editorial articles daily, usually of varying length, with the first article being the longest and thus considered the most prominent.

Pakistan being a former British colony, the country's editorial journalism appears to be strongly influenced by the British editorial writing model, which includes daily three editorials of varying lengths with up to 500–600 words and the first article, usually the longest, indicating prominence (Kamboh et al., 2022). Universally, and in Pakistan, the structure of a typical editorial resembles the one described by Pimentel et al. (2021, para. 27):

> In their first part, editorials briefly describe a particular event or issue so that it can be contextualized (and widely known events are more quickly presented than others). In the second part, the newspaper develops its opinion on the subject to evaluate what is good or bad, wrong or right, especially concerning the actions of the institutions or agents involved. In this same section, the principles, positions, and ideologies shaping the journalistic organization's opinion are put into action. The third and last part brings the conclusions, embracing the explanations that form the basis of the editorial opinion and communicating a recommendation or a summon to take action.

Previous research assigns a powerful role to editorial journalism in terms of having multidimensional effects. Firmstone (2019) mentions that newspaper editorials have a strong influence on readers, the internal news agendas and coverage of newspapers, the external news agendas and coverage in other news media, and political or policy agendas. However, despite carrying such a strong impact, newspaper editorials have historically been ignored by mass media researchers. To this perspective, Elyazale (2014, p. 22) adds:

> Editorials seem to be neglected in media discourse research compared to the abundant work conducted on other newspaper texts, especially the news. One of the important reasons for the necessity to consider editorials in research is based on their familiarity. According to Van Dijk (1996), they 'are probably the widest circulating forms of opinion discourse'.

Ansary and Babaii (2005, p. 271), while further confirming the gap, state that 'print journalism opinion discourse has been and still is considered by many a neglected genre, especially if it is compared with the abundant existing work on other newspaper text types'. Moreover, studies reported that newspaper editorials have scarcely been examined for advocating environmental and climate change issues across the globe (Das, 2019, p. 65).

Editorial journalism has a close connection with advocacy journalism. In a bid to highlight some commonalities between the two, Firmstone (2019) notes that both editorial and advocacy writeups are subjective in nature and are intended to influence social or political change. Though editorial journalism is considered a specific form of advocacy journalism, it is rarely theorized or empirically researched as such. Though some studies underline a few differences between editorial and advocacy journalism as well, this chapter considers editorial writers as advocate-journalists because the term advocacy journalism is not as clearly defined in the scholarship as other media studies concepts such as gatekeeping or agenda-setting. Moreover, Pakistani editorial writers have historically been known for taking the role of advocate-journalists right from the period of the Indian Independence movement from British rule (Kamboh et al., 2022).

Based on the above discussion, we argue that climate and environmental ethics issues can be best contended through newspaper editorials. These issues are usually underrepresented in the mainstream media of the Majority World (Das, 2019; Kamboh & Yousaf, 2020, Kamboh et al., 2022) and need some special journalistic treatment (e.g. editorials) to have better influence on public policy debates. Taking due account of this significance, it would be worthwhile to explore the extent to which Pakistani advocate-journalists gave editorial coverage to environmental and climate ethics issues versus all other types of issues.

2 Regional and Local Climate and Environmental Ethics Issues in Pakistan

Proponents of climate justice contend that the impacts of climate change are not borne equally or fairly between rich and poor nations. Countries in the richer Minority World are responsible for emitting more greenhouse gases (GHGs), while poorer Majority World nations – particularly in places with low emissions – are the first to suffer and get the worst hit (Almeida, 2019). However, it has been observed that the affected societies in the Majority World not only are the victims of excessive GHG emissions from the Minority World, but also have to bear the brunt of several types of climate and environmental ethics violations from a number of surrounding countries in their respective regions. While adding to this perspective, Fuhr (2021) contends that, like Minority World countries, the Majority World's emissions are highly concentrated. For instance, contrary to 120 other Majority World countries that emit only 22%, India and China alone account for some 60%, and the top ten countries account for some 78% of all Majority World emissions (Fuhr 2021).

In this chapter, we argue that as citizens of a Majority World country, the people of Pakistan are facing six types of climate and environmental ethics violations from a number of neighbouring countries in the South Asia and wider region:

1. Owing to its commitments at the 2015 United Nations Climate Change Conference (COP21) in Paris to help reduce global GHG emissions, China abandoned coal-based energy projects in the country. However, later on, its companies exported and reinstalled the dirtiest of the dirty energy plants in Pakistan (Watts, 2019).
2. Despite being a signatory of the Paris Agreement, Qatar – the world's largest exporter of liquefied natural gas (LNG) – has signed a multi-billion-dollar LNG deal with Pakistan in 2016 ('Qatar Emerges', 2019).
3. In spite of the scientific revelations that the livestock sector is a major contributor to global GHG emissions (Grossi et al., 2019), China has heavily invested in the agri-livestock sector in Pakistan to meet its domestic meat demands ('Pakistan For', 2021).
4. Owing to the burning of crop stubble by farmers; the burning of coal, petrol, diesel, gas, and biomass by industry; and emissions from coal power plants and vehicles (Basu, 2019), India is reported to be responsible for spreading transboundary smog to the neighbourhood countries, including Pakistan (Abas et al., 2019; David & Ravishankara, 2019). Concerning this case, it is important to note that Article 2C of DEPCC highlights to 'seek and promote transnational cooperation before deploying new technologies that may have negative transnational impacts' (UNESCO, 2017, p. 130). The poor compliance with this particular Article is evident from the fact that the letter sent by the chief minister of Pakistan's Punjab province to his Indian counterpart in November 2017 for the creation of a regional cooperation arrangement to tackle the issue of smog and other forms of environmental pollution did not receive a response from India (Shi, 2018) or a reminder from Pakistan.
5. Being included in the 'red list' of endangered species by International Union for the Conservation of Nature (IUCN), the hunting of Houbara Bustards is prohibited in Pakistan. However, every year the country's foreign office grants special hunting licences to please Royal Gulf hunters from the Middle East region as the cornerstone of its foreign policy (Khan, 2016).

6. Lastly, on account of loopholes in illegal wildlife trade laws and their lax implementation in China, poaching and trafficking of pangolin from Africa and Asia, including Pakistan, have eventually made it the world's most smuggled and critically endangered mammal (Ingram et al., 2019; Mahmood et al., 2019).

It is, therefore, important to expand upon the conventional debates of the nascent concepts of climate justice and climate ethics in terms of assigning greater obligations to the Minority World countries for reducing GHGs and paying heed to said poor compliance with climate ethics by Majority World countries as well. Moreover, we further contend that, like climate ethics scholarship, environmental ethics debates also need to be expanded. First, besides discussing environmental ethics issues between 'privileged' and 'disadvantaged' groups within a country (Hein & Dünckmann, 2020), this scholarship should also include cases of breach of fundamental environmental obligations across different countries and regions. For instance, questions should be raised about the spread of Indian, Chinese, and Indonesian smog to the neighbouring countries (David & Ravishankara, 2019; Huang, 2017; Yeung, 2019) and eventually seizing everybody's equal right to a good environment. Similarly, there is a dire need to voice concerns of Majority World citizens about hunting expeditions and their negative impact on wildlife conservation efforts in poor African and Asian countries ('Eighty Birds', 2017; Malik, 2014). The brutal murder of a Pakistani wildlife activist, Nazim Jokhio, and its widespread media coverage acted as a catalyst to highlight this issue in 2022 (Jaferii, 2022). Second, environmental justice and ethics debates need to be expanded and consolidated beyond environmental pollution and should include all other factors that cause damage to a country's ecological balance, including wildlife conservation and biodiversity loss issues.

Besides that, on the local front, Majority World countries, including Pakistan, are the victims of climate and environmental ethics violations by a number of actors from within. A few examples are listed below.

1. Due to its increased reliance on fossil fuels to fulfil domestic energy needs, annual GHG emissions in Pakistan have rapidly increased over the past few decades (Tiseo, 2020).
2. Despite having the second highest share of agriculture and livestock sector (44.8%) in the national GHG emissions (Mir et al., 2017), authorities are committed to give top priority to the dairy industry in the country's national development plan ('Government Committed', 2019).
3. As opposed to scientists' warnings that 'humans alter the climate by emitting greenhouse gases' (Cohen, 2010, p. 163), local religious leaders and Islamist militant groups are creating hurdles in population control policy-making efforts in Pakistan ('Pakistan Birth', 2017). It is important to mention here that DEPCC's Article 7(d) asks decision-makers to 'build effective mechanisms

to strengthen the interface between science and policy to ensure a strong knowledge base in decision-making' (United Nations Educational, Scientific and Cultural Organization, 2017, p. 131).

4. Despite experts' advice to mitigate GHG emissions through forestry activities (Law et al., 2018), timber and land mafias have terrifyingly reduced Pakistan's forest cover to below 2% (Faiza et al., 2017).

5. A recent UN report revealed that air pollution caused the world's highest number of deaths in Pakistan during the year 2015 while identifying transport, industry, and agriculture sectors as main contributors to the deteriorating air quality in the country ('FAO Report', 2019). However, the Government of Pakistan is paying no heed to making polices to effectively control agriculture, transport, and industrial particulate matter (PM) emissions.

6. Contrary to the guidelines of DEPCC's Article 4(c), which stipulates that 'it is important for all to take measures to safeguard and protect Earth's terrestrial and marine ecosystems, for present and future generations' (UNESCO, 2017, p. 130), human activities are changing the ocean's chemistry, destroying habitats, and killing marine life in Pakistan. For instance, mangrove forests are considered as one of the most efficient methods of offsetting GHG emissions (Alongi, 2012). However, coastal pollution by industrial and urban wastewater, overfishing, and overharvesting by local communities are greatly contributing to rapid degradation of coastal ecosystems in the country (Mukhtar & Hannan, 2012).

7. In the face of the fact that the plastic industry is the second largest source of industrial GHG emissions (Shield, 2019), Pakistan has the highest percentage of ill-managed plastic in South Asia (Mukheed & Alisha, 2020), attributable to community malpractices and poor waste management (Akmal & Jamil 2021). Regarding this, DEPCC's Article 6(a) underscores the need to 'promote the implementation of the United Nations 2030 Agenda for Sustainable Development and its SDGs, especially by adopting sustainable patterns of consumption, production and waste management' (UNESCO, 2017, p. 131).

8. Though wildlife conservation helps in maintaining a balance in ecosystems, a number of scientific studies reveal that largely due to poaching, smuggling, and illicit trade at the hands of local hunters, many endangered wildlife species are going extinct in the country (Mahmood et al., 2019; Ullah et al., 2020). In the name of infrastructural expansion, local authorities are paying no heed to increasing pollution and biodiversity loss in the country ('Lahore Without', 2016).

For these reasons, there is a strong need to go beyond the classic notion of climate justice that assigns rights and responsibilities only to the nation-states to 'either be protected from the effects of climate change, or to take action to reduce emissions or support adaptation' (Bulkeley et al., 2014, p. 31). Hence, we argue that along with sovereign states, the role of certain other social institutions, including local politics, economics, religion, and various community malpractices, should also be analysed and consequently held accountable if they are found to have been involved in contributing to GHG emissions or negatively impacting people's right to a good environment. For this purpose, the press' watchdog role is vital, and it is worthwhile to examine if advocate-journalists bring the aforementioned cases of poor compliance of climate and environmental ethics to the attention of local authorities. Moreover, it is necessary to analyse the editorial priorities of climate and environmental issues because at times editorial writers give more space to comparatively less important issues, and vice versa.

3 Method

We used the quantitative content analysis method to answer both research questions because it has long been recommended as a useful approach to measure and compare the frequency of media content (Neuman, 2005). To proceed with this method, we followed Wimmer's (2014) defined procedures. In the first place, we selected editorial contents of two Urdu language dailies – *Jang* and *Dunya* – and two English language dailies – *Dawn* and *The Nation* – as sample content. We chose these dailies because they are the most influential and the largest mainstream Urdu and English language newspapers of the country. English dailies carry a much smaller readership than their Urdu counterparts, yet the former possess more influence over the policy agenda (Kamboh & Yousaf, 2020).

We counted 'one entire writeup' from the editorial page of the selected newspapers as a unit of analysis. We made this choice because of the importance of editorials as an advocacy platform to raise voice for marginalised parts of society, their influence on the public policy-making process, and their significance to present the official policy of a newspaper. In Pakistan, editorials are published on a specifically devoted space on the editorial page and are published without the name of the author(s).

We purposively selected two different time periods for published editorials: (1) 1 July 2015 to 30 June 2016, and (2) 19 August 2018 to 18 July 2019. The first was selected because 2015–16 was when climate change effects were evidently noticeable in Pakistan. For instance, the country faced the highest death toll in the world due to air pollution in 2015 ('FAO Report', 2019). The country's largest city Karachi was hit by its worst ever heat wave which killed around 1300 people ('Deadliest', 2015). A new environmental phenomenon, that is, transboundary winter smog across India and Pakistan, started making human life miserable. Pakistan started installing dirty energy plants with support from China and Qatar. The country suffered from a fifth consecutive year of disastrous flooding caused by torrential rainfall and glacial lake outbursts killing dozens of people. In the name of infrastructural development, the big cities started to be turned into concrete jungles by removing trees ('Lahore Without', 2016). The historic COP21 took place in late 2015, and the Paris Agreement was signed the following year. The second time period was selected because on 19 August

2018, chairman of Pakistan Tehreek-e-Insaf (PTI), Imran Khan, took the oath as new prime minister of Pakistan after winning the general elections in July 2018. Throughout Khan's political career and during the election campaign, he reaffirmed his commitment to make a clean and green Pakistan (Khan, 2017). As soon as he formally assumed the office of the country's chief executive, he started launching several climate change adaptation and mitigation programmes in the country, including Ten Billion Tsunami Tree Project, Clean Green Champion Programme, and Clean Energy Initiatives ('60 pc', 2020). Given the fact that all these initiatives and their launching ceremonies received high editorial attention (Weaver & Elliott, 1985), it is important to examine the extent of editorial journalism's support in terms of discussing the merits and demerits of the said initiatives. We then constructed and operationalized the following two major content categories (A and B) and two subcategories (A1 and A2) to place relevant units of analysis into them, which eventually would help us answer both the RQs.

4 Category A: Climate and Environmental Ethics Issues

The present study is based on the DEPCC, which 'sets out a shortlist of the globally agreed ethical principles that should guide decision-making and policy-making at all levels and help mobilize people to address climate change' (UNESCO, 2017, para. 3). All such editorials highlighting any violation of already defined climate and environmental ethics were placed in this category. To further operationalize this abstract and multidisciplinary concept, we divided this category into two subcategories: local (A1) and regional (A2) climate and environmental issues. We defined these subcategories as follows:

4.1 A1: Local Climate and Environmental Ethics

The editorials that criticized any local eco-destructive human activity or threat and eventually assigned rights and responsibilities in relation to climate change and environmental degradation to various stakeholders in Pakistan were placed here. For instance:

1. An editorial reporting the contribution of fossil fuels to national GHG emissions was tagged as a *fossil fuel consumption threat.*
2. The editorials that highlighted the role of increased size of livestock or extravagant use of synthetic fertilizers and pesticides in domestic GHG emissions were identified as an *agri-livestock threat.*
3. The editorials that highlighted the population–climate change nexus in Pakistan were marked as a *population–climate change threat* (Yousaf et al., 2022).
4. The editorials that emphasized the importance of reforestation or afforestation in the country for absorbing CO_2 and other gases from the atmosphere were considered as *deforestation and land use change and forestry (LUCF) threat.*

5. The editorials advocating mitigation of solid waste and crops residue burning by local communities, ill-maintained public/private transport, and industrial emissions, for example, brick kilns without zigzag technology (Abubakar, 2020) which hinder the availability of fresh air to all, were tagged as an *air pollution threat*.
6. The editorials advocating human-triggered abuses to mangrove ecosystems were labelled as *mangrove forests threat*.
7. The editorials highlighting the menace of plastic bags and plastic waste mismanagement in the country were tagged as *plastic pollution threat*.
9. The editorials advocating to rescue rapidly dwindling wildlife species at the hands of local hunters or owing to the loss of habitat were marked as *conservation threats*.
10. The editorials that reported the failure of government authorities to implement eco-friendly policies (regarding the ban on plastic bags, sustainable transportation, deforestation, clean energy sources, to name a few) were tagged as *official inability threats*.

4.2 A2: Regional Climate and Environmental Ethics

This subcategory includes editorials that mention various environmental and climate change threats or ethical violations to Pakistan by neighbouring or nearby countries in the South Asia and Middle East regions. This subcategory was further expanded to include:

1. Any editorial that criticized the installation of coal and nuclear power plants under the China Pakistan Economic Corridor (CPEC) agreement was labelled as *CPEC dirty energy plants threat*.
2. Editorials warning about the environmental risks of LNG-based power production were marked as *Pak-Qatar LNG deal threat*.
3. An editorial highlighting the livestock-related GHG emission risks stemming from joint ventures (JVs) between Pakistan and China was labelled as a *Pak-China livestock JVs threat*.
4. Air pollution from fossil fuels and the burning of crop residue are two major reasons of smog between India and Pakistan (Basu, 2021; FAO Report, 2019). The editorials that suggested minimizing winter smog between India and Pakistan by establishing regional cooperation were tagged as *transboundary smog threat*.
5. An editorial opposing the issuance of any endangered wildlife species hunting permit to a Middle Eastern dignitary was labelled as *Gulf hunting threat*.
6. The editorials highlighting how hunting, trading, and transport of all terrestrial wild animals used for human consumption in China are posing an extinction threat to pangolins and other endangered species in Pakistan were labelled as *Chinese wet markets threat*.

5   Category B: Other Issues

In this main content category, we included all editorials that did not fit any of the above-mentioned categories or subcategories. This operationalization helped us to create a codebook, especially with two subcategories, A1 and A2 – with 15 relevant priority areas (issues) needed to be highlighted by selected newspapers' editorialists to influence public policy-making process around environmental issues in Pakistan. This purposeful instrument then eventually operated as a coding sheet for two trained coders, who later on undertook manifest coding, which involves putting the observable and countable surface data of the sample content into a relevant category (Lawrence, 2005). Following this guideline, each unit of analysis was placed into a relevant category or subcategory by the coders. We trained them to place a unit of analysis that had at least one relevant sentence conforming to the operationalized definition of that category or subcategory. In the end, we counted the coded content from all categories or subcategories. By applying descriptive statistics, the resultant numerical value helped us answer the set RQs.

6   Findings and Discussion

This study was primarily aimed at measuring the extent of editorial coverage given to climate and environmental ethics issues vis-à-vis other issues (RQ1). For this purpose, this study analysed a total of 7973 editorials. To consider the coverage of climate-related issues as a yardstick, we selected two different time periods.

The data in Table 10.1 indicate that only 59 (1.5%) editorials were published to advocate cases of local and regional climate and environmental ethics during the first time period, while 3844 (98.5%) were written on all other types of issues (e.g. politics, foreign affairs, show business, sports, crime, and international issues). Interestingly, during the second time period (Table 10.2), when an

TABLE 10.1 Comparison between category A and B issues for first time period (2015–16) (percentage coverage of both category issues and the average number of editorials per newspaper issue)

Newspaper issues	A: Climate and environmental ethics issues				B: Other issues	Total
	Local	Regional	Percentage	AVG*		
Jang (360)	09	00	9 (0.23)	0.02	1021 (26.2)	1030 (26.4)
Dunya (360)	08	00	8 (0.20)	0.02	711 (18.2)	719 (18.4)
Nation (360)	11	09	20 (0.51)	0.05	1057 (27.1)	1077 (27.6)
Dawn (360)	17	05	22 (0.56)	0.06	1055 (27.0)	1077 (27.6)
Total (1440)	45	14	59 (1.5)	0.04	3844 (98.5)	3903 (100)

* Average no. of editorials per newspaper issue.

TABLE 10.2 Comparison between category A and B issues for second time period (2018–19) (percentage coverage of both category issues and the average number of editorials per newspaper issue)

Newspaper issues	A: Climate and environmental ethics issues				B: Other issues	Total
	Local	Regional	Percentage	AVG*		
Jang	19	01	20 (0.49)	0.06	1033 (25.4)	1053 (17.0)
Dunya	25	01	26 (0.64)	0.07	852 (21.0)	878 (17.2)
Nation	29	05	34 (0.84)	0.10	1034 (25.4)	1068 (17.9)
Dawn	35	06	41 (1.01)	0.11	1030 (25.3)	1071 (17.9)
Total	108	13	121 (2.9)	0.08	3949 (97.1)	4070 (100)

* Average no. of editorials per newspaper issue.

environment-friendly prime minister came into power with a strong green agenda (Khan, 2017), the editorial coverage of climate and environment-related issues nearly doubled – to 121 (2.9%). This increase is attributable to the media agenda-building factor, which is referred to as who sets the agenda for the media (Weaver & Elliott, 1985). Though there are many factors that eventually shape the media agenda such as journalistic routines, news values, organizational culture, media ownership, financial constraints, and ideological values in media, public relations (PR) activities (e.g. Facebook pages, tweets, YouTube videos, press releases, press conferences, and political ads) are considered the most influential ones (Curtin, 1999). Kiousis et al. (2006, p. 267) found that 'public relations impact anywhere from 25% to 80% of news content'. Based on the findings revealed in Table 10.2, it can be safely remarked that PR tactics and the status of a news source are the most influential factors determining a media organization's news and editorial agenda. In other words, the higher a news source' political status or access to PR facilities, the stronger its agenda-building impact.

However, despite having this sharp increase during the second time period, the overall results still reveal that the selected newspapers gave insufficient editorial coverage to climate change and environmental issues. As evidence to support this claim, we present the findings of a recent worldwide study that compares mainstream newspaper coverage of climate change issues. In their enquiry, Barkemeyer et al. (2017) reported that selected Chinese, Nepali, Filipino, Thai, and Pakistani newspapers scarcely published an average number of 1.01, 0.31, 0.20, 0.54, and 0.16 articles per newspaper issue on climate-related issues, respectively. In the case of Pakistan, the findings of our study – with an average number of 0.04 and 0.08 editorials per newspaper issue on climate change issues (Tables 10.1 and 10.2) – reveal that the situation has not changed much despite the global concern on climate-related issues has grown significantly between 2008 and 2019.

While answering the first part of the second research question on local climate ethics issues, the data in Tables 10.3 and 10.4 show that a total of 72 and 179 editorials were published to expose local threats to climate change during the first and second time periods, respectively. The results further reveal that the editorialists' advocacy priorities were highly skewed. To substantiate this claim, we compared the sectoral contribution of national GHG emissions with editorial priorities of local climate and environmental ethics issues. For instance, according to the latest national GHG inventory of Pakistan (Mir et al., 2017), the highest national GHG contributor is the fossil fuel-based energy sector with a 45.9% share that conversely received the sixth (11.1%) and fifth (18.5%) least editorial attention during the first and second time periods, respectively (see Tables 10.3 and 10.4). It was followed by the agriculture and livestock sector that accounted for 44.8% GHG share (Mir et al., 2017), which received the least editorial priority during both of the time periods. In contrast, deforestation and LUCF issues that have just a 2.6% share (Mir et al., 2017) received the highest number of supportive mentions during both of the time periods. Moreover, despite strong scientific evidence regarding the role of rapid population growth in environmental degradation and GHG emissions in the country (Yousaf et al., 2022; Zaman et al., 2011), the population–climate change threat received almost negligible editorialist support. Interestingly, the editorial coverage of air pollution and plastic threats tangibly increased during the second time period from 15.5% to 37% and from 6.6% to 19.4%, respectively, owing to the fact that the PTI government had taken strong initiatives to counter both of these environmental issues (Constable, 2019).

Similarly, with respect to the other part of the second research question, the results reveal that the editorial priorities of regional climate ethics issues were also jumbled and flawed (see Tables 10.5 and 10.6). For instance, during the first time period, an arguably lesser ethical violation, that is, the hunting of houbara bustard, received the highest editorial criticism (64.3%). Contrary to this, an arguably more pressing ethical issue, that is, the establishment of CPEC coal power projects (being the dirtiest of the dirty energy plants) comparatively received the least editorial criticism (28.6%). Moreover, despite the agri-livestock sector being the second highest contributor to national GHG emissions (Mir et al., 2017), climate change risks associated with Pak-China livestock joint ventures were entirely ignored by the selected newspapers' editorial writers during both of the time periods. Similarly, despite having an extinction threat to many endangered wildlife species in the country, the threat from Chinese wet markets was not discussed at all by any of the selected newspaper editorialists. Last but not least, despite experts' warning that natural gas-based energy solutions are as harmful to the environment as coal (Morton, 2019), Qatar LNG projects received almost negligible criticism by the editorialists during both time periods.

TABLE 10.3 Percentage of supportive mentions of local climate ethics issues by selected dailies (2015–16)

Newspaper	Editorials	1	2	3	4	5	6	7	8	9	Total SM*
Jang	09	01 (11.1)**	02 (22.2)	03 (33.3)	08 (88.9)	02 (22.2)	03 (33.3)	01 (11.1)	01 (11.1)	00 (00)	21 (29.2)
Dunya	08	02 (25.0)	01 (12.5)	00 (00)	06 (75.0)	02 (25.0)	02 (25.0)	00 (00)	01 (12.5)	02 (25.0)	16 (22.2)
Nation	11	01 (09.1)	00 (00)	00 (00)	04 (36.4)	01 (09.1)	01 (09.1)	02 (18.2)	03 (27.3)	03 (27.3)	15 (20.8)
Dawn	17	01 (05.9)	00 (00)	00 (00)	07 (41.2)	02 (11.8)	05 (29.4)	00 (00)	01 (05.9)	04 (23.5)	20 (27.8)
Total	45	05 (11.1)	03 (6.6)	03 (6.6)	25 (55.5)	07 (15.5)	11 (24.4)	03 (6.6)	06 (13.3)	09 (20.0)	72 (100)

1, fossil fuel; 2, agri-livestock; 3, population; 4, deforestation; 5, air pollution; 6, mangrove forests; 7, plastic; 8, conservation; 9, official
* Supportive mentions.
** No. of supportive mentions (percentage).

TABLE 10.4 Percentage of supportive mentions of local ethics issues by selected dailies (2018–19)

Newspaper	Editorials	1	2	3	4	5	6	7	8	9	Total SM*
Jang	19	02 (10.5)**	00 (00)	00 (00)	13 (68.4)	05 (26.3)	03 (15.8)	2 (10.5)	04 (21.0)	02 (10.5)	31 (17.3)
Dunya	25	04 (16.0)	00 (00)	00 (00)	14 (56.0)	12 (48.0)	02 (08.0)	6 (24.0)	03 (12.0)	02 (08.0)	43 (24.0)
Nation	29	06 (20.7)	00 (00)	03 (10.3)	12 (41.4)	15 (51.7)	02 (06.9)	04 (13.8)	07 (24.1)	03 (10.3)	52 (29.1)
Dawn	35	08 (22.8)	02 (05.7)	04 (11.4)	04 (11.4)	08 (22.8)	07 (20.0)	09 (25.7)	07 (20.0)	04 (11.4)	53 (29.6)
Total	108	20 (18.5)	02 (1.8)	07 (6.5)	43 (39.8)	40 (37.0)	14 (13.0)	21 (19.4)	21 (19.4)	11 (10.2)	179 (100)

1, fossil fuel; 2, agri-livestock; 3, population; 4, deforestation; 5, air pollution; 6, mangrove forests; 7, plastic; 8, conservation; 9, official
* Supportive mentions.
** No. of supportive mentions (percentage).

TABLE 10.5 Percentage of supportive mentions to regional ethics issues by selected dailies (2015–16)

Daily	Editorials	CPEC dirty energy	Qatar LNG	Pak-China livestock JVs	Transboundary winter smog	Gulf hunters	Chinese wet markets	Total
Jang	00	00 (00)*	00 (00)	00 (00)	00 (00)	00 (00)	00 (00)	00 (00)
Dunya	00	00 (00)	00 (00)	00 (00)	00 (00)	00 (00)	00 (00)	00 (00)
Nation	09	03 (33.3)	01 (11.1)	00 (00)	00 (00)	05 (55.5)	00 (00)	09 (64.3)
Dawn	05	01 (20.0)	00 (00)	00 (00)	00 (00)	04 (80.0)	00 (00)	05 (35.7)
Total	14	04 (28.6)	01 (07.1)	00 (00)	00 (00)	09 (64.3)	00 (00)	14 (100)

* No. of supportive mentions (percentage).

TABLE 10.6 Percentage of supportive mentions to regional ethics issues by selected dailies (2018–19)

Daily	Editorials	CPEC dirty energy	Qatar LNG	Pak-China livestock JVs	Transboundary winter smog	Gulf hunters	Chinese wet markets	Total
Jang	01	00 (00) *	00 (00)	00 (00)	01 (100)	00 (00)	00 (00)	01 (07.7)
Dunya	01	00 (00)	00 (00)	00 (00)	01 (100)	00 (00)	00 (00)	01 (07.7)
Nation	05	00 (00)	00 (00)	00 (00)	04 (80)	01 (20)	00 (00)	05 (38.5)
Dawn	06	04 (66.6)	00 (00)	00 (00)	02 (33.3)	00 (00)	00 (00)	06 (46.1)
Total	13	04 (30.8)	00 (00)	00 (00)	08 (61.5)	01 (07.7)	00 (00)	13 (100)

* No. of supportive mentions (percentage).

10.7 Policy Implications: The Roadmap to Adequate and Effective Climate Ethics Issues Coverage

The evidence from Pakistani editorial writing practices points towards a number of factors, including editorial inattention and incapability, that have contributed to the inadequate coverage and muddled editorial priorities regarding climate and environmental ethics issues. Behind these editorial flaws, both state and mass media are non-adherent to different articles of DEPCC. First, contrary to Article 12, which binds both the state of Pakistan and local mass media to 'promote awareness regarding climate change and the best practices for responding to it, through strengthening social dialogue, and communication by the media' (UNESCO, 2017, p. 132), local advocate-journalists are insufficiently highlighting poor compliance to climate and environmental ethics issues. Similarly, Article 4(4) asks for similar support from media and states while revealing that 'states and other pertinent actors should facilitate and encourage public awareness, and participation in decision-making and actions by making access to information and knowledge on climate change' (UNESCO, 2017, p. 130).

Second, in contrast to Article 7(4C), which prompts both stakeholders to 'promote accurate communication on climate change based on peer-reviewed scientific research, including the broadest promulgation of science in the media and other forms of communication' (UNESCO, 2017, p. 131), there has been no proper communication of the scientific knowledge on how to tackle climate and environmental changes to the media practitioners and eventually to the audience. Due to this gap, editorialists and the journalist community reflect muddled priorities to climate and environmental ethics issues. The above-revealed poor compliance to different DEPCC guidelines suggests that Pakistani advocate-journalists are equally responsible for the present climate change woes of the country and hence – instead of being part of the solution – have become part of the problem.

As a way out, we suggest that newspaper editors and civic advocacy groups (1) explore scientific studies on environmental issues conducted by the local scientific community, (2) organize reporting and editorial staff training workshops on how to use the findings of such studies to effectively influence environmental policy-making process, (3) ensure necessary amendments in relevant journalism course contents for future journalists (Kamboh, 2019), and, above all, (4) persuade editorial writers to provide enough space to climate change and environmental degradation issues.

References

'60 pc of Pakistan's energy will be "clean" by 2030: PM Imran'. *Dawn*, 12 December 2020. www.dawn.com/news/1595373.

Abas, Naeem, Muhammad Shoaib Saleem, Esmat Kalair, and Nasrullah Khan. 2019. 'Cooperative control of regional transboundary air pollutants'. *Environmental Systems Research* 8 (1): 1–14. https://doi.org/10.1186/s40068-019-0138-0.

Abubakar, Muhammad, S. 2020. 'Pakistan embraces zig-zag technology-led brick kilns to fight air pollution'. *Dawn*, 7 May. www.dawn.com/news/1552655.

Akmal, Tanzila, and Faisal Jamil. 2021. 'Testing the role of waste management and environmental quality on health indicators using structural equation modeling in Pakistan'. *International Journal of Environmental Research and Public Health* 18 (8): 4193. https://doi.org/10.3390/ijerph18084193.

Almeida, Paul. 2019. 'Climate justice and sustained transnational mobilization'. *Globalizations* 16 (7): 973–979. https://doi.org/10.1080/14747731.2019.1651518.

Alongi, Daniel M. 2012. 'Carbon sequestration in mangrove forests'. *Carbon Management* 3 (3): 313–322. https://doi.org/10.4155/cmt.12.20.

Ansary, Hasan, and Esmat Babaii. 2005. 'The generic integrity of newspaper editorials: A systemic functional perspective'. *RELC Journal* 36 (3): 271–295. https://doi.org/10.1177/0033688205060051.

Barkemeyer, Ralf, Frank Figge, Andreas Hoepner, Diane Holt, Johannes Marcelus Kraak, and Pei-Shan Yu. 2017. 'Media coverage of climate change: An international comparison'. *Environment and Planning. C, Politics and Space* 35 (6): 1029–1054. https://doi.org/10.1177/0263774X16680818.

Basu, Mausumi. 2019. 'The great smog of Delhi'. *Lung India* 36 (3): 239. https://doi.org/10.4103/lungindia.lungindia_363_18.

Bulkeley, Harriet, Gareth A. S. Edwards, and Sara Fuller. 2014. 'Contesting climate justice in the city: Examining politics and practice in urban climate change experiments'. *Global Environmental Change* 25 (1): 31–40. https://doi.org/10.1016/j.gloenvcha.2014.01.009.

Caney, Simon. 2021. 'Climate justice'. In *The Stanford Encyclopedia of Philosophy*, ed. Edward N. Zalta. Stanford University Press. https://plato.stanford.edu/archives/win2021/entries/justice-climate.

Cohen, Joel E. 2010. 'Population and climate change'. *Proceedings of the American Philosophical Society* 154 (2): 158–182. www.jstor.org/stable/41000096.

Constable, P. 2019. 'Pakistan moves to ban single-use plastic bags: The health of 200 million people is at stake'. *Washington Post*, 13 August. www.washingtonpost.com/world/asia_pacific/pakistan-moves-to-ban-single-useplastic-bags-the-health-of-200-million-people-is-at-stake/2019/08/12/6c7641ca-bc23-11e9-b873-63ace636af08_story.html.

Curtin, Patricia A. 1999. 'Reevaluating public relations information subsidies: Market-driven journalism and agenda-building theory and practice'. *Journal of Public Relations Research* 11 (1): 53–90. https://doi.org/10.1207/s1532754xjprr1101_03.

Das, Jahnnabi. 2019. *Reporting Climate Change in the Global North and South: Journalism in Australia and Bangladesh*. Routledge.

David, Liji M., and A. R. Ravishankara. 2019. 'Boundary layer ozone across the Indian subcontinent: Who influences whom?' *Geophysical Research Letters* 46 (16): 10008–10014. https://doi.org/10.1029/2019GL082416.

'Deadliest, week-long Pakistan heatwave kills 1,233 in Karachi'. *DW*, 28 June 2015. www.dw.com/en/deadliest-week-long-pakistan-heatwave-kills-1233-in-karachi/a-18545483.

Eckstein, D., M. L. Hutfils, and M. Winges. 2018. 'Global climate risk index 2019: Who suffers most from extreme weather events? Weather-related loss events in 2017 and 1998 to 2017'. https://germanwatch.org/en/16046.

'Eighty birds of prey take flight – On jet to Jeddah'. *BBC*, 31 January 2017. www.bbc.com/news/world-middle-east-38809355.

Elyazale, Nabila. 2014. 'Characteristics of newspaper editorials: "Chouftchouf" in "Almassae" Moroccan newspaper as a case study'. *New Media and Mass Communication* 32 (1): 21–43.

Faiza, N., J. Weiguo, Y. Aijun, and S. Wenxing. 2017. 'Giant deforestation leads to drastic eco-environmental devastating effects since 2000: A case study of Pakistan'. *Journal of Animal and Plant Sciences* 27 (4): 1366–1376.

'FAO report analyses the causes of smog in Punjab focusing on agriculture'. *FAO*, 5 February 2019. www.fao.org/pakistan/news/detail-events/en/c/1179183.

Firmstone, J. 2019. 'Editorial journalism and newspapers' editorial opinions'. In *Oxford Research Encyclopedia of Communication*, ed. J. F. Nussbaum. (pp. 1–24). Oxford University Press. https://doi.org/10.1093/acrefore/9780190228613.013.803.

Fuhr, Harald. 2021. 'The rise of the Global South and the rise in carbon emissions'. *Third World Quarterly* 42 (11): 2724–2746. https://doi.org/10.1080/01436597.2021.1954901.

'Government committed to developing livestock sector on modern lines'. *Business Recorder*, 7 February 2019. https://fp.brecorder.com/2019/02/20190207445241.

Graham, Sonia. 2015. 'Climate change, ethics and human security'. *Ethics, Policy & Environment* 18 (1): 112–115. https://doi.org/10.1080/21550085.2015.1016966.

Grossi, Giampiero, Pietro Goglio, Andrea Vitali, and Adrian G. Williams. 2019. 'Livestock and climate change: Impact of livestock on climate and mitigation strategies'. *Animal Frontiers: The Review Magazine of Animal Agriculture* 9 (1): 69–76. https://doi.org/10.1093/af/vfy034.

Hein, Jonas, and Florian Dünckmann. 2020. 'Narratives and practices of environmental justice'. *Journal of the Geographical Society of Berlin* 151 (2–3): 59–66. https://doi.org/10.12854/erde-2020-524.

Huang, E. 2017. 'China's neighbors are getting a whiff of its terrible pollution'. *Quartz*, 8 January. https://qz.com/879718/chinas-neighbors-are-getting-a-whiff-of-its-terrible-pollution.

Ingram, Daniel J., Drew T. Cronin, Daniel W. S. Challender, Dana M. Venditti, and Mary K. Gonder. 2019. 'Characterising trafficking and trade of pangolins in the Gulf of Guinea'. *Global Ecology and Conservation* 17: e00576. https://doi.org/10.1016/j.gecco.2019.e00576.

Jaferii, A. M. 2022. 'Nazim Jokhio murder case: With great power, comes the ability to bend he criminal justice system to your will'. *Dawn*, 18 April. www.dawn.com/news/1685132.

Jorgic, Drazen. 2019. 'Qatar emerges as front-runner for long-term LNG deal for Pakistan'. *Reuters*, 26 April. www.reuters.com/article/pakistan-gas-lng/qatar-emerges-as-front-runner-for-long-term-lng-deal-for-pakistan-idUSL5N2264ZN

Juliet, Ifechukwu. 2021. Climate justice and reproductive related issues/challenges: Why you should care as a feminist. *Global Environment Commons*, 7 July. https://gdc.unicef.org/resource/climate-justice-and-reproductive-related-issueschallenges-why-you-should-care-feminist?fbclid=IwAR151ePNR3KvCEFGmRiok7TXKxW-1QAbWZLPiDZDfMQpBdeEfv6JbXPczag.

Kamboh, Shafiq A. 2019. 'Missing links in practical journalism in developing Pakistan'. In *Journalism and Journalism Education in Developing Countries*, ed. B. Dernbach and B. Illg (pp. 157–168). Manipal Universal Press.

Kamboh, Shafiq A., Muhammad Ittefaq, and Muhammad Yousaf. 2022. 'Editorial journalism and environmental issues in the Majority World'. *International Journal of Communication* 16 (23): 2646–2668.

Kamboh, Shafiq A., and Muhammad Yousaf. 2020. 'Human development and advocacy journalism: Analysis of low editorial coverage in Pakistan'. *Development Policy Review* 38 (5): 646–663. https://doi.org/10.1111/dpr.12443.

Khan, Ilyas M. 2016. 'Houbara bustards: Pakistan hunting ban sparks political row'. *BBC News*, 14 December. www.bbc.com/news/world-asia-38301815.

Khan, Rania S. 2017. 'Pakistan cricket hero aims to turn green tide to political tsunami'. *Reuters*, 2 September. www.reuters.com/article/pakistan politics-environment-idUKL5N1LT3UZ.

Kiousis, S., M. Mitrook, X. Wu, and T. Seltzer. 2006. First- and second-level agenda-building and agenda-setting effects: Exploring the linkages among candidate news releases, media coverage, and public opinion during the 2002 Florida gubernatorial election. *Journal of Public Relations Research* 18 (3): 265–285. https://doi.org/10.1207/s1532754xjprr1803_4.

'Lahore without trees'. *Dawn*, 4 May 2016. www.dawn.com/news/1256107.

Law, Beverly E., Tara W. Hudiburg, Logan T. Berner, Jeffrey J. Kent, Polly C. Buotte, Mark E. Harmon, and Oregon State University. 2018. 'Land use strategies to mitigate climate change in carbon dense temperate forests'. *Proceedings of the National Academy of Sciences* 115 (14): 3663–3668. https://doi.org/10.1073/pnas.1720064115.

Lawrence, Neuman W. 2005. *Social Research Methods: Quantitative and Qualitative Approaches.* Allyn and Bacon.

Mahmood, Tariq, Faraz Akrim, Nausheen Irshad, Riaz Hussain, Hira Fatima, Shaista Andleeb, and Ayesha Aihetasham. 2019. 'Distribution and illegal killing of the endangered Indian pangolin Manis Crassicaudata on the Potohar Plateau, Pakistan'. *Oryx* 53 (1): 159–164. https://doi.org/10.1017/S0030605317000023.

Malik, N. 2014. 'Rich Gulf Arabs using Tanzania as a playground? Someone opened the gate'. *The Guardian*, 17 November. www.theguardian.com/commentisfree/2014/nov/17/rich-gulf-arabs-tanzania-hunters-masai-rights.

Mir, Kaleem Anwar, Pallav Purohit, and Shahbaz Mehmood. 2017. 'Sectoral assessment of greenhouse gas emissions in Pakistan'. *Environmental Science and Pollution Research International* 24 (35): 27345–27355. https://doi.org/10.1007/s11356-017-0354-y.

Morton, A. 2019. 'Booming LNG industry could be as bad for climate as coal, experts warn'. *The Guardian*, 2 July. www.theguardian.com/environment/2019/jul/03/boom ing-lng- industry-could-be-as-bad-for-climate-as-coal-experts-warn.

Mukheed, Muhammad, and Alisha Khan. 2020. 'Plastic pollution in Pakistan: Environmental and health implications'. *Journal of Pollution Effects & Control* 8 (4): 251–258. https://doi.org/10.35248/2375-4397.20.8.251.

Mukhtar, Irum, and Abdul Hannan. 2012. 'Constrains on mangrove forests and conservation projects in Pakistan'. *Journal of Coastal Conservation* 16 (1): 51–62. https://doi.org/10.1007/s11852-011-0168-x.

O'Brien, K., A. L. S. Clair, and B. Kristoffersen, eds. 2010. 'The framing of climate change: Why it matters'. In *Climate Change, Ethics and Human Security.* (pp. 3–22). Cambridge University Press.

'Pakistan birth rate a "disaster in the making" as population passes 207 million'. *Independent*, 10 September 2017. www.independent.co.uk/news/world/asia/pakistan-population-muslim-birth rate-census-disaster-poverty-million-a7938816.html.

'Pakistan for JVs with China in livestock sector, SEZs'. *News International*, 16 May 2021. www.thenews.com.pk/print/835034.

Pimentel, Pablo Silva, Francisco Paulo Jamil Marques, and Deivison Henrique Freitas Santos. 2021. 'The structure, production routines, and political functions of editorials in contemporary journalism'. *Atlantic Journal of Communication.* https://doi.org/10.1080/15456870.2021.1931218.

Shi, Kevin. 2018. 'Pollution knows no borders – So India, Pakistan would do well to fight it together'. *Scroll*, 22 November. https://scroll.in/article/901421/pollution-knows-no-borders-and-india-pakistan-would-do-well-to-fight-it-together.

Shield, Charli. 2019. 'The plastic crisis isn't just ugly – It's fueling global warming'. *DW*, 15 May. www.dw.com/en/a-48730321.

Tiseo, Ian. 2020. 'Carbon dioxide emissions from fossil fuel and industrial purposes in Pakistan from 1970 to 2019'. *Statista*, 12 November. www.statista.com/statistics/486054.

Ullah, Zaib, Inayat Ullah, Ikram Ullah, Sajid Mahmood, and Zafar Iqbal. 2020. 'Poaching of Asiatic black bear: Evidence from Siran and Kaghan valleys, Pakistan'. *Global Ecology and Conservation* 24: e01351. https://doi.org/10.1016/j.gecco.2020.e01351.

UNESCO. n.d. 'Declaration of ethical principles in relation to climate change'. https://en.unesco.org/themes/ethics-science-and-technology/ethical-Principles.

UNESCO. 2017. 'Annex III declaration of ethical principles in relation to climate change'. https://unesdoc.unesco.org/ark:/48223/pf0000260889.page=127.

Watts, J. 2019. 'Belt and Road summit puts spotlight on Chinese coal funding'. *The Guardian*, 25 May. www.theguardian.com/world/2019/apr/25/belt-and-road-summit-puts-spotlight-on-chinese-coal-funding.

Weaver, David, and Swanzy Nimley Elliott. 1985. 'Who sets the agenda for the media? A study of local agenda-building'. *Journalism Quarterly* 62 (1): 87. https://doi.org/10.1177/107769908506200113.

Wimmer, Roger D., and Joseph R. Dominick. 2013. *Mass Media Research: An Introduction*. Cengage Learning.

Yeung, Jessie. 2019. 'Indonesian forests are burning, and Malaysia and Singapore are choking on the fumes'. *CNN*, 11 September. https://edition.cnn.com/2019/09/11/asia/malaysia-singapore-pollution-intl-hnk/index.html.

Yousaf, Usman Saleem, Farhan Ali, Babar Aziz, and Saima Sarwar. 2022. 'What causes environmental degradation in Pakistan? Embossing the role of fossil fuel energy consumption in the view of ecological footprint'. *Environmental Science and Pollution Research International* 29 (22): 33106–33116. https://doi.org/10.1007/s11356-021-17895-4.

Zaman, K., I. Ali Shah, and M. Mushtaq Khan. 2011. 'Exploring the link between poverty-pollution-population (3Ps)'. *Journal of Economics and Sustainable Development* 2 (1112): 27–51

11

SOCIOECOLOGICAL ENTANGLEMENTS, INVASIVE ECOLOGY, AND CLIMATE INJUSTICE

A Story of Cape Town, South Africa

Grace D. O'Donovan

On 18 April 2021, fires burned across Table Mountain National Park in Cape Town, South Africa, spreading so quickly and ferociously through the brittle *fynbos* that they destroyed significant parts of historical buildings on the slopes of Devil's Peak and its surrounds. The University of Cape Town's Jagger Library fell victim to the fire, as well as an irreplaceable collection of ancient African Studies archives and the historic Mostert's Mill, as it raged on for days (Burke 2021). The fire also spread quickly to flammable species of pine (some invasive, some non-invasive, and some protected by heritage legislation) and nearby non-Indigenous palms with high fuel loads. The disaster made global headlines, not necessarily because of the fire in-and-of-itself but because of destruction and incalculable loss of historic university buildings and artefacts (Goldbaum and de Greef 2021). Local news sources quickly reported that the fire was started by displaced people living on the slopes of Table Mountain who had left their fire unattended, provoking a testy debate about 'homelessness' and poverty in the city (Marvin 2021). Yet, the fire threatened not only 'the disadvantaged' but also 'the historically-advantaged' as well as 'heritage' infrastructure (van Wilgen and van Wilgen-Bredenkamp 2021).

In the month of the Table Mountain fire, Cape Town also battled unprecedented winter rainfall and flooding. A coalition of residents of Western Cape informal settlements, including those from Langa, Delft, Mfuleni and eThembeni, known as the Intlungu yaseMatyotyombeni Movement (IYM), gathered to protest the poor drainage, inadequate housing, and lack of the city's disaster relief services and ability to manage water and sanitation within informal settlements. In nearby Kraaifontein, COVID-19 migrants and pandemic-displaced families, branded as 'land invaders', appealed to the city's disaster risk management (DRM) to initiate disaster relief for flooding of their newly erected informal shelters on 'illegally

DOI: 10.4324/9781003214021-12

occupied' land, as they condemned the insensitivity of not being entitled to service provision in compensation for their 'forced' resettlement.

In the Western Cape, 'moments of crisis' (Lund 2016) reveal a nexus of underlying challenges that demands reflection on these nested complexities. As points of disruption, they bring vulnerabilities to light within the political landscape of South Africa that have become endemic and invisibilized through their normalization in everyday life. They reveal the ways in which political spatialities of language, power, race, and experience come together, but also open up new focused conversations about possible ways of disentangling and challenging these vexed problems. For example, the veld fires that burned across the City of Cape Town's Table Mountain National Park in 2000 gave rise to renewed focus on invasive ecology, water conservation, and existing sociopolitical legacies of apartheid. In 2017/2018, the drought across the Western Cape became an exemplar for vulnerabilities around food and water insecurity, precarious livelihoods, biodiversity loss, urban densification, and burgeoning inequality now faced by many metropolitan cities around the world (Joubert and Ziervogel 2019). The 2021 fire has reignited conversations around invasive ecology as well as urban densification, crime/violence, poor governance, displacement, and 'homelessness' in the city. Moreover, the global COVID-19 pandemic since 2019 has compounded existing stresses to reveal a 'new' type of displacement-disenfranchisement amongst other political, environmental, and socioeconomic effects.

Intricately interconnected to this assemblage of socioecological crises is the looming threat and global emergency of climate change, with impacts on diverse local contexts in ways that, together, foreground the increasing frequency and extremes of disaster risks and impacts. At the heart of socioecological challenges is (inter)relationships. Increasingly, as we examine the impact of climate change and socioecological crises in the Majority World, the ways in which the social, ecological, political, and economic are mutually informing and intertwined cannot be overstated. Yet, the differentiated experiences of climate change impacts on the Majority World compared with those of the Minority World explain the extent not only of the inequality between them but also of the intertwined instabilities of socioecological and political systems that represent our planetary existence within the Anthropocene. Thus, the Majority World experiences of climate change give emphasis to the extent of burgeoning, intersectional oppressions and the fragilities of global justice.

This chapter contributes to closing a gap in the literature by attending more closely to the ways in which 'the sociopolitical' and 'the ecological' are intricately intertwined, drawing on critical climate justice scholarship, and paying attention to the theme of 'invasiveness' from sociological, decolonial, and ecological perspectives, bringing these together productively to attend to resonances. Cape Town acts as a case study for the ways in which these resonances are realized. Here, climate change acts as a catalyst for socioecological disaster, fuelling conditions of 'crisis' (Gray 2019). An increased frequency of fire breakouts, as an example, causes suffering to 'displaced others' living in contexts of deep informality, paralleling fire effects on displaced Indigenous flora and biodiversity loss. In both instances, 'invasiveness' has created the conditions for disaster with respect to fire and other socioecological disasters.

The rest of the chapter is given over to discussions that foster arguments around socioecological complexities in the Majority World from different entry points. The first section, 'Historicizing South Africa's Climate Challenges', contextualizes the socioecological challenges being faced in South Africa, leading to 'Invasive Ecologies in Cape Town', which not only defines the importance of this concept in context but also provides an ecological metaphor for understanding how the settler paradigm and (post)apartheid legacies have created the conditions for and exacerbated socioecological disasters and climate change effects. Following, 'Ecosophy and Political Ecology' offers a deeper theoretical-philosophical undertaking to situate arguments in terms of both decolonial thought and the politico-ecological impacts on Majority World contexts of Cape Town. 'Climate Justice Complexities: The Case of Cape Town' draws, in part, on extracts of an interview with a leading South African ecologist, David Le Maitre, towards my doctoral research project[1] and illustrates how climate justice scholarship and political ecology is useful in framing an understanding of invasive socioecology in Cape Town. The conclusion ties arguments together, summing up key concepts and tropes. The arguments in this chapter seek to offer a nuanced, conceptual perspective to the deleterious effects of climate injustices on the Majority World, with Cape Town as exemplar, and highlight how intimately the social, political, historical, economic, and ecological are intertwined and mutually informing in situated context.

This chapter draws its transdisciplinary influence from literature within the fields of political science (specifically political ecology), environmental humanities, climate justice scholarship, and critical (Mignolo and Vazquez 2017) space theory. It embraces invasive ecology metaphors to forward an argument on socioecological complexity and 'thinks with' decolonial scholarship (Mignolo 2017) in this undertaking. Section 1 opens with climate justice scholarship and provides a historicizing view to contextualize the arguments that follow it.

1 Historicizing South Africa's Climate Challenges

Southern Africa is experiencing consistent temperature rises at approximately twice the average rate of the rest of the world (Diop 2014). Economists have outlined the dire ways in which an increase in global temperature of just 2°C will impact the country, with severe water scarcity occurring more frequently at approximately every 100 years (Diop 2014). Just one standard deviation change in climate towards either heat or increased rainfall patterns, as witnessed in the April 2021 fires, sees a 4% rise in interpersonal violence and a 14% rise in intergroup conflict (Hsiang, Burke and Miguel 2013). In South Africa, the effects of average temperature rise are already present in polluted streams and water scarcity – the urban disconnect to our relationship with water (von Zeil 2011). Biodiversity loss, lack of food sovereignty, insecure livelihoods, and poor health of disaffected populations are also exacerbated by climate change effects. Disenfranchised populations are further precaritized by ongoing racialism; economic, spatialized urban, cultural, and educational apartheid; and lack of opportunity, security, and welfare. Residents' daily realities of living in a country grappling with climate change, structural inequalities, biodiversity loss, socioeconomic vulnerability, (un)sustainable livelihoods, 'state capture' and cronyism, racialized tensions, historical political precarities, and shadow/illicit economies (Preiser et al. 2017) signify an assemblage of mutually-informing challenges. These 'moments of crisis' grow in frequency and scale as climate change increasingly seeps into 'the political', its effects multiplying in complexity as they ripple across these stresses.

South Africa's hard-earned, post-apartheid constitution,[2] ratified in 1996, is still viewed as politically innovative and pioneering. Yet, the legacy of apartheid still dominates country narratives (Ziervogel et al. 20110) to which the lived realities of everyday life bear witness. Nevertheless, within the policy domain, South Africa's constitution emphasizes non-racial suffrage, equality, and fairness for people previously disadvantaged by apartheid, and formatively links South Africa's *natural* heritage with *national* heritage (Clarke 2002). Thus, it enshrines an edict of 'conservation' within its Bill of Rights as being equally important to the rights of citizens:

Everyone has the right
    a. to an environment that is not harmful to their health or well-being; and
    b. to have the environment protected, for the benefit of present and future generations, through reasonable legislative and other measures that
        i. prevent pollution and ecological degradation;
        ii promote conservation; and
        iii secure ecologically sustainable development and use of natural resources while promoting justifiable economic and social development.[3]

By framing this edict within terms of 'sustainable economic development', post-apartheid national policy prioritizes labour and poverty reduction alongside conservation (Clarke 2002). In 1995, the national Working for Water Programme emerged as the largest joint poverty relief and conservation programme ever instituted (Neely 2010), with scientific practice and ecological research addressing the country's growing biodiversity and water challenges fuelled by the deleterious spread of invasive or alien species. Table Mountain National Park is an area that contributes to the fynbos biome of the Cape Floral Kingdom as a UNESCO World Heritage Site and is one of the most biodiverse regions of the world. The fires that burned across Table Mountain National Park in 2000 increased the tempo in response to this crisis and saw a swift and decisive challenge to invasives. With biomass-heavy invasive species, such as the Hakea (*H. sericea*) and the Australian Acacia (A. saligna and A. cyclops) lining the urban perimeter of Cape Town, renewed attention has been paid to the impact on disaffected populations and the wider ecology. By establishing a direct link, policymakers could now incorporate a challenge to invasive species into wider socioeconomic development pathways, and through the Working for Water Programme, invasive ecology has sat on the frontline of conservation response, connecting biodiversity, water, and poverty with a constitutional mandate to empower and support the most marginalized (Clarke 2002).

Despite global praise for this labour-intensive, invasive clearing programme, legitimated by the National Water Act (1998) to redress historical inequalities, the integrated water resources management (IWRM) (Schreiner 2013) is still marked by 'geographies of difference' (Power 2003) and injustice. The 1998 Act succeeded the Water Act, No. 54 of 1956 that was promulgated during the 'first phase of apartheid' (1948–60) (Tempelhoff 2017). The 1956 legislation deliberately outlined the distribution of water along racialized lines, guided by the principles of a 'water-rich Europe' (Schreiner 2013) instead of the water-scarce region of South Africa, reinforcing what could be viewed as 'resource colonialism'. In 2015, in recognition of the disproportionately adverse effects of climate change on sub-Saharan Africa, and the Western Cape particularly, the City of Cape Town earned recognition by the C40 Cities Climate Leadership Group for their Water Conservation and Water Demand Management (WC/WDM). This initiative succeeded in reducing water demand and loss by 30%, despite an over 30% population increase in Cape Town since 2001, a surge attributed to climate and poverty migration (Tempelhoff 2017). Despite these efforts, in 2018, with Cape Town on the brink of becoming the first major metropolitan city globally to run out of municipal water, the world debut of the country's turbulent 'story of water', and the role invasive ecology plays, brought renewed awareness of how climate and socioecological disasters are experienced by those living in informality. The effects on disaffected populations reflect how spatialities and inequalities underscored by politico-ecological 'crisis' are intertwined. This context of the politics of water highlights the precarities of a planet burdened by the Anthropocene (Barlow 2013) while also revealing the differential effects on residents' livelihoods and welfare (Preiser et al. 2017).

Countering a settled 'deficit view', South Africa also reflects a history of political resistance and solidarity groupings (Nagy 2002), borne from a long history of oppression and anti-apartheid resistance. While South Africa's post-apartheid constitution of 1996 offers amongst the greatest human rights and constitutional freedoms in the world (Nagy 2002), contradictorily, deeply racialized inequalities and conflicts remain. These tensions are often realized in the form of oppositional ideological views and xenophobia, framed by a notion of 'the African other'. In this way, racially rooted post-apartheid inequalities 'invade' the lived experiences of everyday life. Here, entrenched maladministration, abuse of power, cronyism, and localized global capitalism deleteriously impact people and land (Ziervogel et al. 2010). While South Africa is considered the most unequal country in the world (World Bank 2022), these complexities are underscored by competing imaginaries of contradiction and possibility, characterized by geographies of difference, unequal relations, cycles of vulnerability (Misselhorn 2005), and multiple socio-material, sociopolitical, and socioecological dimensions of inequality (Ziervogel et al. 2010). These ecopolitical realities invoke a discussion on 'invasive ecology' (Neely 2010) to assert ecological and metaphorical explanations of South Africa's historical, political, and sociocultural story.

2   Invasive Ecologies in Cape Town

Invasive ecology is conceptually relevant in the South African context in respect of environmental impacts, but can also be appropriated metaphorically to give historical meaning to settler politics and effects on people and land. Nevertheless, some contextualizing is first needed to understand 'invasive ecology' in accordance with its ecological meanings.

Renewed attention to invasive ecology in South Africa has arisen out of the country's environmental crises since the 1930s, and specifically in its role within the wider politics of water, which has increasingly become a focus of both the media and climate scholarship. The country's ongoing struggles with drought and flooding have come to overlay existing structural inequalities, reflected in increasing human precarity in a climate-vulnerable region. The recent National State of Disaster, declared by South African president Cyril Ramaphosa in April 2022 regarding the devastation caused by flooding in Kwazulu-Natal province, is indicative of these cross-cutting challenges between climate change, poor governance, and poverty (Ramaphosa 2022). Likewise, biodiversity loss of the unique Cape Floral Kingdom (discussed below) works alongside failing water infrastructure, political tensions, and poor resource management. These are coupled with rising population and increased demand for food and sustainable water resources, compounded by a global 'cost of living crisis' (Erasmus 2022). Yet, the role of alien vegetation and invasive species in exacerbating the 2015–18 crisis has been a lesser-known contributing factor (Wild 2018). As noted earlier, invasive alien species have been considered a significant factor in the Cape Town fires in 2000, causing 8000 hectares of natural veld and fragile fynbos to burn along the Cape Peninsula, while also revealing a nexus of concerns over poverty, conservation, labour, and post-apartheid socioecological vulnerabilities (Neely 2010). The Table Mountain Fire of February 2021 serves as a cyclical repeat of the same entrenched, socioecologically intertwined circumstances that led to the devastation and misery it created.

To recognize the role played by invasive species in socioecological disasters, to which climate change has contributed, there is a need to understand how invasive species have come to be so pervasive in South Africa. Historically, many invasive species were cultivated intentionally to manipulate ecosystems for the forestry or agricultural industries. Others have crossed South Africa's borders accidentally through migration, trade, or tourism (Potgieter et al. 2020). In 2017, the South African National Biodiversity Institute (SANBI) published a first-of-its-kind report on invasive species monitoring, noting that despite increased regulation, the numbers of invasive species are growing (van Wilgen and Wilson 2018). The South African National Environmental Management Biodiversity Act's Alien and Invasive Species Regulations list 556 invasive vegetation, although actual numbers are closer to 775. 107 species are classed as 'major threats' to biodiversity and/or human well-being (Van Wilgen and Wilson 2018).

The Cape Floral Kingdom is one of the most biodiverse regions of the world and boasts extremely high levels of endemism in its fynbos biome. As many as 68% of all plant species worldwide can be found within this single kingdom (Helme & Trinder-Smith 2006). Despite this abundance of beauty and life, 1700 species are listed as critically threatened, in part due to Cape Town's urban expansion and high levels of migration as well as its precarious location on two specific zones of endemicity. The 2017 SANBI report notes that 'biological invasions account for 25% of the reduction in South African biodiversity seen to date' (Helme & Trinder-Smith 2006, 19). Apart from invasive species, the largest threats to biodiversity in the Cape are agricultural fertilizers and crops, unseasonal fires, and deepening effects of climate change. It is important to note that the fynbos biome actually does *rely* on seasonal, low-level, fire-triggered regeneration (Neely 2010), if these fires are not too frequent. Increased fire frequency would lead to species destruction rather than regeneration, the risk of which has increased substantially, while average temperatures rise and unregulated urbanization intensifies (Neely 2010). The seeds of some invasive species, such as Australian acacia, are active pyrophytes, where fire heat breaks open their hard dormant shells (Holmes 1997). Others, such as Pinus, are passive pyrophytes, being fire-resistant and outliving their native neighbours by crowding into empty spaces left behind by these disturbed ecosystems. This inherent hardiness allows the invasive species to flourish, while their Indigenous neighbours fail to regrow (Neely 2010). Floral composition and biodiversity survive in a delicate balance, as the frequency, intensity, and (un)seasonality of changing weather patterns grow increasingly more unpredictable under climate change. The Prosopis genus of trees has also overtaken the growth of local plants, and the water hyacinth (*Eichhornia crassipes*) can be found in rivers and dams across the country. Other known invasives include the Australian hakea, Eucalyptus, and Acacia wattles. Around 30% of the water supply delivered by the 44 dams in the Western Cape is used by invasive species (Le Maitre et al. 2020).

The historical introduction of alien species through (un)intentional means played a significant yet globally underreported role in the Cape Town water crisis, where alien vegetation used over 100 million litres per day of the Western Cape Water Supply System (WCWSS), or 20% of Cape Town's daily usage (Slingsby & Botha 2018). The webs of influence across governance structures – such as the

National Department of Environmental Affairs' Working for Water Programme, the National Department of Water and Sanitation, and CapeNature, together with others – form a tenuous conservation bond. The affiliation suffers from a lack of joint cooperation, the prioritization of small-term projects with low cost labour at the expense of long-term projects, disagreement over long-term rehabilitation plans for clearing the Cape's catchment areas of alien vegetation, and the favouring of resources used for industries that may reap greater economic rewards, such as forestry (Slingsby & Botha 2018). In an interview, ecologist David Le Maitre (2022) noted that together these governance structures form a typical example of a complex system, and that their oft-disjointed work reflects the erratic ways in which catchment experiments have been funded. It has also shed light on how cities forecast climate change for adaptation purposes. Resource planning has traditionally relied on historical records to plan for the future, and yet the flux in statistical rainfall distribution caused by inconsistent climate change-related weather patterns has meant that water resource planning is no longer adequate. Apart from biodiversity loss and increasing drought, contributing factors to water shortage include urbanization and growing resource use. It also implies unemployment for those whose livelihoods are precariously dependent on industries such as forestry and water resource management. There are health concerns too for those suffering from a lack of clean water and proper sanitation facilities. Disparities between water resource use by private land owners and distribution to agricultural industries, and those living in deep informality, are stark (Le Maitre 2022). Underscoring this is a growing gap between the priorities of existing governance policies and the well-being of citizens. The result is a costly entanglement of competing challenges. Alien vegetation currently costs South Africa approximately R6.5 billion annually in addition to the R1.5 billion it spends on managing the challenge. Yet, this financial cost does not account for the stresses experienced by those facing increases in drought, or indeed flooding, within cities and the instability of livelihoods caused by biodiversity loss and water insecurity. Compounding these disasters is catalytic climate change as an overarching challenge, developing an increasingly unsustainable environment for all, but especially the most disenfranchised.

Robbins introduced the concept of 'sociobiological networks', drawing on the principles of Actor-Network Theory (ANT), to view 'invasion' as occurring when biological and social actors engage, 'each influencing the other as well as propelling the process of invasion forward' (Robbins in Neely 2010, 873). Here, competing historical, global/local, geopolitical, and eco-material crises, namely the political, ecological, economic, and social fragilities that dominate the post-apartheid landscape, can come to be viewed as invasive in the daily lives of the most vulnerable, instead of the species in-and-of-themselves. Invasive ecology or, in this sense, invasive socioecology, which can be understood in terms of its colonial or settler paradigm effects, can be deployed as a politico-spatial lens in this context to consider the roles of human and non-human. The narratives surrounding the story of 'land invaders' in Kraaifontein, for example, can in this way be reinterpreted in relation to post-apartheid legacies that still 'invade' their lives and in their experiences of climate change. Spatialities striated by experiences of socioecological disasters, induced by climate change, tell a story of invasion, devastation, and the effects of Euromodernism and legacies of colonialism. The settler paradigm can speak as much to a notion of political invasion in South Africa as it can to ecological invasiveness.

The only difference to this state of affairs might be the question of who the 'invader' or 'alien' might be. For the displaced 'vagrants' whose unattended fire caused the Table Mountain disaster in February 2021, or those deemed 'land invaders' as a consequence of their political and material displacement, could it be that Euromodernism and settler paradigms are, instead, the 'alien invaders' that have set in place a cataclysm of socioecological events?

3 Ecosophy and Political Ecology

Presenting the 'complexity' argument of socioecological crisis relies in part on Forsyth and Walker's (2008) demonstration that Global North-led climate discourse often depends on an oversimplification of complex environmental challenges (Bassett and Peimer 2015) in order to promote simple management processes (Turner 1993). The 'steady state' narrative 'helps to justify the "restoration" of and intervention in an ecological process that is assumed to be preventing the return of a "normal state"' (Bassett and Peimer 2015, 160). This is why political ecology, in the Minority World, can learn from the environmental crisis discourse in the Global South and even work towards a symmetry of practice (Robbins 2002). This draws on McCarthy's work in blurring the lines between geographies of distance (North and South) within political ecology, where he specifically insists that there is 'nothing about the epistemology, methodology, philosophy or politics of [Global South] political ecology that bars its deployment in other contexts' (2002, 1510). Turning this gaze towards a preoccupation with both 'local' political ecology and central institutions of power (Scheper-Hughes 1995) means that the 'documents, practices and logics of national and supranational organizations are as crucial as the discourses and practices activated in local resistance' (Frank et al. 2000 in Robbins 2002, 1511). As Robbins (2002) mentions, Brodt (1999), for example, specifically critiqued a 'Western science vs Indigenous knowledge' dichotomy through her studies of local knowledge in India, where tree farmers' ways of knowing were integrated through co-production principles with those held by agricultural officials. Leach and Fairhead's (2000) formative rehistoricizing of forest savanna landscapes in West Africa also emphasized socioecological relations and global/local positionalities. Instead of focusing on North–South dichotomies of environmental crisis narratives without considering historically marginalized forms of Indigenous knowledge and epistemicide (de Sousa Santos 2014), we can follow Stengers' (2005, 186) 'ecology of practice', which 'aims at the construction of new "practical identities" for practices [and] new possibilities for them to be present … it thus does not approach practices as they are, but as they may become'.

Considering climate justice and ecological research through a complexity lens might mean inviting in a post-humanist 'ecosophy' (Naess 1989) along the lines of Bateson's (1972) transversal relations that produce assemblages within/through the way we live. Braidotti and Hlavajova (2018, 131) argue that instead of starting from:

> *oppositions* between mind and body, man and animal, man and nature, nature and culture, technology and earth, but instead from *relations*, [ecosophy will] allow us to analyse the crises of today in a completely different way, if only because the role of ecological thinking as such now changes from critical (oppositional) to affirmative (mutual coexistensive): … it searches for ways to be interwoven with the movements and the swerves of today.

This ecosophy brings together environmental, social, and mental ecologies through their intertwining relations (Guattari 1989). For Bassett and Peimer (2015, 162), this means paying attention to how the material world influences social action and, within the context of natural resource management, how the biophysicalities of resources '"resists", "assists" or "redirects" political economic prerogatives'. For Pickering (2010, 7), this looks like a 'dance of agency' that is 'embedded in a decentered and open-ended becoming of the human and non-human'. It also calls for radical decolonizing of our thinking and being (Mignolo 2017) and demands a less dichotomous view of the world that would reinforce essentialisms (Santos 2010) and prevent new possibilities for research and thought (Fabricio 2017).

In South Africa, an example of this socio-material, post-humanist 'dance' can be witnessed in the national Working for Water Programme (Neely 2010). In parallel, it means that concerns about fire in the Western Cape are also concerns about 'conservation' and biodiversity; invasive vegetation; water scarcity, drought, or flooding; homelessness and displacement; poverty, inequality, and socioeconomic precarity; and climate vulnerability. Reinterpreting the human and the non-human as intertwined offers opportunities to 'reclaim resilience' in the Majority World; and through Kuhlken's (1999) idea of incendiarism as 'a global ecological weapon of political resistance' (Robbins 2002, 1510), we can even imagine our natural world protesting the conditions wrought on it by the Anthropocene. This ecosophy brings to light a posthuman approach to how we interpret meanings of 'crisis', especially in light of climate change, and indeed a '(post)COVID-19 world'. As Guattari (2000, 29) stated, 'now more than ever, nature cannot be separated from culture; in order to comprehend the interactions between ecosystems, the mechanosphere and the social and individual Universes of reference, we must learn to think "transversally"'. Understanding, then, that life is a 'vortex of shared precariousness and unchosen proximities' (Cohen 2015,

107), the discussion that follows, through a case study of Cape Town, considers the ways in which climate justice scholarship and decolonial, Indigenous approaches are crucial to negotiating the complex challenges of climate (in)justice in South Africa today.

4 Climate Justice Complexities: The Case of Cape Town

Climate justice scholarship draws attention to how climate-related socioecological disasters are violating the 'right to be free from climate change, its related impacts and other forms of ecological destruction' (Cohen 2015, 365). South Africa's 2020 Climate Justice Charter – the result of a struggle-driven participatory process that emerged from joint action between networks involved in the South African Food Sovereignty Campaign – offers an example of how consultative, localized responses to complex challenges of systemic climate injustice are crucial to the movement (COPAC 2020). Aimed at facilitating deep democratic and decolonial just transition, the charter specifically outlines reforms to redress socioecological crises facing the country. Formative to the climate justice movement is the understanding that these issues are deeply entangled. The interconnections that bind these crises are realized in particular vulnerabilities in the Majority World, such as in South Africa, implying that Indigenous onto-epistemologies and decolonial thought help to articulate tensions and inform activism around climate change injustices that impact local environments and communities worldwide (Makondo and Thomas 2018). Euromodernist and instrumentalist discourses dominate structural climate narratives within spatialities of power. The oft-heard terms of 'mitigation', 'adaptation', and 'resilience' (Stewart et al. 2019), and even the notion of 'crisis' itself (Norval 1994) that emanate from Global North/Minority World supernational organizations (e.g. in the form of the UN Sustainable Development Goals) can often neatly bypass how deeply interconnected political, economic, social, and ecological stresses can be, especially for the most marginalized (Adelman 2017). Often relying on solutionist and reductionist approaches, the dominance of these scientistic voices in the climate change and conservationist landscapes poses a number of significant threats, the preeminent being that governance procedures and policies remain rooted in northern-dominated or 'Western' scientific knowledge forms and often tend to be unreflective of the situated challenges and circumstances facing local disenfranchised populations in the Majority World (Agrawal 1995). The concept of socioecological 'resilience' itself arose out of a counterrevolution to the prevalence of neoliberal environmental policies, but has since been co-opted by 'neo-liberal strategies of control' (Nelson 2014). Here, language within policy speak may be canalized to assign meaning, highlight particular messages, and legitimize certain agendas for the Majority World. Indigenous, decolonial, and ecofeminist responses seek to 'reclaim' a notion of resilience in the Majority World by promoting an intersectional, complexity-driven approach to challenges of the Anthropocene (Nelson 2014).

Climate justice complexity in political ecology and the need for consultative, decolonial approaches, such as those offered by the Climate Justice Charter, can be demonstrated in a specific, local example of the relationship between invasive species and the residents of the Cape Flats, an area outside of Cape Town characterized by informal settlements. As one of the many consequences of the impact the COVID-19 pandemic has had in exacerbating existing social, economic, political, and environmental 'precarities', loss of livelihoods has led to a rise in 'displaced' communities. Many have migrated from the African National Congress (ANC)-run Eastern Cape (South Africa's poorest province), in particular, to the Democratic Alliance (DA)-run Western Cape in the hopes of pursuing greater employment opportunities as well as conditions with 'better' infrastructure (Le Maitre 2022). The 2019 *Municipal Financial Stability Report*, published by Ratings Afrika, noted that the DA-run Western Cape is considered the 'best'-run province for its ability to maintain and deliver services and infrastructure. Large-scale intranational migration, exacerbating the swift expansion of informal housing, has also occurred more broadly over time, in part due to a political push to increase voting support for the DA (Le Maitre 2022). Despite the increasing droughts in the Eastern Cape and the higher 'hopes' of employment in the Western Cape, many migrants regard themselves as temporary. By living in informality, some migrants eschew paying for formal healthcare, electricity, or water infrastructure. Challenging Global North-led instrumental views of poverty, migration, and livelihoods, a bottom-up, localized approach can offer perspectives on 'scales of complexity' that account for situated socioeconomic and environmental precarities. Cape Town also possesses an excess of groundwater, often in low-lying areas characterized by growing informal settlements. Where groundwater-prevalent areas are subject to nearby wastewater run-off, what prevails is poor sanitation and biological contamination from informal settlements that use groundwater to compensate for water shortages. Where high winter rainfall has caused the groundwater table to rise above the surface, the Cape Flats has been subject to flooding like Kraaifontein. Early settlers struggled to manage the natural vegetation on this land, but in the hopes of being able to farm and supply livestock to the Dutch East India Company, and latterly the British, they sought to turn the land through controlled burning (Le Maitre 2022). The land became mobile sand, and no vegetation could survive. From 1870 onwards, in an attempt to address the poor land quality, Cape Town's rubbish was distributed across the sand to prevent further movement, and invasive Acacia seeds were sown throughout. Acacia thickets still exist along the Cape Flats today, and local residents have a complex, multifaceted relationship with the invasive species.

Coarsely dense Acacia thickets present a security risk for residents of the Cape Flats, specifically women, children, and the most marginalized who are the predominant targets of South Africa's high rates of gender-based violence. These invasive species have become ecospatial sites of crime, and residents are forced to travel long distances around the thickets to ensure their safety. The Working for Water Programme has also historically employed residents of informal settlements, but poor managerial oversight has often impacted the efficacy of their invasive clearing programmes. Other impacts are social: clearing teams undertaking labour-intensive work that camp out for long periods often see a higher rate of pregnancies, leading to escalated rates of turnover in employment, increased stated need to deliver continuous training, and lower levels of female employed staff. Thus, the quality of clearing programmes suffers, a lack of opportunity remains entrenched, and the rate of growth of invasive species still outstrips clearing. Many clearing teams are underpaid, affecting their efficacy. Hard-to-reach locations make access to invasive species difficult and this also contributes to overgrowth. Further, if an area is deemed 'uncleared', the teams do not receive compensation. Current data-recording systems only show the state of vegetation and the state of invasion in an area and lack detail regarding how effectively an area has been cleared (Le Maitre 2022). Consequently, to ensure compensation of those employed in clearing programmes, managers often approve contracts as having been fulfilled, even when not. The result is that uncleared invasive species continue to grow.

The settler paradigm thus can be seen to operate as a web, from destructive, racialized ecological legacies; entrenched sociospatial oppressions and violence from apartheid; lack of post-apartheid security; lack of access to opportunity, education, and well-being; and intranational migration within South Africa. Demonstrating a precarious relationship between livelihoods, climate justice, and invasive ecology, Indigenous, ecofeminist, and decolonial perspectives on local political ecology are required to appropriately grapple with socioecological entanglements to enable the 'deep', 'just' transition for which the Climate Justice Charter calls for. Cape Town offers a unique case: in terms of its vulnerability to climate change due to its reliance on cold-weather frontal systems and its increasing levels of drought; its high levels of invasive species that exacerbate socioecological and politico-economic precarities; but also as a site exemplar for the ways in which apartheid legacies continue to pervade residents' lives through an invasive socioecology. Municipal workers have often worked contrarily to city's policies, staking out plots, illegally, for informal migrants from the Eastern Cape on which to build housing. For Le Maitre (2022), 'apartheid taught people to disrespect the law, and that's the legacy'. These 'sociobiological networks' of complexity continue to 'influence' and 'propel forward' these 'invasions', exacerbating settler paradigm effects and compounding the way climate change disproportionately affects the disenfranchised.

5 Conclusion

The effects of socioecological disasters related to climate change have differentially impacted the disenfranchised in the Majority World, who have borne the brunt of these impacts. Marginalized populations of Cape Town exemplify this reality. In the Western Cape, these moments of crisis reveal a nexus of underlying challenges that gives a pause for reflection on the interconnectedness of these socioecological complexities. As points of disruption, they bring vulnerabilities to light that have become endemic within the political landscape of South Africa, 'invisibilized' through their normalization in everyday life. These points of disruption reveal how political spatialities of language, power, race, and experience come together, but also how they are entangled within increasing climate change effects and socioecological disaster risks.

Cape Town acts as an important case whereby the social, political, ecological, and economic are complexly intertwined, and where legacies and spatialities of apartheid and settler politics have indelibly produced conditions that have led to increased socioecological disasters and suffering. Political ecology offers some important tropes in coming to understand such entanglements, and invasive ecology offers a further entry point to drawing parallels, metaphorically, between ecological devastation and the kinds of devastation borne from the violences of apartheid and Euromodernism. A question therefore could be raised around the 'land invaders' in Kraaifontein, or those living in the Cape Flats alongside invasive species planted during colonial times, as to whether their experiences could be reinterpreted and reimagined in relation to post-apartheid legacies that still 'invade' their lived experiences under climate change. The spatialities striated by experiences of socioecological disaster risks induced or exacerbated by climate change tell a story of invasion, devastation, and the effects of structural inequality, Euromodernism, and legacies of colonialism. In effect, the settler paradigm can speak as much to a notion of political invasion in South Africa as it can to ecological invasiveness.

These spatial nodes therefore occupy a space in Cape Town's 'overstory' and 'understory', an above-and-below-ground web of interrelated complexities. Here, I argue, the concept of invasive species is not only a situated account of challenges affecting disenfranchised residents of the Western Cape today, but also acts as a metaphor for thinking through how climate change and socioecological vulnerabilities are intimately intertwined with human precarity and the history of structural inequality in South Africa.

Notes

1 These interview extracts are drawn from my ESRC-funded doctoral research project, undertaken at the University of Edinburgh. The research is taking critical socioecological and eco-materialist approaches to examining food, water, and inequality as experienced by disenfranchised residents in the Western Cape, South Africa.
2 The 1996 South African constitution is a collective consequence of South Africa's transition to democracy in 1994.
3 The Constitution of the Republic of South Africa, No. 108 of 1996, G 17678, 18 December 1996, c.2, n.24.

References

Adelman, S. (2017). The sustainable development goals, Anthropocentrism and neoliberalism. In Duncan French and Louis Kotzé (eds.), *Global Goals: Law, Theory and Implementation*. Cheltenham: Edward Elgar, pp. 15–40.

Agrawal, A. (1995). Dismantling the divide between Indigenous and scientific knowledge. *Development and Change*, 26(3), 413–439.

Barlow, M. (2013). *Blue Future: Protecting Water for People and the Planet Forever.* Toronto: House of Asansi Press.

Bassett, T. & Peimer, A. (2015). Political ecological perspectives on socioecological relations. *Natures Sciences Sociétés*, 23, 157–165. https://doi.org/10.1051/nss/2015029.

Bateson, G. (1972). *Steps to an Ecology of the Mind.* Chicago: University of Chicago Press.

Braidotti, R. & Hlavajova, M. (2018). *Posthuman Glossary.* London: Bloomsbury.

Brodt, S. (1999). Interactions of formal and informal knowledge systems in village-based tree management in India. *Agriculture and Human Values*, 16, 355–363.

Burke, J. (2021). Cape Town fires: Police investigate causes after library damaged. www.theguardian.com/world/2021/apr/19/cape-town-fire-table-mountain-evacuate-university-jagger-library.

Clarke, J. (2002). *Coming Back to Earth: South Africa's Changing Environment.* Jacana: Houghton.

Cohen, J. J. & Duckert, L. (eds). (2015). *Elemental Ecocriticism: Thinking with Earth, Air, Water and Fire.* Minneapolis: University of Minnesota Press.

Cooperative and Policy Alternative Center. (2019). Climate Justice Charter. https://safcei.org/wp-content/uploads/2019/11/Climate-Justice-Charter-Draft_Nov-2019.pdf.

de Sousa Santos, B. (2010). *Descolonizar el saber, reivnetar el poder.* Montevideu: Ediciones Trilce

de Sousa Santos, B. (2014). *Epistemologies of the South: Justice against Epistemicide.* London: Paradigm.

Diop, M. (2014). Listen more closely to Africa's voice on climate change. www.worldbank.org/en/news/opinion/2014/09/22/op-ed-listen-more-closely-to-africas-voice-on-climate-change.

Erasmus, D. (2022, 26 January). SA's poor people getting poorer as prices climb. www.businesslive.co.za/bd/economy/2022-01-26-sas-poor-people-are-getting-poorer-as-prices-climb.

Fabricio, B. F. (2017). Processos de ensino-aprendizagem, educação linguística e decolonialidade. In F. Zolin-Vesz (ed.), *Linguagens e descolonialidades: práticas linguageiras e produção de (des)colonialidade no mundo contemporâneo*, vol 2. Campinas: Pontes, pp. 15–38.

Forsyth, T. & Walker, A. (2008). *Forest Guardians and Forest Destroyers: The Politics of Environmental Knowledge in Northern Thailand.* Seattle: University of Washington Press.

Frank, D. J., Hironaka, A., & Schofer, E. (2000). The nation-state and the natural environment over the twentieth century. *American Sociological Review*, 65, 96–116.

Goldbaum, C. & de Greef, K. (2021). Wildfire deals hard blow to South Africa's archives. www.nytimes.com/2021/04/19/world/africa/cape-town-table-mountain-fire.html.

Gray, E. (2019). Satellite data record shows climate change's impact on fires. https://climate.nasa.gov/news/2912/satellite-data-record-shows-climate-changes-impact-on-fires.

Guattari, F. (2000). *The Three Ecologies.* London: The Athlone Press.

Helme, N. & Trinder-Smith, T. (2006). The endemic flora of the Cape Peninsula, South Africa. *South African Journal of Botany*, 72(2), 205–210.

Holmes, P. (1997). Diversity, composition and guild structure relationships between soil-stored seed banks and mature vegetation in alien plant-invaded South African fynbos shrublands. *Plant Ecology*, 133, 107–122. https://doi.org/10.1023/A:1009734026612.

Hsiang, S. M., Marshall, B., & Edward, M. (2013). Quantifying the influence of climate on human conflict. *Science*. http://emiguel.econ.berkeley.edu/research/quantifying-the-influence-of-climate-on-human-conflict.

Joubert, L. & Ziervogel, G. (2019). One city's response to a record breaking drought. In *African Centre for Cities* Day Zero. Cape Town: African Centre for Cities, pp. 1–52.

Kuhlken, R. (1999). Settin' the woods on fire: Rural incendiarism as protest. *Geographical Review*, 89, 343–363.

Leach, M. & Fairhead, J. (2000). Fashioned forest pasts, occluded histories? International environmental analysis in West African locales. *Development and Change*, 31(1), 35–59.

Le Maitre, D. (2022). Interview on invasive ecologies in the Western Cape, with Grace D. O'Donovan, March.

Le Maitre, D. C., Blignaut, J., Clulow, A. et al. (2020). Impacts of plant invasions on terrestrial water flows in South Africa. In B. W. van Wilgen, J. Measey, D. M. Richardson, J. R. Wilson, & T. A. Zengeya (eds.), *Biological Invasions in South Africa*. Berlin: Springer, pp. 429–456. https://doi.org/10. 1007/978-3-030-32394-3_15.

Lund, C. (2016). Rule and rupture: State formation through the production of property and citizenship. *Development and Change*, 47(6), 1199–1228.

Makondo, C. & Thomas, D. (2018). Climate change adaptation: Linking Indigenous knowledge with western science for effective adaptation. *Environmental Science & Policy*, 88, 83–91.

Marvin, C. (2021). Why the Cape Town fire has thrust the homeless into the spotlight. www.news24.com/news24/southafrica/news/why-the-cape-town-fire-has-thrust-the-homeless-into-the-spotlight-20210422.

McCarthy, J. (2002). First World political ecology; lessons from the Wise Use movement" Environment and Planning A 34 1281-1302

Mignolo, W. & Vazquez, R. (2017). Pedagogía y (de)colonialidad. In C. Walsh (ed.), *Pedagogías decoloniales: prácticas insurgentes de resistir, (re)existir y (re)vivir*. Educador: Abya-Yala, Tomo II, pp. 489–508.

Misselhorn, A. (2005). What drives food insecurity in southern Africa? A meta-analysis of household economy studies. *Global Environmental Change*, 15(1), 33–43.

Naess, A. (1989). From ecology to ecosophy, from science to wisdom. *World Futures*, 27(2), 185–190.

Nagy, R. (2002). Reconciliation in post-commission South Africa: Thick and thin accounts of solidarity. *Canadian Journal of Political Science / Revue Canadienne De Science Politique*, 35(2), 323–346. www.jstor.org/stable/3233430.

Neely, A. (2010). 'Blame it on the weeds': Politics, poverty and ecology in the new South Africa. *Journal of Southern African Studies*, 36(4), 869–887.

Nelson, S. H. (2014). Resilience and the neoliberal counter-revolution: From ecologies of control to production of the common. *Resilience*, 2(1), 1–17, https://doi.org/10.1080/21693293.2014.872456.

Norval, A. J. (1994). The dichotomization of political space and the crisis of Apartheid discourse. In S. S. Nagel (ed.), *African Development and Public Policy. Policy Studies Organization Series*. London: Palgrave Macmillan, pp. 128–154. https://doi.org/10.1007/978-1-349-23355-7_6.

Pickering, A. (2010). *The Mangle of Practice: Time, Agency, and Science*. Chicago: University of Chicago Press.

Potgieter, L. J., Douwes, E., Gaertner, M., Measey, J., Paap, T., & Richardson, D. M. (2020). Biological invasions in South Africa's urban ecosystems: Patterns, processes, impacts, and management. In B. van Wilgen, J. Measey, D. Richardson, J. Wilson, & T. Zengeya (eds.), *Biological Invasions in South Africa. Invading Nature*, vol. 14. Cham: Springer, pp. 275–309. https://doi.org/10.1007/978-3-030-32394-3_11.

Power, M. (2003). *Rethinking Development Geographies* (1st ed.). London: Routledge. https://doi.org/10.4324/9780203006184.

Preiser, R., Pereira, L., Biggs, O., Drimie, S., Metlerkamp, L., Hamman, M., Maciejewski, K., & Cloete, D. (2017). *Guidance for Resilience in the Anthropocene.* Stellenbosch: University of Stellenbosch, Centre for Complex Transitions. www0.sun.ac.za/cst/wp-content/uploads/2017/03/CST-GRAID-2016-Report- FINAL-1.pdf.

Robbins, P. (2002). Obstacles to a First World Political Ecology? Looking near without looking up. Environment and Planning. 34, pp. 1509–1513. [Online]. Available at: DOI:10.1068/a34217

Scheper-Hughes, N. (1995). The primacy of the ethical: propositions for a militant anthropology. *Current Anthropology,* 36, 409–440.

Schreiner, B. (2013). Viewpoint – Why has the South African National Water Act been so difficult to implement? *Water Alternatives,* 6(2), 239–245.

Slingsby, J. & Botha, M. (2018). Aliens are greatest threat to Cape Town water security. www.groundup.org.za/article/aliens-are-greatest-threat-cape-towns-water-security/.

South African Government. (2022, 18 April). President Cyril Ramaphosa: Declaration of a national state of disaster to respond to widespread flooding. www.gov.za/speec hes/president-cyril-ramaphosa-declaration-national-state-disaster-respond-widespr ead-flooding.

Stengers, I. (2005). Introductory notes on an ecology of practices. *Cultural Studies Review,* 11(1), 183–196.

Stewart, R., Dayal, H., Langer, L. et al. (2019). The evidence ecosystem in South Africa: Growing resilience and institutionalisation of evidence use. *Palgrave Communications,* 5, 90. https://doi.org/10.1057/s41599-019-0303-0.

Tempelhoff, J. (2017). The Water Act, No. 54 of 1956 and the first phase of apartheid in South Africa (1948–1960). *Water History,* 9, 189–213. https://doi.org/10.1007/s12 685-016-0181-y.

The National Water Act of South Africa (1998).

Turner, M. (1993). Overstocking the range: A critical analysis of the environmental science of Sahelian pastoralism. *Economic Geography,* 69(4), 402–421.

van Wilgen, B. & van Wilgen-Bredenkamp, N. (2021). The Table Mountain fire: What we can learn from the main drivers of wildfires. www.sun.ac.za/english/Lists/news/DispForm.aspx?ID=8195.

van Wilgen, B. W. & Wilson, J. R. (eds.) (2018). *The Status of Biological Invasions and Their Management in South Africa in 2017.* Stellenbosch: South African National Biodiversity Institute, Kirstenbosch and DST-NRF Centre of Excellence for Invasion Biology.

von Zeil, C. (2011). Reclaiming Camissa: Tedex Talks Cape Town. www.youtube.com/watch?v=Z9022ydUiRg.

Wild S (2018) South Africa's invasive species guzzle precious water and cost US $450 million a year.Nature 563: 164-166. https://doi.org/10.1038/d41586-018-07286-0

World Bank (2022). Press Release: New World Bank Report Assesses Sources of Inequality in Five Countries in Southern Africa. [https://www.worldbank.org/en/news/press-release/2022/03/09/new-world-bank-report-assesses-sources-of-inequality-in-five-countries-in-southern-africa]

Ziervogel, G., Shale, M., & Du, M. (2010). Climate change adaptation in a developing country context: A case of urban water supply in Cape Town. *Climate and Development,* 2(2), 94–110.

12

RESISTING NARRATIVES OF FUTURE FORECLOSURE

Rethinking Adaptation and Resilience in Favour of Climate Justice in the Maldives

Africa Bauzà Garcia-Arcicollar

1 Introduction

A well-known impact of climate change is sea level rise. The International Panel on Climate Change (IPCC) predicts that by 2100, global mean sea level could rise from 0.26 m to 0.55 m in the best-case scenario, and from 0.52 m to 0.98 m in the worst-case scenario (IPCC 2013, 1140). This threatens the existence, or at least inhabitability, of a group of Small Island Developing States (SIDS), including the Republic of the Maldives in the Indian Ocean. On the basis of these predictions, the future of some island nations has been declared and treated as finite (Barnett 2017; Felli & Castree 2012). The foreclosure of island futures illustrates what Hulme (2011, 247) refers to as climate reductionism, which he defines as 'a form of analysis and prediction in which climate is first extracted from the matrix of interdependencies that shape human life within the physical world. Once isolated, climate is then elevated to the role of dominant predictor variable'. Hulme (2011) further elaborates this by reflecting on the hegemony that the predictive natural and biological sciences hold towards visions of the future. He argues that through system models, these disciplines have a wider epistemological reach than the social sciences or humanities. As a result, it is tempting to base visions of the future on this type of knowledge. Applied to the context of SIDS, climate predictions suggest islands might become submerged, and so islanders will have to move. Operating under such a deterministic future, however, leads to reductive understandings of adaptation and resilience, which fail to take into account justice considerations appropriately. This chapter positions itself against future foreclosure, instead it attempts to hold the future open and explore within it alternative emerging understandings of adaptation and resilience by asking: what meanings might adaptation and resilience adopt if we conceive

DOI: 10.4324/9781003214021-13

of the possibility of a future on the islands? In attempting to answer this question, the chapter is structured as follows. First, it looks at the meanings that adaptation and resilience adopt under a foreclosed future for the islands. Second, it reviews a range of reasons why it might be important to resist future foreclosure and discusses the methodology of this research. Third, it presents islanders' alternative future visions, where adaptation and resilience are understood differently. The chapter closes by redefining adaptation and resilience in a way that foregrounds islanders' right to remain at home.

2 Adaptation, Resilience, and Climate-Related Migration from SIDS

Adaptation and resilience are two key concepts in climate change literature, which have been interpreted and utilized in a number of ways, with key questions revolving around the scope and limits of what it means to adapt and what it means to be resilient (Barnett et al. 2015). As introduced above, the basis of this chapter is that narratives of future foreclosure, where the end of life on the islands is assumed, lead to predetermined conceptualizations of adaptation and resilience. The flexibility of interpretation and meaning of these concepts is reduced to fit one particular vision of the future. With predictions suggesting that small islands' inhabitability is only a matter of time, and thus that their populations have no choice but to move or to drown, the question for the islanders becomes whether 'to stay or to go', 'to live or to die' (Methmann & Oels 2015, 63). When the only option is to move, international migration becomes the only alternative to drowning, and consequently conceptualizations of adaptation and resilience become tailored to this particular end. International migration ceases to be a problem and starts to be defined as a potential adaptive response (Black 2001; Foresight 2011; Tacoli 2009). What emerges in this context is an idea of migration as a powerful adaptation strategy for populations encountering environmental and climate challenges (Bardsley & Hugo 2010; McLeman & Smit 2006).

McLeman (2014) argues that thinking about adaptation in relation to mobility is consistent with the way in which migration has evolved within the IPCC and the United Nations Framework Convention on Climate Change (UNFCCC). The very first IPCC Assessment Report released in 1990 spoke of out-migration as a solution for islanders (IPPC & Houghton 1990). Twenty years later, the Cancun Agreement (UNFCCC 2010) was the first to mention migration in the context of adaptation, where, in order to increase action on adaptation, parties were invited to enhance understanding, coordination, and cooperation in relation to climate change-induced displacement, migration, and planned relocation.

In the Cancun Agreement (2010), the international community explicitly places migration within adaptation policy-making. Against a backdrop where adaptation is conceived in relation to mobility, the meaning of resilience is also tailored and reduced to the particular vision of the future where the islands are uninhabitable. The context in which resilience needs to be developed is one in which in order to adapt, one ought to move. Therefore, early conceptualizations of resilience, such as engineering resilience and ecological resilience which in the context of climate change and SIDS would focus on maintaining the islands as homelands, are no longer applicable (Holling 1973). Instead, and because the end of life on the islands is imminent, a third conceptualization of resilience as socioecological or transformational is adopted, to be achieved through abandoning one's place of residence and migrating to a new location (Methmann & Oels 2015). I do not mean to reduce the meaning of socioecological resilience to migration here, nor adhere to the recognized problematic dichotomy between nature vs. socioecological approaches to resilience (Cote & Nightingale 2012). Instead, I reflect on what is left there for resilience to mean in the particular case of SIDS where the future of the islands has already been declared finite. I critique here its 'transformational' path which is thought to occur when the current system is 'untenable or undesirable' (Cretney 2014, 630).

Under this particular context driven by a 'modelling culture that is preoccupied with determining ecological outcomes' (Cote & Nightingale 2012, 481), human mobility is seen as a way to build resilience and increase adaptive capacities for affected populations (Gemenne & Blocher 2017). This is understood in terms of having a set of skills and abilities that will make the international migration story a successful one. If affected populations develop skills and abilities that are useful in a potential host country, they will increase their chances of being employed and reduce the difficulties one encounters when arriving somewhere new. Bettini (2014) explains the idea of the resilient climate migrant as a docile and mobile individual committed to the international labour market by developing the skills and capacities required by it. Therefore, it is in achieving this form of resilience that migration can be thought of as a successful adaptation strategy as opposed to an apocalyptic scene of masses feeling their sinking countries in desperation. By developing a set of skills, and thus their adaptive capacity, individuals become resilient against the prospect of climate change. As such, they are prepared and well equipped to embark on a future elsewhere.

An example where these conceptualizations of adaptation and resilience in the context of island futures and climate-related migration are adopted is the Foresight report on migration and global environmental change produced by the UK's Government Office for Science (Foresight 2011). The authors of the report present international migration as bringing both opportunities and challenges, arguing that it is the most effective way of allowing individuals and communities to diversify income and build resilience. Therefore, they insist on the need to create channels for voluntary migration. Scholars have proposed a range of benefits of migration for the migrants as well as the original and the host communities, such as the diversification of income or a chance for professional and personal development through acquiring new skills while migrating (Warner 2012). Another example of how the Foresight report authors argue migration could benefit a community is by an individual migrating to obtain money and goods that could be sent back to the place of origin. This, the authors claim, would additionally benefit host countries where there might be demographic deficits and labour shortages (Foresight 2011).

Following the report's rationale, it seems, at least initially, that thinking about adaptation and resilience in terms of movement and the ability to move makes sense in terms of finding an effective way of responding to the challenges that climate change presents to SIDS. However, migration does not always result in upward social mobility or improved well-being. Instead, it can lead to a sense of loss, hopelessness, and alienation, with negative impacts on the migrants' health and well-being. Even if it results in some improvements, emerging scholarship suggests a more accurate way of representing migration: as a trading of vulnerabilities (Cundill et al. 2021; Thomas & Benjamin 2020; Tschakert et al. 2017). Therefore, instead of assuming the inevitability of movement and portraying it as an 'unfortunate but acceptable' (McNamara & Gibson 2009, 482) response to climate change, it is important to hold a comprehensive understanding of migration and engage with the justice dimensions of climate displacement in order to better define what fair understandings of adaptation and resilience might look like.

Scholars have raised a number of justice concerns in relation to climate displacement. De Shalit (2011), for example, recognizes that environmental displacement involves the loss of the crucial functioning of having a sense of place. Further, he stresses the significance of the permanence of this loss in what he calls 'harsh cases' of environmental displacement due to the impossibility of return, like would be the case of SIDS. Heyward (2014) discusses the loss of 'traditional ways of life' in these states as an injustice alongside losing their status as a distinct and self-governing, self-determining political community with control over their own affairs. Zellentin (2015) acknowledges it is not only the physical homeland that will be lost but also the social structures and cultural communities. These arguments demonstrate why climate-related migration and displacement might constitute an injustice.

It is essential that conceptualizations of adaptation and resilience are aligned to these justice concerns and not only address efficiency priorities. However, there is a recognized tension between justice and effectiveness when it comes to adaptation. As Byskov et al. (2019) recognize, often the most effective adaptation options might not necessarily be the most ethically justifiable. They use the example of populations who might be forced to be displaced for their own safety. While this would be effective from a technocratic perspective, it is problematic from the ethical perspective.

My argument here is that framing adaptation and resilience of SIDS in terms of unavoidable movement and refusing to engage with island futures shift attention away from important questions of justice. In particular, it dismisses justice implications of displacement and obfuscates islanders' rights to remain at home, on the one hand, and the international community's responsibility to ensure such a right, on the other hand. It is thus important to question narratives of future foreclosure and of the inevitability of SIDS sinking and becoming uninhabitable. This chapter positions itself against such future foreclosure, instead attempting to hold the future open and explore within it alternative understandings of adaptation and resilience. The driving question is: what meanings might adaptation and resilience acquire if we conceive of the possibility of a future on the islands?

Despite scientific predictions, there are at least two reasons to resist future foreclosure. First, although there is relative confidence in the global mean sea level rise trajectory, less information is available about the changes in regional sea level rise. The IPCC *AR5* report stated that it is very likely that in the twenty-first century and beyond, sea level change will have a strong regional pattern, with some places experiencing significant differences from the global mean change. In the Indian Ocean, the estimated sea level rise is around 1.5 mm/year (UN–Habitat 2015). Second, and as Hulme highlights, there is more to the future than climate (Hulme 2011). Scientific predictions do not account for the potential of human societies to find solutions. To assume the future of the island is finite does not take into account the possibility of a human response that might impact the outcome of climate change, nor does it seriously consider today's islanders' ancestral past. As Barnett puts its (2017, 10):

> Atoll peoples voyaged in canoes across vast distances and settled on narrow sandy islands with no soils and surface water. Some atoll islands have been settled for well over two thousand years, with no 'collapse' in the populations living on these most marginal of terrestrial environments despite climate extremes, colonisation, blackbirding, world war and dramatic changes in economic and political conditions.

Instead of foreclosing the future of entire island nations and their populations, Baldwin (2014, 526) suggests that we can think of the future as 'ours to make' and invent it the way we want, as opposed to preparing for a future the experts tell us to expect. This remaining uncertainty in relation to regional sea level rise, coupled with the potential of island communities to respond to climate change impacts in their own ways, is sufficient to challenge the idea of future foreclosure to hold the future open and redefine meanings of adaptation and resilience in ways that work alongside key concerns of climate justice.

In order to do so, this chapter draws on a methodological approach that relies on creative and participatory methods. The data presented in this chapter was generated in Male, the capital of the Maldives, in 2018 and 2019. The chapter draws from two sources of data: participatory workshops at the Maldives National University (MNU), and key informant interviews. The workshops were conducted with students of the BA in environmental management programme. The first workshop was with the third-year group of 12 students, and the second with the first-year group of 8 students. The central activity of the workshop was to conduct group-based drawings of the future of the Maldives. In the first workshop, groups of four students each were formed. In the second workshop, there were two groups of three students each, and one group with two students. The students were asked to draw the future of the Maldives as they imagined it. Then, each of the groups explained their visions and were asked questions by other groups. At the end, there was a general discussion about what the future of the Maldives might look like and what would be important if the prospect of climate-related displacement became true. To finish, students were given some time to write an individual reflection on the topic. One of the motivations to do this was to ensure that everyone had an opportunity to say something and state their opinion on the matter. In addition, in-depth qualitative interviews were conducted in Male with individuals working in research, policy, and academic circles in the area of climate change and environment. They included people working at the MNU, the Ministry of Environment, NGOs, and independent researchers and artists. I followed a purposive sampling strategy in order to select information-rich cases and include a variety of perspectives. Section 2.1 presents the findings that resulted from the workshops and key informant interviews.

2.1   The Future 'Here', in the Maldives

It was through hope and possibility, as Baldwin (2014) suggests, that islanders lived with the uncertainty of future predictions. Instead of giving up on a future on their islands, participants conveyed a strong sense of faith and confidence in the islands' futures. Through the workshops at MNU and key informant interviews, participants imagined a future where island communities thrived alongside their natural environments. The diversity of opinions was perhaps not in whether the country would survive or not, but in the ways in which guaranteeing such a future should be pursued. While some believed in the sustainability of the 'island concept', others placed their confidence in the human capacity to design ways to continue to inhabit the islands with projects such as land reclamation.

Those that had faith in the former believed that the islands have a natural capacity to adapt to changing conditions, as long as this is respected and not interfered with in excess. A key informant who had completed his PhD on the adaptive capacity of the Maldivian islands and was now working in Ministry of Environment in Male shared his reflections on the islands' capacity to self-repair and maintain:

> The islands have been in existence even with much harsher conditions, because they are allowed to self-repair and self-maintain … Adaptive capacity for me means that those biophysical limits should not be interfered [with], it should be kept as natural as possible and we should try to build with nature, and to live with nature, not against it.

The interviewee was referring to the islands' inherent dynamism, where erosion and accretion become regular island-building processes. The unconsolidated nature of beaches and the reversal of monsoon from north-east to south-west creates a dynamic and unstable environment, where land changes in the islands are normal (Kench & Brander 2006; Shaig 2008). Having witnessed such changes for generations, islanders held on to the potential that the islands have to self-repair and self-maintain as a desirable way to adapt to climate change. Another way participants thought of the islands' natural adaptive capacity was through mangroves. This was reflected in one of the drawings from the workshop conducted at MNU (Figure 12.1) where the students imagined the future of the Maldives. As the authors of the drawing described: 'Also here is the mangrove, mangrove protection and healthy reefs. I don't see any sea walls in the future, because we will be able to protect our mangroves and our reefs will be healthy.' In the drawing, mangroves surround the coastline of the island, and this is described as the preferred way to ensure protection against environmental challenges. A key informant working at MNU reflected on the potential and importance of mangrove systems: 'Mangroves are one of the most resilient ecosystems, they are able to move with the tides, if the tide is extending into the land, the mangrove can extend into the tide and cover it up, so this provides a lot of protection to the island.' The focus was placed on supporting island ecosystems as a way of adapting

FIGURE 12.1 Future vision A. Drawing by participants.

to climate change. This was also illustrated by sustainable fishing practices, sources of renewable energy, greenery, and healthy marine life and then accompanied by a representation of a developed island with a university, hospital, and recycling points (Figure 12.1).

Despite the preference for this type of natural adaptation, other participants recognized that there are many potential barriers to this process occurring naturally. Instead, some participants grounded their confidence, or lack of fear over the disappearance of the islands, in the changes that humans can make to ensure their survival. An example of this is land reclamation or land-raising.

The former, land reclamation, refers to the construction of ground where water once had been (Grydehøj 2015). The latter, land raising, involves the artificial raising of whole islands to appropriate heights to cope with future sea level rise (Brown et al. 2019). A second drawing from the workshops reflects this, relying not on the protection by mangroves but on the islands being elevated: 'We have even elevated the land. One of the ways of coastal adaptation, so to adapt to rising sea level.'

Figure 12.2 shows an island with a higher elevation relative to the sea level, which the authors describe as a coastal adaptation measure. The island has a few buildings and plenty of greenery with a big sign that reads: 'Plastic is Banned in the Maldives.' The sea has a floating solar panel, a sea plane, and shark fins to illustrate the existence of marine wildlife. Land reclamation and elevation have been used in the Maldives since the 1970s for both public and private purposes

FIGURE 12.2 Future vision B. Drawing by participants.

(Naylor 2015). For example, the island of Hulhumale was built on a two-metre-high platform, which is higher than the average Maldivian island in relation to sea level rise. Another example is the island of Th. Vilufushi. After the 2004 Indian Ocean Tsunami made the island uninhabitable, it was also rebuilt according to Gayoom's 'Safer Island Strategy'; its land area was tripled, with its mean elevation raised from 1 to 1.4 m above sea level (Naylor 2015). A key interviewee working for an environmental NGO in Male reflected on this:

> Hulumale', when they reclaimed [it], actually, they reclaimed one meter [higher]. They actually took the extra cost and reclaimed one meter above our normal sea level. Like at that time, the predictions were, there were some predictions done at that time, there will be one meter rise in sea level.

Land reclamation and elevation have been, to a certain extent, popular measures adopted in the Maldives to address a range of challenges, amongst which environmental or climate change is one. However, as illustrated above, concern has been raised regarding their lack of consideration for the environment. The key informant who worked at MNU shared: 'What we do now is we reclaim as much as we can and we tend to remove those shallow areas as well, so that is not really good, it's really bad coastal engineering.' Aware of the negative impacts of land reclamation on the environment, a more recent project in the Maldives also plays with the idea of creating inhabitable land, although this time, instead of relying on the reclamation of land, it relies on creating floating land. In March 2021, the Maldives Floating City (MFC) project was inaugurated as the first true floating island city, characterized as a 'futuristic dreamscape finally poised to become reality' (TTM 2021). In partnership with Dutch Docklands, the Government of the Maldives launched the project as a new way of increasing resilience and conceptualized it as a 'next-generation sea-level-rise-proof urban development' without the negative implications of land reclamation. As described on the project website:

> Maldives Floating City is the first development of a new era in which Maldivians return to the water with resilient eco-friendly floating projects. The city has a nature-based structure of roads and water canals resembling the beautiful and efficient way in which real brain coral is organised. The idea of having brain coral as the leading concept is that the goal of living with nature and learning to improve and respect natural coral is at the heart of the development, which leads to new knowledge emphasising the responsibility Maldives takes as centre for coral protection in the world.

The floating structure is planned to start the construction stage in 2022 in a lagoon near Male and consist of a mixture of housing, resources, restaurants, and shops, as well as a hospital, a school, and a government building. The main forms of land-based movement are expected to be walking and biking on natural white sand roads, with cars not being allowed. In this way, the planned floating city resembles the vision that islanders shared regarding an 'eco-city' and the development of local islands. In describing their drawing of a future 'eco-city', the group shared:

> We have an ecocity here with trees, there is sustainable development and for neighbourhoods we have trees, playgrounds, solar panels are installed on the roof of buildings and this is the bridge, these are the windmills in the sea, we have lanes for people who are walking, jogging and cycling, and also people are using public transport. People will not smoke on the road, they will be very healthy, and people will be outside a lot. Every neighbourhood will have a park, very green, so kids can go outside and exercise or play around. Basically, an island life in a city. We put this sign here to show that people will be educated and care for the environment and they will want this sustainable development.

The development of local islands was imagined along similar lines, as the group described, and presented as a chosen alternative to international migration. In describing Figure 12.4, the group shared:

> In our future, we do not see us moving to another country or anything, we see our island communities developing. So, what we have tried to represent here is a, yes, as an island nation, we are very vulnerable, we are prone to wave action as you can see here, but if everyone is willing to work together, they can actually protect the islands and still have all of the development. You have your renewable energy, you have your pristine reefs, if, the big if being here that you all work together. It's your island, so it's your responsibility.

At the centre of the drawing, there is an island with colourful houses in front of a jungle area with palm trees and sources of renewable energy. At the front, there is a line of people holding hands to represent a sense of unity and togetherness in facing the vulnerability of the islands represented by a wave rising behind the island. The ocean in the foreground is filled with healthy and colourful marine life (Figure 12.4).

These visions of the future are not oblivious to the vulnerability of the Maldives to climate change impacts or to the threat it poses towards the very habitability and existence of the islands. However, they chose to resist a future where that is the only possible scenario. Instead, through joint efforts and collaborations, they imagine a future where the islands and their populations thrive. These future visions of the islands were characterized by a range of 'everyday' and 'general long-term' changes. When speaking about what would characterize the islands in terms of everyday life, participants spoke about having lanes for walking or jogging (Figures 12.1 and 12.3), more playgrounds so that every neighbourhood

FIGURE 12.3 Future vision C. Drawing by participants.

can enjoy some outside space (Figure 12.3), or widespread greenery such as trees by the roadsides in Male or in the islands to increase shaded spaces (Figures 12.1–12.4). Public transport was also reflected in one of the drawings in an attempt to minimize the use of private vehicles (Figure 12.3). A series of other elements were identified to increase the chances for a future on the islands. These focused on sustainability and development, including local islands having a jetty as opposed to a harbour (Figure 12.4), practising sustainable fishing (Figure 12.1) and relying on renewable energy (Figures 12.2 and 12.3). Altogether, these changes would, according to research participants, lead to healthy and pristine reefs, allowing islanders to live in harmony with wildlife and their environments.

Workshop participants spoke about these changes taking place in local islands as a way of increasing their quality of life while responding to the threat climate change poses for the Maldives. Key informants spoke about developing regional hubs instead, implying that it was not viable to ensure each inhabited island in the Maldives could be adapted and made resilient to climate change impacts. However, workshop participants wished for the development of adaptation and resilience measures in the Maldives to go beyond the Greater Male area. They explained that so far the majority of sea protection measures had been built in the Greater Male region, while most other atoll islands had not received the same level of investment. An example of this is the $60 million seawall built around the entire island of Male after the floods of 1987, which was funded by the Japanese government under former president Gayoom's regime (Hamilton 2008).

FIGURE 12.4   Future vision D. Drawing by participants.

Therefore, participants and interviewees suggested that a way of adapting and increasing resilience was to develop local islands and create regional 'hubs'. The key informant who worked at the environmental NGO put it in the following way:

> I think there should be regions and, in each region, a bigger settlement where all services and protection measures are built in. I think that is something we should do. So, as I said, then you have to think about how to compensate to [*sic*] those who don't live on those islands, but as an adaptation thing, that is something we should do.

The creation of these bigger settlements in different regions with adequate protection measures was conceived as a way of adapting to climate change and increasing resilience. In 2008, Bluepeace, another Maldivian environmentalist group, made a similar suggestion (Bluepeace Maldives 2008). The proposed project consisted of constructing three-metre-high islands throughout the atolls as both centres for development and as settlements that are more resilient against sea level rise. This would follow the model of a contingency-adapted island where despite the islands sitting three metres higher, buildings would still be constructed on stilts in order to reduce the force of floods and strong storm events. Bluepeace

Maldives suggest that there should be at least seven contingency-adapted islands in seven different regions across the archipelago, although they indicate that indeed all the resort islands in the Maldives could be adapted as water villages or boat houses. Importantly, this proposal was explicitly framed as an alternative to a planned relocation of the Maldivian population, and therefore stating a preference for finding on-site adaptation solutions (Bluepeace Maldives 2008).

Therefore, alongside a narrative of foreclosure, where a future on the islands has been declared impossible, there are a range of visions from different groups and perspectives that are committed to pursuing a future in the Maldives. Through the islands' own natural adaptive capacity, processes of land reclamation and elevation, and the creation of regional hubs and floating inhabitable structures, affected populations shed light on alternative futures for their islands. When such futures are made possible, the limitations for understanding adaptation and resilience dissipate, creating space to reconceptualize and redefine these key concepts. Section 3 turns to exploring the meanings adaptation and resilience adopt under a narrative where a future on the islands is possible.

3 Revisiting Conceptualizations of Adaptation and Resilience

The chapter established that under a narrative where the future of the islands is foreclosed, adaptation and resilience take on reductive and specific meanings. The only alternative to drowning, after the islands become submerged, is to migrate internationally. Thus, adaptation and resilience begin to be conceptualized in relation to mobility. In order to adapt, one ought to move and be resilient, one ought to develop a set of skills that will aid their success in a new country. Although this rationale seems to serve policymaking concerns around effectiveness (Foresight 2011), it obscures important considerations of justice in failing to prioritize the right that islanders have to remain at home and the international community's responsibility to ensure such a right. The chapter established that it is important to re-examine possible meanings of adaptation and resilience which are aligned with the concerns of climate justice.

Drawing on participatory and creative methodologies, this chapter has explored alternative future imaginings for the islands in an attempt to rethink how one might conceptualize adaptation and resilience. The drawings from the workshops, alongside proposed projects by the government and other organizations, illustrate a range of possible futures for the Maldives, where international migration is avoided. Instead, throughout the visions of the future developed above, adaptation and resilience were understood in a range of other ways. For some, to have adaptive capacity meant not to interfere with the islands' natural processes and capacity to self-repair and self-maintain. For others, adaptation and resilience were to be achieved through land reclamation and land elevation projects, prioritizing a number of regional areas with appropriate measures to reduce levels of vulnerability against climate change impacts. Others have proposed the creation of floating structures resilient to climate change impacts where islanders could live. Often, these were stated explicitly as alternatives to having to migrate, which has consistently been opposed by affected populations (McNamara & Gibson 2009). Thus, unlike previous conceptualizations where resilience and adaptive capacity were understood in relation to mobility and meant developing a set of skills to increase chances of successful migration, both adaptation and resilience were, for islanders, understood in terms of the physical. As a key informant, who worked for an NGO called Plastic Noon Gotheh, illustrates:

> For me, I think very much in [terms of] the physical. Because the real basis of our existence itself would be sort of, we don't know what would happen, so I really think on the physical and see how we can build settlements, where we can live [in] 20, 30 years and at that time [undefined], they were saying 50 years on, and now we're there. And at the time, in UNDP, we were not talking so much on adaptation. At that time, we were talking on mitigation, and I remember they're saying, we need to do this project, because 50 years on, I don't want to be sitting in UNDP and planning evacuation of Maldivian[s] from that. But for me, it's more of a physical settlement where people can live, and a liveable environment. Even now, we don't have water, we too use technology, so basically for me, it's more of an existential thing, where do we exist? Do we adapt, and how? So physically, it's looking like better planning, better drainage, where there will be flooding, and for me it's, it's always at the back of my mind, where will we live? Where would we go? And that's the kind of thing that I think of when I say resilience.

It is thus these physical and infrastructural elements that lie at the core of islanders' conceptualizations of adaptation and resilience. These coincide with conceptualizations of resilience introduced above, namely engineering resilience (resilience as maintenance) and ecological resilience (resilience as adaptation) (Holling 1973), as opposed to socioecological (or transformational) resilience, where resilience might have meant the abandonment of the settlement and having to migrate elsewhere (Methmann & Oels 2015). None of the participants spoke about adaptation and resilience in terms of preparing for a future elsewhere, whereby they had to become 'adaptable' or 'employable' in a potential new host country. This point sheds light on an important disjuncture between the way policymakers think of these concepts and the way they are interpreted on the islands.

4   Conclusion

In an attempt to understand the way islanders think about their futures and, in doing so, deconstruct future foreclosure, this chapter has examined islanders' hopes, aspirations, and imaginings about the future. Through examining these narratives and visions, it has resisted the ways in which the future is being 'predetermined' (Hulme 2011). Further, it has responded to critiques of global narratives of climate change which privilege technical science and exclude lay knowledge (Nightingale et al. 2020), and elaborated on emerging counternarratives on climate change and small islands (Kelman 2018).

Holding the future open has allowed for alternative understandings of adaptation and resilience to be firmly grounded in the continuation of life on the islands. In this way, these conceptualizations are able to prioritize questions of climate justice. They are grounded in islanders' right to remain home and accommodate their ways of life and future visions. In rethinking adaptation and resilience in this light, this chapter further demonstrated a disjuncture between academic and policy framings of the climate change and migration nexus and the way these themes are understood in the islands and reflected on their implications for just climate futures. In doing so, it further strengthens the position first shared by former president Gayoom in 1998 and consistently maintained by people from SIDS ever since – that they do not wish to drown nor become environmental refugees and that what they wish to do is to 'stand up and fight' (Gayoom 1998) and expect the international community to do their share. It seems, therefore, that if we are concerned with justice or with the experiences of those who might 'have to migrate', we cannot speak about migration as a solution to climate change; a future where islanders are involuntarily displaced from their homelands is not a just future. Therefore, prioritizing climate justice requires that we reconsider the possibilities of a future *here* and take a braver, more creative approach to on-site adaptation.

References

Baldwin, A. (2014). Pluralising climate change and migration: An argument in favour of open futures. *Geography Compass*, 8(8), 516–528. https://doi.org/10.1111/gec3.12145.

Bardsley, D. K. & Hugo, G. J. (2010). Migration and climate change: Examining thresholds of change to guide effective adaptation decision-making. *Population and Environment*, 32(2), 238–262. https://doi.org/10.1007/s11111-010-0126-9.

Barnett, J. O. (2017). The dilemmas of normalising losses from climate change: Towards hope for Pacific atoll countries. *Asia Pacific Viewpoint*, 58(1), 3–13. https://doi.org/10.1111/apv.12153.

Barnett, Jon, Evans, L. S., Gross, C., Kiem, A. S., Kingsford, R. T., Palutikof, J. P., … Smithers, S. G. (2015). From barriers to limits to climate change adaptation: Path dependency and the speed of change. *Ecology and Society*, 20(3). https://doi.org/10.5751/ES-07698-200305.

Bettini, G. (2014). Climate migration as an adaption strategy: De-securitizing climate-induced migration or making the unruly governable? *Critical Studies on Security*, 2(2), 180–195. https://doi.org/10.1080/21624887.2014.909225.

Black, R. (2001, 7 February). Environmental refugees: Myth or reality? www.unhcr.org/research/working/3ae6a0d00/environmental-refugees-myth-reality-richard-black.html

Bluepeace Maldives. (2008). New homeland elsewhere is not the only solution for climate victims of the Maldives. http://www.bluepeacemaldives.org/blog/climate-change/new-homeland-not-only-solution-climate-victims-maldives.

Brown, S., Wadey, M. P., Nicholls, R. J., Shareef, A., Khaleel, Z., Hinkel, J., … McCabe, M. V. (2019). Land raising as a solution to sea-level rise: An analysis of coastal flooding on an artificial island in the Maldives. *Journal of Flood Risk Management, 13*(August), 1–18. https://doi.org/10.1111/jfr3.12567.

Byskov, M. F., Hyams, K., Satyal, P., Anguelovski, I., Benjamin, L., Blackburn, S., … Venn, A. (2019). Climate and development: An agenda for ethics and justice in adaptation to climate change. *Climate and Development, 13*(1). https://doi.org/10.1080/17565 529.2019.1700774

Cote, M. & Nightingale, A. (2012). Resilience thinking meets social theory: Situating social change in socio-ecological systems (SES) research. *Progress in Human Geography, 36*(4), 475–489.

Cretney, R. (2014). Resilience for whom? Emerging critical geographies of socio-ecological resilience. *Geography Compass, 8*(9), 627–640.

Cundill, G. et al. (2021). Toward a climate mobilities research agenda: Intersectionality, immobility, and policy responses. *Global Environmental Change, 69*, 1–7

De Shalit, A. (2011). Climate change refugees, compensation, and rectification. *The Monist, 94*(3), 310–328.

Felli, R. & Castree, N. (2012). Neoliberalising adaptation to environmental change: Foresight or foreclosure? *Environment and Planning A, 44*(1), 1–4. https://doi.org/10.1068/a44680.

Foresight. (2011). *Migration and Global Environmental Change.* The Government Office for Science.

Gayoom, M. A. (1998). International assistance can save our peoples. In *The Maldives: A Nation in Peril.* Ministry of Planning Human Resources and Environment.

Gemenne, F. & Blocher, J. (2017). How can migration serve adaptation to climate change? Challenges to fleshing out a policy ideal. *Geographical Journal, 183*(4), 336–347. https://doi.org/10.1111/geoj.12205.

Grydehøj, A. (2015). Making ground, losing space: Land reclamation and urban public space in island cities. *Urban Island Studies, 1*, 96–117.

Hamilton, J. (2008). Maldives builds barriers to global warming. www.npr.org/templates/story/story.php?storyId=18425626?storyId=18425626&t=1588344501003.

Heyward, C. (2014). Climate change as cultural injustice. In *New Waves in Global Justice.* Palgrave Macmillan, pp. 149–150.

Holling, C. S. (1973). Resilience and stability of ecological systems. *Annual Review of Ecology and Systematics, 4*, 1–23.

Hulme, M. (2011). Reducing the future to climate: A story of climate determinism and reductionism. *Osiris, 26*(1), 245–266. https://doi.org/10.1086/661274.

IPCC & Houghton, J. T. (1990). *IPCC First Assessment Report.* WMO.

IPCC. (2013). *Climate Change 2013: The Physical Science Basis. Contribution of Working Group I to the Fifth Assessment Report of the Intergovernmental Panel on Climate Change.* https://doi.org/10.1016/B978-0-12-409548-9.10820-6.

Kelman, I. (2018). Islandness within climate change narratives of small island developing states (SIDS). *Island Studies Journal, 13*(1), 149–166. https://doi.org/10.24043/isj.52.

Kench, P. S. & Brander, R. W. (2006). Response of reef island shorelines to seasonal climate oscillations: South Maalhosmadulu atoll, Maldives. *Journal of Geophysical Research: Earth Surface, 111*(1), 1–12. https://doi.org/10.1029/2005JF000323.

McLeman, R. (2014). *Climate Change and Human Migration. Past Experiences, Future Challenges.* Cambridge University Press.

McLeman, R. & Smit, B. (2006). Migration as an adaptation to climate change. *Climatic Change, 76*(1–2), 31–53. https://doi.org/10.1007/s10584-005-9000-7.

McNamara, K. E. & Gibson, C. (2009). 'We do not want to leave our land': Pacific ambassadors at the United Nations resist the category of 'climate refugees'. *Geoforum, 40*(3), 475–483. https://doi.org/10.1016/j.geoforum.2009.03.006.

Methmann, C. & Oels, A. (2015). From 'fearing' to 'empowering' climate refugees: Governing climate-induced migration in the name of resilience. *Security Dialogue, 46*(1), 51–68. http://files/209/Methmann et Oels 2015.pdf.

Naylor, A. K. (2015). Island morphology, reef resources, and development paths in the Maldives. *Progress in Physical Geography, 39*(6), 728–749. https://doi.org/10.1177/0309133315598269.

Nightingale, A. J. et al. (2020). Beyond technical fixes: Climate solutions and the great derangement. *Climate and Development, 12*(4), 343–352. https://doi.org/10.1080/17565529.2019.1624495.

Shaig, A. (2008). *Settlement Planning for Natural Hazard Resilience in Small Island States: The Population and Development Consolidation Approach.* James Cook University.

Tacoli, C. (2009). Crisis or adaptation? Migration and climate change in a context of high mobility. *Environment and Urbanization, 21*(2), 513–525. https://doi.org/10.1177/0956247809342182.

Thomas, A. & Benjamin, L. (2020). Non-economic loss and damage: Lessons from displacement in the Caribbean. *Climate Policy, 20*(6), 715–728. https://doi.org/10.1080/14693062.2019.1640105.

Tschakert, P. et al. (2017). Climate change and loss, as if people mattered: Values, places, and experiences. *Wiley Interdisciplinary Reviews: Climate Change, 8*(5), 1–19. https://doi.org/10.1002/wcc.476.

TTM. (2021). Introducing the world's first true floating island city – Maldives Floating City. www.traveltrademaldives.com/introducing-the-worlds-first-true-floating-island-city-maldives-floating-city.

UN-Habitat. (2015). Urbanization and climate change in small island developing states. https://sustainabledevelopment.un.org/content/documents/2169(UN-Habitat, 2015) SIDS_Urbanization.pdf.

UNFCCC. (2010). Report of the Conference of the Parties on its sixteenth session, held in Cancun from 29 November to 10 December 2010: Addendum. https://digitallibrary.un.org/record/708999?ln=fr

Warner, K. (2012). Human migration and displacement in the context of adaptation to climate change: The Cancun Adaptation Framework and potential for future action. Environment and Planning C: Government and *Policy, 30*(6), 1061–1077. https://doi.org/10.1068/c1209j.

Zellentin, A. (2015). Climate justice, small island developing states & cultural loss. *Climatic Change, 133*(3), 491–498. https://doi.org/10.1007/s10584-015-1410-6.

CONCLUSION

Towards Justice in Climate Justice Research – Feedback from Chapter Contributors

Michael Mikulewicz, Kavya Michael, and Neil J. W. Crawford

Climate Justice in the Majority World reflects a deliberate commitment on the part of the editors and the authors to foreground the diverse ways through which people make sense of their relationships with the climate and their environments. As such, the book captures multiple epistemologies and ontologies of people who experience climate change and environmental challenges first-hand, often in conjunction with the aftermaths of colonization, neoliberal economic development models, increasing socioeconomic inequalities, and general lack of recognition for their own lived realities. We have attempted to anchor this collection within an intersectional perspective, emphasizing the complex nature of today's socioecological crises. At the same time, we are also calling for more climate justice scholarship *from* (rather than *on*) the Majority World, a step necessary for overcoming the hegemony of mainstream climate knowledge frames rooted in the Minority World. We have deliberately chosen the term 'Majority World' – as opposed to 'Global South', 'developing countries' or 'Third World' – to challenge perspectives that view the world through the prism of economic, military, and technological power rather than its humanity and (bio)diversity. By doing so, we sought to underscore the need for epistemic justice in the climate justice literature that, despite its commitment to equity, still in many ways elevates certain bodies of knowledge over others (for a detailed discussion, see Ndlovu-Gatsheni 2021). The chapters in this book also interrogate diverse movements of resistance, protest, and community-based action happening in the Majority World, testifying to the need to overcome the barriers people with local knowledge and lived experiences of climate change face when trying to influence climate-related decision-making in their locales.

That said, like any collection of its kind, *Climate Justice in the Majority World* has its limitations, beyond the ones mentioned in the Introduction. Despite numbering just 12 chapters, the book is quite broad in scope and as such cannot hope to

DOI: 10.4324/9781003214021-14

capture the breadth and depth of climate justice issues, struggles, contestations, and policymaking efforts taking place in the extremely diverse Majority World. For instance, while countries like Bangladesh and India are the focus of two chapters each, most of the African continent and the Middle East do not feature, at all. This was not due to a desire to omit specific places or issues, but rather a result of the way in which contributions were collected and of ever-present limitations of space and resources. Topic-wise, most chapters pertain to case studies in individual countries. Meanwhile, more comparative or multi-site studies are required to learn from the struggles, resistances, and policy challenges across locations and to foster meaningful alliances between them. Lira Luz Benites Lazaro et al.'s chapter offers a welcome regional overview of climate justice issues in Latin America; similar studies on other world regions are needed. Moreover, the book does not focus sufficiently on many of the international mechanisms, structures, organizations, and global policies that govern national and local climate change policy and action across the Majority World. A notable exception within the collection, Jessica Omukuti and Aidan O'Sullivan's chapter constitutes a much-needed critical analysis of the Green Climate Fund, with more investigations of this kind required to uncover the inequalities and injustices embedded within global climate finance and governance structures. Despite these limitations, our hope is that the book provides a snapshot of climate justice issues in some parts of the Majority World at the time of writing, and that in the future it will be joined by many more collections that explore additional aspects of climate justice across the vast, diverse, and dynamic Majority World.

1   Challenges Encountered by Climate Justice Scholars from the Majority World

Alongside systemic and more abstract issues like epistemic injustice that we mentioned in the Introduction, one cannot forget that there exist a number of more practical challenges that Majority World scholars face on a regular basis in their work. Having realized the diversity of these challenges following multiple conversations on this topic with our colleagues throughout the years, we decided to reach out to chapter contributors and ask them to share their views on the issues faced by Majority World scholars, with most (but not all) survey respondents self-identifying as such. To this end, we distributed a short questionnaire to all chapter leads and co-authors, in which we sought to capture specific examples of such obstacles. We also shared a draft of this concluding chapter with all authors to ensure we managed to reflect their views to a satisfactory standard and seek additional feedback from others. The resulting analysis, while rudimentary, has revealed a number of interrelated challenges, presented here in no particular order. These include inequality between Majority and Minority World scholars, the former's restricted access to knowledge and technology, publishing difficulties, systemic pressures, development challenges, and repression of climate justice scholars and activists.

One of the key obstacles identified by the authors was the fundamental inequality that often dominates the relationship between scholars in the Minority and Majority Worlds. One of the causes was seen in the preconceptions by Minority World academics about the 'academic canon', in the words of one of the authors, and who is allowed to shape it or comment on it (Carrión and Acosta 2020). Relatedly, it was mentioned, partners from the Majority World are often tasked with carrying out fieldwork and collecting data for case studies, while theorizing is reserved for their Minority World partners. This is yet another reason why 'Western' theories and bodies of knowledge dominate in studies on the Majority World, including those pertaining to climate change. Authors also pointed out the often extractive relationship between Minority World academics and Majority World activists and NGOs, whereby the former find themselves in a privileged position relative to the latter due to financial, technological, linguistic, and other advantages. Not without implications is the fact that communication within such partnerships is, more often than not, carried out in the language spoken by Minority World partners rather than the other way round. In general, linguistic issues were seen as a major obstacle for creating equitable partnerships.

Next, a key identified challenge was that of access. According to our contributors, Majority World scholars frequently find themselves shut out of the global knowledge production systems, an exclusion which takes a myriad of forms. Our authors indicated what they described as insufficient baseline knowledge on climate justice in their countries or at their institutions. This is related, among other factors, to the suggested lack of 'serious' academic work on climate justice and poor access to specialist knowledge on this topic, in general. In addition, some authors pointed out low credibility and reliability of various kinds of data at the national and subnational levels, including data compiled by scientists, governments, and advocacy groups.

A crucial issue which transcends national contexts is the widely recognized difficulty in accessing scholarly publications by Majority World researchers and activists. Due to lack of funding, university libraries in Majority World countries frequently lack access to scientific databases or key journal subscriptions, which makes it very challenging to keep up with theoretical and empirical advancements in many disciplines, whether in natural or social sciences. The price point of academic books and e-books, such as this very one, is often beyond reach for scholars and activists based in the Majority World. This, however, is not to suggest that this issue is experienced equally by all universities in the Majority World. After all, they are subject to prestige and funding hierarchies just like their counterparts in the Minority World, with some impacted more than others. Still, Majority World scholars are frequently forced to rely on open-access works, which – while growing in number – still represent only a small proportion of published scholarship. Another concern is overreliance on the more accessible journals that may not meet strict academic peer review standards, which may result in lower-quality or less-respected publications. A related issue is limited access to technology, including specialist software and equipment such as geographic information systems (GIS), remote sensing, or statistical analysis software – a notorious issue experienced by our research partners in the Majority World. This technological exclusion translates into limited specialist knowledge and skill base, which only perpetuates the reliance on Minority World experts and expertise. Finally, the exclusion of Majority World scholars and activists can take a more tangible form – that of visa restrictions that prevent many from entering spaces where key partnerships are forged and the bulk of climate knowledge is produced. Obtaining visas to attend conferences and other networking events or undertaking visiting appointments are both a time-consuming and expensive process which is often out of reach for Majority World-based academics.

Another challenge mentioned by the authors was the difficulty in publishing in established journals and other venues. There is certainly a wide scope for improvement in the level of support offered to Majority World scholars by academic journal editors, who wrongly assume the former are familiar with the publication process and have access to editorial assistance, including English language support (which is often a paid service). Indeed, the language barrier in the world of publishing cannot be overstated, as non-Anglophone scholars must ensure not only that their work meets strict academic standards, but also that it complies with uncompromising publication language requirements. The fact that writing incentives and writing support services at their home institutions are usually very limited makes this hurdle even worse. These practical challenges, authors said, affect their creativity and frequently translate into lack of high-impact references for their work, further reinforcing their exclusion from global academic and knowledge production systems.

A common theme we identified can be described as systemic pressures, often caused by the neoliberalization of higher education. Authors mentioned low pay and high levels of disorganization at their home institutions, alongside increasing time pressures and workloads, all leading to overwork and burnout. Another systemic issue relates to intersecting inequalities in workplaces, which cut across gender, ethnicity, sexuality, and other categories of difference. Ethnic diversity and discrimination within the Majority World may impose intersectional inequities on some climate justice scholars and the populations they study and partner with. It is evident from the responses that women academics – much like in the Minority World – face more barriers in their work than their male colleagues due to sexism, patriarchy, and misogyny as well as gendered social responsibilities, time pressures which weigh especially heavily on those with families, and mentorship or representation demands in universities where few women have yet been hired. One reflection of the masculinist nature of academia (similar to the Minority World) is the relatively low prioritization of community-based research so essential for climate justice knowledge production and the concomitant lack of interest in – and funding for – climate justice-related work among key stakeholders, including governments, media, and many academic circles. There are very few Majority

World research centres focused explicitly on climate justice issues, a reflection of the general lack of spaces where the basics of climate justice can be communicated to the general public and decision-makers, either by scholars or by those directly affected by the uneven impacts of climate change.

Crucially, Majority World-based scholars and activists often find themselves impacted directly by climate change, a rare issue for their colleagues in the Minority World. This is related to broader development challenges that affect the former's work and their lives in general. Respondents mentioned poor healthcare and related health issues, poor infrastructure and education systems, power outages, limited internet access, transportation challenges, and experiencing poverty. Authors also pointed out the political and security challenges which come in the form of pressure or even repression of outspoken scholars of climate justice and social justice at the hands of authoritarian governments. Shocking violence – such as the assassination of Indigenous environmental activists Berta Cáceres (2016, Honduras), Sergio Rojas Ortiz (2019, Costa Rica), Ari Uru-Eu-Wau-Wau (2020, Brazil), Fikile Ntshangase (2020, South Africa), or journalist Nazim Jokhio (2021, Pakistan) along with thousands of other climate justice activists – clearly demonstrates the very real risks faced by climate or environmental advocates in the Majority World (Temper et al. 2020; Feng, Mildenberger, and Stokes 2020; Jaferii 2022). Needless to say, a climate of repression and fear of this kind can make scholars and activists wary of sharing their work with the public or forging cross-sectoral and international alliances with like-minded collectives and colleagues.

As can be seen from these important insights, our contributors' challenges can be both symptomatic of epistemological injustices (e.g. perceptions of non-Western expertise as less scientific or otherwise inferior), and simultaneously working to perpetuate them (e.g. through lack of access to knowledge dissemination channels, such as journal subscriptions). Some challenges are unique to the country or discipline in which our contributors work, but others broadly reflect the difficulties faced by Minority World academics, such as systemic pressures on higher education in terms of time and financial resources. For all these reasons, it is not a secret that Majority World scholars find themselves at a disadvantage when interacting with the overwhelmingly Anglophone and increasingly standardized global knowledge production systems, in most cases controlled by Minority World-based actors. These include publishers, journal editors and reviewers, national scientific bodies, foundations, governments, among others.

2 Towards Decolonizing Climate Justice Research

Alongside examples of challenges faced by Majority World authors, we asked our contributors about what needs to change to address the issues they identified. A first step, and in our view a perfectly achievable one, is building more horizontal and democratic partnerships between Majority and Minority World scholars. Authors had multiple suggestions as to how this should be achieved. For example, they noted that Minority World scholars should engage more with their in-country and regional partners to better understand local contexts. Oftentimes (and this is true in the Minority World as well), academics are too detached from the lived realities of the places they are researching, which prevents research partnerships from reaching their full potential. Minority World academics, it was also argued, should explore 'non-Western works' as part of their academic training and progression. Relatedly, some authors argued for the preferential inclusion of Majority World scholarship on course reading lists to ensure students in wealthier countries are exposed to diverse knowledges throughout their academic journeys. Another suggestion was ensuring fair co-authorship of written outputs such as academic articles. Too often, it was noted, these publications are spearheaded by Minority World partners, with little incentive or assistance offered to Majority World partners in the process.

There is also a need for more trans- and interdisciplinary collaborations and knowledge sharing that would increase the visibility of Majority World scholars in academia. Here, authors mentioned equitable 'North–South' collaborations, including between academics and activists, as well as partnerships between scholars in the Majority World, or 'South–South' collaboration. Emphasis was placed on the need to disseminate success stories and grassroots initiatives in the Majority World to highlight local agency, resourcefulness, and creativity. It was also highlighted that there was a need for more spaces like this book, which 'talks about us and about our issues'. Authors suggested additional support for Minority World-funded exchange programmes at public and academic institutions with 'concrete capacity building actions', for instance, training on publishing in the more selective or reputable peer-reviewed journals. Even short visits enabling Majority World scholars to present their ideas and discuss them with new audiences in the Minority World would go a long way in increasing the former's visibility. Some authors pointed out that knowledge sharing of this kind does not need to be resource-intensive, citing low-cost mapping information sharing through the EJAtlas as an example (Temper, del Bene, and Martinez-Alier 2015). The CLACSOnetwork and its open-access libraries are another example of an initiative that seeks to overcome the financial and institutional barriers related to access to knowledge, in this case within Latin America and the Caribbean region (CLACSO n.d.). Similarly, there is a need for more academic scholarships for students from the Majority World, both at home and abroad. In addition to funding from Minority World governments, this could be buttressed by mobilizing national diasporas for creating academic and political linkages and bridges of support. Domestically, it was suggested that national governments and other institutions in the Majority World should allocate research funding to climate justice issues specifically. This could include the establishment of national climate justice research agendas and funding frameworks. In general, there was consensus that more funding streams specifically for Majority World scholars are needed. It is of crucial importance that calls for research funding offer these scholars an opportunity to *lead* projects as principal/lead investigators rather than playing an auxiliary role as 'in-country partners'. Relatedly, travel budgets for Majority World scholars should be increased, at the cost of their partners' travel budgets, if necessary, and visa regulations relaxed to facilitate international research, teaching, and conference visits. Minority World universities should take the lead in pressing their governments for this, as a matter of academic freedom.

Authors had a number of suggestions for overcoming the language barriers inherent in most international collaborations. First, they suggested that Minority World academics should learn the languages of the places they study, rather than relying on the command of their own language by Majority World partners. Second, they proposed that calls for papers and presentations should be advertised in more than just one language, and that online publications should always be available in multiple languages (translating the abstract into different languages being insufficient). This might require a considerable investment of financial resources into translation services by the publishers and academic departments – an investment that we believe is both feasible and essential, and could benefit from the rapidly improving closed captioning and auto-translation technologies. Third, in the context of academic review, language issues present in manuscripts submitted by Majority World-based authors should be overlooked at the first review stage, since they are fairly easy to remedy, or become of secondary importance relative to the academic content when reviewing such submissions. At the very least, free language services should be offered to authors who require support of this kind. Relatedly, it was suggested that academic journals should be more flexible when it comes to the expected writing styles and manuscript structures, which are seen as overly rigid and inaccessible to authors not trained in conventional academic writing. In reviewing the chapters for this book, we sought to follow this advice ourselves.

Finally, authors put emphasis on the need to ensure protection of academic and personal freedom for climate justice scholars and activists in the Majority World. This can be facilitated via collegial academic societies and networks, acting within larger Minority World academic associations. Certain contributors to this book have experienced theft and offloading (being forced off a flight) (Vaughan and Vidal 2015), likely due to their outspoken advocacy for climate justice. Majority World-based scholars' freedom to communicate the findings of their studies to wider audiences, both at home and abroad, needs to be guaranteed.

The scope of changes to how global academia works is as wide as it is complex. After all, in the words of one of our authors, 'decolonizing isn't a box-ticking exercise', as it touches on most aspects of what it means to be a researcher. Change is urgently needed in how research is funded, how international partnerships operate, how national governments see climate justice issues and those who study them, how research is published, and how scholars across the world communicate. Despite the daunting scope of transformation that needs to take place, these changes are necessary not just in climate justice and climate change studies but throughout the entire world of research and academia.

Alongside these insightful suggestions by chapter contributors, the editors would like to highlight two more. First, and related to the epistemic injustice we talked about here and in the Introduction to the book, the perception among funders and decision-makers of Minority World knowledge on climate change – and any other topic for that matter – as more precise, correct, or otherwise superior must end. In our view, one of the reasons for the current paucity of successful approaches to solving the climate crisis and the accompanying injustice issues stems from the suppression – or the 'epistemicide', in the words of Boaventura de Sousa Santos (2016) – of perspectives that fall outside of the realm of market and technological solutions and that seek to evade control of Minority World actors. This is particularly relevant for the UNFCCC process, which has long been criticized for being unreceptive to Majority World interests and perspectives. Relatedly, it is worth highlighting that Majority World scholars themselves may perceive scholarship from their own countries as not serious or credible enough, which arguably constitutes another form of epistemic injustice. Fricker (2007) calls this 'hermeneutical injustice', whereby a structural identity prejudice leads to lack of trust and credibility in one's own knowledge practices. Second, researchers and activists alike should avoid perpetuating vulnerabilizing imaginaries of people living in the Majority World, and instead focus on their agency in responding to the climate crisis. There is no shortage of positive stories testifying to the creativity, resourcefulness, and solidarity of people in the Majority World, yet the arenas of policy, academia, and the media notoriously rely on discourses that stage them as hopeless victims of floods, droughts, and other climate impacts or as passive recipients of climate aid. One of the key strengths of climate justice scholarship is its ability to foreground local agency and how successful it can be in improving lives and livelihoods, often without help from external donors and aid providers (Mersha 2017; Tokar and Gilbertson 2020; Garland 2015; Khalil et al. 2020).

We hope that the contents of this book have managed to convey the essential contribution Majority World scholarship and activism makes to the fight for global climate justice, and how important it is to continue decolonizing research, policy, and action in this field. As mentioned at the beginning of this concluding chapter, we see this book as a modest contribution to the efforts to foreground the diversity of thinking about climate change and its associated inequities. Moving forward, we urge research funders and other key decision-makers to follow the decolonizing strategies suggested by the authors in this book and call for further discussion on the need to decolonize our field of research and practice with colleagues from Minority and Majority Worlds alike.

References

Carrión, Andrea, and María Elena Acosta, eds. 2020. *Investigación aplicada sobre cambio climático: Aportes para ciudades de América Latina.* Quite: FLACSO Ecuador.

CLACSO. n.d. 'Open Access to Knowledge'. Consejo Latinoamericano de Ciencias Sociales. www.clacso.org/en/open-access-to-knowledge.

de Sousa Santos, Boaventura. 2016. *Epistemologies of the South: Justice against Epistemicide*. Routledge.

Feng, Jeff, Matto Mildenberger, and Leah C. Stokes. 2020. 'Inhumane Environments: Global Violence against Environmental Justice Activists as a Human Rights Violation'. In *A Research Agenda for Human Rights*, edited by Michael Stohl and Alison Brysk, 141–53. Edward Elgar Publishing. https://doi.org/10.4337/9781788973083.00014.

Fricker, Miranda. 2007. *Epistemic Injustice: Power and the Ethics of Knowing*. Oxford University Press.

Garland, Allison, ed. 2015. *Urban Opportunities: Perspectives on Climate Change, Resilience, Inclusion, and the Informal Economy*. Wilson Center. www.wilsoncenter.org/publication/urban-opportunities-perspectives-climate-change-resilience-and-inclusion.

Jaferii, Abdul Moiz. 2022. 'Nazim Jokhio Murder Case: With Great Power, Comes the Ability to Bend the Criminal Justice System to Your Will'. www.dawn.com/news/1685132.

Khalil, Momtaj Bintay, Brent C. Jacobs, Kylie McKenna, and Natasha Kuruppu. 2020. 'Female Contribution to Grassroots Innovation for Climate Change Adaptation in Bangladesh'. *Climate and Development* 12 (7): 664–76. https://doi.org/10.1080/17565 529.2019.1676188.

Mersha, Sara. 2017. 'Black Lives and Climate Justice: Courage and Power in Defending Communities and Mother Earth'. *Third World Quarterly*, November, 1–14. https://doi.org/10.1080/01436597.2017.1368385.

Ndlovu-Gatsheni, Sabelo. 2021. 'Epistemic Injustice'. In *Knowledge for the Anthropocene*, edited by Francisco Carrillo and Günter Koch, 167–77. Edward Elgar Publishing. https://doi.org/10.4337/9781800884298.00026.

Temper, Leah, Sofia Avila, Daniela Del Bene, Jennifer Gobby, Nicolas Kosoy, Philippe Le Billon, Joan Martinez-Alier, et al. 2020. 'Movements Shaping Climate Futures: A Systematic Mapping of Protests against Fossil Fuel and Low-Carbon Energy Projects'. *Environmental Research Letters* 15 (12): 123004. https://doi.org/10.1088/1748-9326/abc197.

Temper, Leah, Daniela del Bene, and Joan Martinez-Alier. 2015. 'Mapping the Frontiers and Front Lines of Global Environmental Justice: The EJAtlas'. *Journal of Political Ecology* 22 (1): 255–78. https://doi.org/10.2458/v22i1.21108.

Tokar, Brian, and Tamra Gilbertson, eds. 2020. *Climate Justice and Community Renewal: Resistance and Grassroots Solutions*. Routledge.

Vaughan, Adam, and John Vidal. 2015. 'Greenpeace India Campaigner Prevented from Travelling to the UK'. *The Guardian*, 12 January. www.theguardian.com/environment/2015/jan/12/greenpeace-india-campaigner-prevented-from-travelling-to-the-uk.

INDEX